PENGUIN BOOKS

# THE FATAL STRAIN

Alan Sipress is the economics editor at *The Washington Post* and a longtime foreign correspondent, based most recently in Southeast Asia. In 2005, a *Post* team he anchored was awarded the Jesse Laventhol Prize for Deadline Writing for coverage of the South Asian tsunami. This is his first book. He lives in Washington, D.C.

Praise for *The Fatal Strain*

## A *Washington Post* Best Book of 2009

"Sipress's book is an important—and highly readable—contribution to our understanding of the all-too-human mistakes that make pandemics possible and, if he is right, inevitable. . . . Weav[es] together the complex and riveting story." —*The Philadelphia Inquirer*

"Masterfully paced, gripping work." —*Seed*

"[Sipress is] a thorough reporter who knows how to tell a story." —Bloomberg.com

"Well written and engaging . . . This story-behind-the-story offers excellent insight into . . . mounting a global effort to contain an outbreak." —*Journal of the American Medical Association*

"Exemplary—and highly frightening—investigative reporting." —*Kirkus Reviews* (starred review)

"Timely, given today's headline-bursting thread of swine flu (H1N1). It is . . . a cautionary tale [as] influenza is about politics." —*Booklist*

"This is a book about much more than a lethal threat from the influenza virus. It's about the fog of war, about reality, about the gap between those who make plans and those who carry them out. And ultimately it's about heroism and determination. There are lessons here for everyone, and in compelling fashion this book drives those lessons home." —John M. Barry, author of *The Great Influenza*

"*The Fatal Strain* reads like a gripping medical mystery novel—only it is not. It is the true story of the scary world of pandemic influenza expertly written by one of the leading flu journalists of our time. Although the H1N1 (swine) influenza pandemic is unfolding before us, we must not take our eye off of avian influenza, as it very well may lead to a deadly 'one-two pandemic punch.' Anyone who cares about what might happen to their loved ones, friends, or colleagues should read this book."　　　—Michael T. Osterholm, PhD, MPH, director, Center for Infectious Disease Research and Policy, University of Minnesota

"Alan Sipress has produced a vivid and enthralling story that could not be more timely. Ever evolving, ever elusive, influenza threatens us on a scale far worse than anything we've yet seen. Sipress pursues the deadly strain through Asia with tenacious energy, revealing the true scale of the danger, and the terrifying inadequacy of our readiness to face it."—Pete Davies, author of *Devil's Flu* and *Inside the Hurricane*

# THE
# FATAL STRAIN

On the Trail of Avian Flu and
the Coming Pandemic

## Alan Sipress

PENGUIN BOOKS

PENGUIN BOOKS

Published by the Penguin Group

Penguin Group (USA) Inc., 375 Hudson Street, New York, New York 10014, U.S.A.

Penguin Group (Canada), 90 Eglinton Avenue East, Suite 700, Toronto, Ontario,
Canada M4P 2Y3 (a division of Pearson Penguin Canada Inc.)

Penguin Books Ltd, 80 Strand, London WC2R 0RL, England

Penguin Ireland, 25 St Stephen's Green, Dublin 2, Ireland (a division of Penguin Books Ltd)

Penguin Group (Australia), 250 Camberwell Road, Camberwell, Victoria 3124, Australia
(a division of Pearson Australia Group Pty Ltd)

Penguin Books India Pvt Ltd, 11 Community Centre, Panchsheel Park,
New Delhi – 110 017, India

Penguin Group (NZ), 67 Apollo Drive, Rosedale, North Shore 0632, New Zealand
(a division of Pearson New Zealand Ltd)

Penguin Books (South Africa) (Pty) Ltd, 24 Sturdee Avenue, Rosebank,
Johannesburg 2196, South Africa

Penguin Books Ltd, Registered Offices:
80 Strand, London WC2R 0RL, England

First published in the United States of America by Viking Penguin, a member of
Penguin Group (USA) Inc. 2009
Published in Penguin Books 2010

1   3   5   7   9   10   8   6   4   2

THE LIBRARY OF CONGRESS HAS CATALOGED THE HARDCOVER EDITION AS FOLLOWS:
Sipress, Alan.
The fatal strain : on the trail of avian flu and the coming pandemic / Alan Sipress.
p.   cm.
Includes bibliographical references and index.
ISBN 978-0-670-02127-7 (hc.)
ISBN 978-0-14-311830-5 (pbk.)
1. Avian influenza—Popular works.   I. Title.
RA644.16S57 2009
636.5'0896203—dc22      2009004191

Printed in the United States of America
Designed by Nancy Resnick
Map by Brenna Maloney
Timeline by Laris Karklis

To Ellen

# The Asian Theater

Hot spots for avian flu outbreaks in birds and the year when the first reported outbreak occurred.

Pearl River Delta 1996

Red River Delta 2003

Central Wetlands: 2003

Mekong Delta 2004

Qinghai Lake

Vladivostok    RUSSIA

CHINA

Beijing

NORTH KOREA

Seoul
SOUTH KOREA    JAPAN

Yellow

CHINA

Shanghai

East China Sea

Brahmaputra

BHUTAN

INDIA

BANGLA.

Yangtze

Hongshui

Guangzhou

Xi    Taipei    Tropic of Cancer

MYANMAR (BURMA)

Chiang Rai

Dong Dang

Hanoi    Thai Binh

Hong Kong    TAIWAN

Irrawaddy

Rangoon

LAOS

Red R.

VIETNAM

Pacific Ocean

THAILAND

Suphan Buri

Mekong

Manilla

Bangkok

Tonle Sap

CAMBODIA

South China Sea

PHILIPPINES

Andaman Sea

Phnom Penh    Ho Chi Minh City

Kampot

BRUNEI

Medan    MALAYSIA

Kubu Sembilang

Kuala Lumpur

MALAYSIA

SINGAPORE

Equator

Sumatra

Borneo

Sulawesi

I  N  D  O  N  E  S  I  A

Jakarta    Pekalongan

Java    Solo    Bali

Dili

E. TIMOR

Indian Ocean

Detail

Asia

Pacific Ocean

Indian Ocean

0    500
MILES

AUSTRALIA

# Avian Flu Timeline

**IN BIRDS**  **IN PEOPLE**

**1996:** Die-off of geese at a farm in Guangdong Province, China. Highly pathogenic avian influenza H5N1 is later found in a sample from a dead goose.

**1996**

**1997**

**May:** Boy in Hong Kong falls fatally ill with what is later confirmed to be the first known human case of H5N1 avian influenza. CDC team investigates.

**November—December:** Seventeen more people in Hong Kong are confirmed to have the virus; five die. CDC team returns to Hong Kong.

**February:** Avian flu resurfaces, striking a Hong Kong family that had recently traveled to mainland China.

**June:** Die-offs in Vietnam and Indonesia. The region's governments refrain from announcing outbreaks until the following year.

**2003**

**October:** Vietnamese children hospitalized with a mysterious respiratory disease, later confirmed as avian flu.

**November:** Die-offs in Thailand.
**December:** Outbreaks reported in South Korea.
**January:** Outbreaks reported in Vietnam, Thailand, Japan, Cambodia and Laos.
**February:** Outbreaks reported in China, Indonesia.

**January:** First human cases confirmed, Vietnam and Thailand; likely episode of human-to-human transmission in wedding party in Thai Binh, Vietnam. Dozens of WHO specialists investigate.

**June-July:** Resurgence of outbreaks in Vietnam, Thailand, Cambodia, China and Indonesia; in Malaysian birds for the first time.

**2004**

**September:** Likely case of human-to-human transmission between mother and daughter in Kamphaeng Phet, Thailand.

**January:** Family clusters in Thai Binh, Vietnam.
**February:** First human case confirmed, Cambodia.

**April:** Thousands of migratory birds begin dying at Qinghai Lake in central China.

**May:** International experts meet to evaluate human cases in Vietnam.

**July-August:** Outbreaks reported in Russia; similar cases in poultry in Kazakhstan and wild birds in Mongolia.

**2005**

**June:** WHO team arrives in Hanoi to check test results indicating that the virus is mutating toward a pandemic strain.

**October:** Outbreaks reported in Turkey, Romania, Croatia, Britain.

**July:** First human case confirmed, Indonesia.

**January:** First human cases confirmed, Turkey; Iraq.

**February 2006:** Outbreaks reported in Egypt; cases begin to spread across West Africa, and Central and Western Europe. More than 30 countries on those continents affected by summer.

**March:** First human cases confirmed, Azerbaijan; Egypt.
**May:** Ginting family cluster in Sumatra, Indonesia.

**2006**

**November:** Indonesian health minister informs WHO that her government will no longer share virus samples.

**January:** First human case confirmed, Nigeria.
**February:** First human case confirmed, Laos.

**April:** Die-offs begin to spread to more countries in West Africa.

**2007**

**June:** Confirmed, 100th human case, Indonesia.

**August:** Confirmed, 100th human case, Vietnam.

**November:** WHO hosts a meeting in a bid to break the impasse over sharing of virus samples.
**December:** First human cases confirmed, Myanmar; Pakistan.

**2008**

**May:** First human case confirmed, Bangladesh.

# Contents

# THE
# FATAL STRAIN

# Prologue

In an underground bunker carved from the soft Swiss hills above Lake Geneva, the daily intelligence briefing they call the morning prayers was beginning. The intel officers had been up since dawn, sifting through reams of electronic communications for any hint of an emerging threat, any anomaly portending disaster. Their customized computer search engine had culled and translated news reports, Web postings, and online rumors in six different languages, and now these were stitched together with official bulletins and tips from informants into a tapestry of microbial peril. Like the Men in Black who work to keep humanity blithely ignorant of the menace to this planet, these officers were engaged in a perpetual, often unpublicized contest with the most accomplished killers known to man. But this was for real.

The officers filed into the command center, a converted movie theater two floors beneath the global headquarters of the World Health Organization (WHO). They passed through doors secured with electronic locks that limit access to all but the privileged few and pulled up black chairs around the blond wood table. There was space for only six. So nearly twenty others—including special operations staff, disease specialists, and a few of the agency's senior brass—lined the edge of the room, some seated, others forced to stand. The chief of the intel section was a stylishly dressed Australian woman, a former senior disease investigator in Canberra who had been seconded to WHO in response to the 2001 anthrax attacks in the United States. She convened the briefing, moving briskly down the list of newly uncovered outbreaks. Some were still unidentified. There were reports and rumblings from Africa, Asia, the Middle East, and the Caribbean. A pair of large plasma screens on the wall flashed maps of these evolving hot spots. A red digital display scrolled through the time in eighteen major cities from Geneva and Washington to Delhi, Hanoi, and Jakarta.

The room was small, even cozy, with recessed lighting and gray carpeting. This was the mezzanine that once housed the movie theater's projector. But if any of the threats developed into a true epidemic, the full war room below, called the Strategic Health Operations Center, would be activated. More commonly known as the SHOC, it offers dedicated satellite communications linking it to 150 agency offices and government ministries around the world. Dispatches from field investigators and doctors in distant hospitals can be beamed directly to the high-definition video wall at the front. A private network instantly transmits data from the remotest of outposts to a series of black planar monitors, which retract with a hum into the surface of the U-shaped table in the middle of the room. The center has its own dedicated power source and computer servers, its own telephone exchange, ventilation, and air-conditioning. In case of a pandemic or a biological weapons attack, the SHOC must go on.

The briefing this morning was confidential. The premature disclosure of a suspected outbreak could sabotage WHO's ability to mount an emergency response. It might cause panic, crippling economies and embarrassing governments.

But given a rare opportunity to sit in on morning prayers, I got a glimpse of humanity's existential conflict with the microbes. Ebola. Monkey pox. Cholera. Plague. Yet of all these and more, it was flu that my hosts feared most. Not ordinary flu, not the kind that sends the sick to bed each winter with chills and a fever, that keeps children home from school and nursing homes on alert. They were watching warily for a novel strain of flu, a new virus against which no one on Earth has natural immunity.

In recent years, such a strain has emerged with the deceptive name of bird flu. All flu viruses in fact emanate from birds. But this particular strain was supposed to stay there. Instead it confounded scientists by repeatedly leaping the species barrier from poultry to people, often with lethal results. Now the agency's officers were searching for any unusual outbreak signaling a sinister mutation in this fatal strain, any omen of a gathering pandemic. A few genetic tweaks and millions could perish.

As an Asia correspondent for the *Washington Post* covering the spread of bird flu, I had learned not to underestimate it. Still, I was surprised by what I saw that morning in the bunker. At the top of the agenda were two reported deaths from avian influenza on the eastern edge of Europe, a continent that had so far been spared human cases. I was required to not identify the specific country but can say that these fatal infections, if confirmed, would have opened a new frontier.

Ultimately, after investigation, the bulletin was dismissed as unsubstantiated. But I was struck by the urgency with which it was handled. The information had originated with Russian medical staff working near the site of the suspected outbreak. From there, the reports had reached U.S. intelligence, which was growing increasingly anxious about the prospect of a flu pandemic. Intelligence agencies worry about its potential to disrupt the world economy, stoke political unrest, and escalate tension among countries vying for scarce medicine and other essentials that would run short. From the U.S. Office of the Director of National Intelligence, the information found its way to WHO headquarters, which in turn directed its European regional office to investigate. Within two days the report was run to ground. The episode underscored for me how gravely the pandemic threat was viewed by those in the know, even as media coverage ebbed and flowed.

By the time I sat in on the briefing, flu specialists had long since concluded we were closer to a global epidemic than we'd been in a generation. Even those who ran the command center, which integrates the best minds and technology the world can muster, conceded that they cannot prevent the inevitable. They didn't know when, and they didn't know whence. But the pandemic was coming.

---

Dr. Michael Ryan has hunted down some of the most virulent biological agents of modern times. He has parachuted into the world's bleakest corners to confront the horrors of Ebola and lesser-known hemorrhagic diseases like Marburg fever, Crimean-Congo fever, and Rift Valley fever—viruses so infectious, so devastating that doctors don biosafety space suits when they encounter these so-called special pathogens. As a member of WHO's special operations team, he has also helped stanch incipient epidemics of meningitis and cholera. He's braved guerrilla insurgencies, minefields, and wars on three continents to reach these outbreaks, dodging drug runners in Afghanistan during the Taliban era and Somali bandits in Kenya. He has crossed battle lines in the Ugandan bush to secretly win the cooperation of the cruel, cultish rebels of the Lord's Resistance Army. "I was shaking in my boots," he recalled. He fractured his spine in multiple places when his car was forced off the road in Kurdistan by an Iraqi military convoy during the rule of Saddam Hussein and careered down an embankment. Yet each time, Ryan's mission has succeeded. This has instilled in him an outsize sense of confidence. Until now.

Ryan is a burly, red-haired Irishman with lively hazel green eyes, full cheeks, and a gap-tooth smile. He grew up playing Gaelic football in what he describes as the "wild and savage" county of Mayo on Ireland's west coast, and he still has the build. His colleagues have called him the bulldozer for the way he muscles aside bureaucratic and logistical obstacles. More often, they call him the Irish traffic cop. After 2001 he took charge of the urgent comings and goings of scores of WHO teams dispatched to the field each year. As the indomitable director of Alert and Response Operations, he assumed overall responsibility for intelligence activities and emergency assistance to stricken countries, as well as the divisions dealing with specific diseases, including special pathogens and influenza. He is quick-talking and can-do, at times barely able to contain his restless intensity. His impassioned monologues cascade from topic to topic. His beefy hands punctuate each exuberant thought. His full, red eyebrows bob, and his chair creaks beneath his shifting frame. Though some agency old-timers initially dismissed Ryan and his crew as a bunch of cowboys, he has brought speed and smarts to fighting disease at a time when disease spreads faster than ever.

"We can do big operations. We *have* done big operations," Ryan put it to me shortly before I sat in on the agency's morning briefing. But influenza tests his optimism. "A containment exercise would be the biggest thing we ever attempted. We need to make that intervention at a higher level of speed, at a higher level of aggression than we might ever have done in the past."

Men like Ryan find it hard to admit the prospect of failure, much less its certainty. But pressed, he acknowledged that flu is so contagious and mercurial that a pandemic strain is destined to break free.

"That's the reality and I absolutely accept that," he finally said. "If you look at history, it's bloody well inevitable."

He moved to the edge of his chair and leaned forward.

"But what if we don't prepare for it? You have to make an honest attempt to try, understanding that by the time we find out about it, we may have lost. What I don't want to do is to try to find my pants in the dark when it does happen."

So Ryan had been doggedly reworking his emergency planning and reinforcing his team. He had brought on board a Canadian contingency expert seasoned in running large emergency exercises and a British specialist in heavy-lift logistics to oversee the delivery of the vast quantities of drugs and medical supplies that would be required by a budding flu epidemic. Ryan was also lining up other UN agencies with aircraft to ferry men and materiel in a crisis.

"You don't prepare to lose. You prepare to win," Ryan insisted. "You get your performance to the level that you think you need, right? You can do nothing about the performance of the other team. The other team is the virus. Its performance is set. So all you can focus on is your own ability to perform. You get your performance to the maximum you can achieve."

At best, that would mean slowing death's inexorable progress.

"If that's too little to stop the virus, fine," he conceded. "I will die in peace."

---

As I neared completion of this book's first edition in 2009, a novel strain broke on our shores. Rarely in public health had an event been so anticipated, yet it came as a surprise. The virus known as swine flu was a strain that global health specialists had not been expecting, that virologists had not been tracking. Most eyes had been turned to the East, the traditional wellspring of influenza viruses. Instead, this flu surfaced in Mexico and over the following months spread to more than two hundred countries and territories on six continents. At WHO, the full strategic operations center was activated.

This book, conceived long before swine flu erupted, is an account of why influenza pandemics are inevitable. The scientific and social forces that produced the global outbreak of swine flu were partly obscured because of its sudden, stealth birth. But they are on display elsewhere, notably in Asia, where humanity has been locked for more than a decade in a contest with flu, its most implacable viral adversary.

For forty years, the world eluded the inevitable. Scholars had urged that a pandemic was overdue. When swine flu broke out, it gave the world an initial fright but the first wave in the spring of 2009 proved exceptionally mild. So did a second later in the year. Whether even more waves are coming remains uncertain as I write now in the spring of 2010. Yet no matter how swine flu ultimately plays out its hand, it offers a warning: even after it's done with us, we're not done. The prospect of another pandemic remains real. And it could be far worse. Nothing about swine flu has altered the cruel dynamic of its avian cousin, which kills more than half of those it infects, or allayed the fears of the disease hunters who stalk it. There's no way to forecast how many lives this fatal strain will take and families it will destroy, how many hospitals and governments could tumble under its siege. The attributes of a novel strain that make global epidemics unavoidable also make its appetite impossible to predict. Yet today we remain closer than we've ever been to a repeat of the Great Influenza of 1918, which cost at least 50 million lives worldwide and killed more Americans than all the wars of the last hundred years combined.

In the first of the book's three parts, we meet the antagonists in this struggle, man and microbe, as they've confronted each other since the waning years of the twentieth century in what America's premier flu hunter, Dr. Keiji Fukuda, called an "accelerated number of near-misses." On one side—or on one team, as Ryan's sporting metaphor would have it—is the World Health Organization. In its own ranks and through external allies like the U.S. Centers for Disease Control and Prevention (CDC), WHO can marshal the world's most brilliant doctors and researchers, intrepid investigators, and sophisticated labs. The agency embodies our faith in rational science and our noblest aspirations for humanity. It may also be all that now separates us from cataclysm.

On the other side is flu. Ubiquitous in nature. Ever changing. Savage. No other virus can match its grim genetics. Given enough chances to remake itself, flu is guaranteed to breed a pandemic strain.

But microbiology alone does not account for the inevitability of pandemic. The virus must be afforded those chances to follow its logic. Today, they are greatest where flu's minute genetic idiosyncrasies meet the vast realities of Asia. The opportunities there are unprecedented and growing.

The second part of the book explores these ground truths. If the struggle to preempt the next great epidemic were waged just outside the doors of WHO's headquarters in Geneva, the flu hunters would face better odds. Switzerland, after all, is a citadel of efficiency, social order, and good government. It is famed for its precision watchmaking and global civil servants. But in the theater of conflict that is Asia, the battlefield is defined by poverty, superstition, unregulated development, and corrupt, parochial politics. Tradition often trumps science. Survival today trumps survival tomorrow. On this unfriendly terrain, the initiative lies not with the flu hunters. It belongs to the backyard chicken farmers, cockfighters, witch doctors, political bosses, and poultry smugglers.

The final part examines the limits inherent in the world's effort to prevent a pandemic strain from emerging and, failing that, snuff one out when it does. Over the centuries, flu has repeatedly preyed on a defenseless world. The first documented global outbreak came in 1580, starting in Asia and spreading west across Africa and Europe to the Americas. Lesser epidemics have occurred since the twelfth century.

Though scholars have yet to glean definitive proof of earlier outbreaks from the historical record, the virus no doubt accounted for a share of the myriad unidentified plagues that have been chronicled as far back as ancient Greece. Modern medicine, dating only to the nineteenth century, has a long way to catch up. The study of molecular genetics, born within the lifetime of many who are still alive, is in its infancy. As astounding as these scientific advances have been, they all too often fall short. Flu researchers working behind the double airtight doors of their biosafety labs, as well as clinicians on the wards, constantly confront the limits of their science, just as flu investigators run into the limits of epidemiological inquiry nearly every time they enter a village. The fight is also bedeviled by a second set of limitations that, while man-made, might as well be equally immutable. These are the constraints of money and mandate.

And with the coming of a pandemic, yet another set of limits threatens to exact a tremendous price. These are the bounds of the emergency planning we've done, the limitations of our downsized health-care system and outsourced drug industry, even potentially of our power supply and food and water. But that's a story history will write.

Now to Asia, where a fatal strain has raised its baleful challenge.

# PART ONE

# CHAPTER ONE

# The Revenge of Begu Ganjang

Dowes Ginting, the most wanted man in Sumatra, lay dying. He had abandoned the Indonesian hospital, where he had seen his family succumb one after another, and fled deep into the mountains, trying to outrun the black magic, seeking refuge in a small, clapboard home beyond the cloud-shrouded peak of the Sinabung volcano. For four straight nights, a witch doctor hovered above him, resisting the evil spell.

The medicine man mumbled the unintelligible words of an incantation. He doused Dowes with cooking oil and massaged his tormented muscles. He placed a hefty chunk of betel nut in the corner of his own mouth and chewed, his lips staining red from the juice of the mild stimulant, and then, with the precision of a surgeon, spit the concoction gob by gob over his patient's head, face, and chest and along his arms and legs. To relieve the surging fever, the witch doctor applied a pasty preparation of pounded rice and ginger called *beras kencur* to the man's damp flesh.

Dowes, a wiry thirty-two-year-old with a mop of dark hair and a whisper of a mustache, seemed at times to improve. His breathing grew less labored and his coughing less wrenching. At these moments, Dowes would hope his affliction was nothing more than the respiratory troubles he sometimes had as a teenager. But other times his condition would suddenly deteriorate. The witch doctor had to repeat the treatment several times in the course of a single night. He even lit some *kemenyan*, a pungent incense made from the balsamic resin of local trees and burned only when battling the most intractable spirits.

When the nighttime therapy was finished, Dowes would retire to a house just next door on a rise overlooking the village of Jandi Meriah. While the witch doctor's home was little more than a shack with a faded white exterior and rickety brown shutters, the neighbors' was more substantial. It had a jolly pink tile facade with a green tile terrace and even a working toilet. It belonged to relatives of Dowes's wife, and for the four days he sheltered there, they looked after him, washing him and cooking his meals. Mostly, he slept. But when he felt ambitious, he sat on the terrace and looked down on the rusting, corrugated metal roofs of the village and the tall stands of coconut palms.

It was to this remote hideaway that Dr. Timothy Uyeki tracked Dowes in May 2006. Uyeki had been dispatched to Indonesia by the World Health Organization after the agency's office in the capital, Jakarta, had put out an urgent call for reinforcements. Influenza specialists suspected that Dowes was carrying a novel flu strain. And he was on the loose.

Since bird flu had first jumped the species barrier from animals to humans in Hong Kong nine years earlier, the strain had infected more than two hundred people, yet it had reassuringly shown little aptitude for spreading from one person to another. Now the outbreak ravaging this one Sumatran family suggested the contrary. It was the largest cluster of confirmed bird flu cases in the world to date and the strongest proof yet that the virus could be passed among people. If the virus that Dowes was harboring had mutated into a more transmissible form, tens of millions of people worldwide could die.

Investigators would ultimately conclude that the Sumatra outbreak was not sparked by an epidemic strain. The toll would be confined to the Gintings. But the outbreak was the closest thing yet to a dry run for an emerging epidemic.

The experience was sobering. This was more than just a stress test of the world's emergency response. The episode, more generally, would cast doubts on some of the biggest bets humanity has been placing, raising profound questions about whether the WHO formula of brilliant scientists, courageous investigators, and modern technology was potent enough to break history's inexorable cycle of pandemics.

In the tight-knit world of highland Sumatra, it did not take long for Uyeki and a pair of Indonesian colleagues to trace the fugitive through his extended family, catching up with Dowes three days after he slipped into Jandi Meriah. Uyeki entered the front room of the hilltop refuge and spied Dowes lying by the window. He was sprawled on a thin, pink mattress against the wall, chest bare, his head propped up on a pair of pillows. His eyes were open. They were dark and sunken, and beneath them hung heavy bags left by the long nights of misery and magic. Above him, the curtains were open. Morning sunlight filled the room.

The patient's wife and several other family members took seats on a red rug along the far wall and eyed their visitors suspiciously. Uyeki sat down by the door. He started by offering his sympathy for the losses they'd already suffered. He explained he was looking for information. He wanted to help. It was a hard sell. Outsiders, even fellow Indonesians, were a rare sight in the village. They were viewed with distrust, even hostility.

Uyeki, an American, was an old hand at this kind of fieldwork. During his career as an epidemiologist with the U.S. Centers for Disease Control and Prevention in Atlanta, he had responded to Ebola in Africa, SARS in Asia, and anthrax attacks in Washington.

But flu was his passion. He had signed on to CDC's influenza branch in 1998 and was instantly dispatched to investigate an unusual summertime outbreak on cruise ships in Alaska that sickened thousands of tourists and tourism workers. In 2002 he headed off with an international team to the island nation of Madagascar, where thirty thousand villagers had come down with a baffling respiratory infection and hundreds had died. Though it was never publicized, some communicable disease specialists at WHO feared that at that very moment a global flu pandemic was hatching, not somewhere in Asia but in that African country's remote rural highlands. Uyeki pitched his tent there and camped for five days in continuous rain, collecting samples from the sick. The test results were ultimately reassuring, revealing that the outbreak was caused by an ordinary seasonal flu strain, exacerbated by Madagascar's crowded conditions, malnutrition, and cold, wet winter.

Uyeki is evangelical about flu. He calls it the Rodney Dangerfield of diseases. He is outraged that for years the virus was unable to garner much attention from researchers and policy makers despite its proven ability to kill tens of thousands of Americans each winter and potentially millions in a pandemic. He is emphatic about this point, as he is about many things. He is perpetually engaged and chronically restless. He keeps his sleeves rolled up and sneakers on his feet. Even into his late forties, this abundant energy, coupled with his intense brown eyes and dark, uneven bangs, lent him a boyish aspect.

Uyeki is also a practicing physician, a trained pediatrician, and, as he sat in that doorway in Sumatra, he understood that he was dealing with more than just the devilish twists of microbiology. He also had to attend to stigma and loss. "These were people who had already suffered a great tragedy," Uyeki later recalled. "This was a family that was trying to grieve. You needed to be empathetic and sympathetic. You feel a little reluctant to keep pushing."

As he spoke with the relatives, Uyeki tried to make eye contact, hoping to win their trust. But out of the corner of his eye he also studied Dowes. The ailing man had pulled himself into a sitting position. There was something odd about his breathing. His chest muscles seemed to be working too hard, too fast. Before long, Uyeki fell silent. He looked down at his watch. Then, for a full minute, he observed Dowes, counting each breath. A healthy man would breathe about fourteen or sixteen times in a minute. But Dowes exceeded that in half the time. Uyeki counted up to thirty-six. He knew then that the patient had pneumonia. It was almost certainly bird flu. But he'd need a specimen of virus to be sure.

Uyeki delicately advised the family that he wanted to take a sample.

"No blood," one relative objected.

"I'm not going to draw blood," Uyeki assured them. His aim was to swab the man's nose and throat.

"No equipment," a relative insisted, making clear they'd be horribly offended if Uyeki donned his protective gear.

It was a terrible demand. This was a monster of a virus that had killed most of those it infected. Flipping a coin would give better odds of survival. It was so deadly that scientists studying the virus were required to wear protective suits and work in special limited-access laboratories behind double doors with reverse ventilation to siphon off contaminated air. In the field, investigators were instructed to wear full outfits, including respirator masks, goggles, surgical caps and gowns, shoe coverings, and gloves. Uyeki usually snapped on three layers of gloves just to be sure. He always carried the complete gear in his backpack. This was going to be a tough negotiation.

"OK," Uyeki offered, "I won't wear a gown and I won't wear goggles." He would compromise. "But I'm going to wear a respirator and gloves." He figured the main hazard would be breathing in the virus or getting it on his bare flesh.

"Nothing," the relative repeated.

"No, this is what I'm going to wear," Uyeki responded, motioning to his face and hands.

They went back and forth. Eventually the family relented. Uyeki slipped on what he considered the minimum protection, a mask and gloves.

Uyeki had wanted one of the Indonesian doctors to help him take samples. But his colleague was petrified, rightfully so. He was just a local country doctor who had never confronted such a cruel and exotic killer. Uyeki tried to reassure him. Uyeki produced another mask and pair of gloves from his backpack and passed them to his reluctant partner, then explained step-by-step what they were about to do.

On Uyeki's instruction, the Indonesian physician lifted a Q-tip–like swab to Dowes's left nostril and gingerly inserted it. Then he handed it back to Uyeki, who carefully snapped off the top, placed it into a tube containing sterile media for transporting virus samples, and screwed on the cap.

Next Uyeki sampled the right nostril himself, repeating the procedure.

Finally Uyeki prepared to take the throat specimen. This would be tougher. He would have to reach into the windpipe with the swab and fish out a glob of mucus. It would be absolutely loaded with virus. The pathogen had already been replicating inside its host for a week, and Dowes would be teeming with infection.

Uyeki squatted beside him and leaned right in. The doctor's eyes were just inches from those of his patient. Uyeki lifted the swab. Then, he carefully inserted it through the open mouth and down Dowes's throat.

The sensation must have tickled. For at that very moment, Dowes coughed. And when he did, it was right in Uyeki's face.

Uyeki didn't blanch. But inside, his stomach dropped. "Oh, this is not good," Uyeki fretted to himself. Despite the mask, most of his face was exposed. His mind raced. He instantly thought about his unprotected eyes. He knew other types of avian flu viruses had been contracted through the eyes. Could the same be true for this strain?

He tried to stay focused. He had long since learned to remain calm no matter what. Countless times in his career he had resuscitated children when one slight miscue or false start might mean their death. This time he stared past his own mortality. Later, after leaving the village, he would urgently dose himself with antiviral drugs, taking four times the usual amount prescribed for prophylaxis, and acknowledge he'd made a mistake. He would promise himself never to forgo goggles again.

But now he had to rein in his rising anxiety. Slowly, he withdrew the swab from the Dowes's mouth and backed away.

Uyeki placed the final, hard-won sample in a tube. With a Magic Marker, he labeled all the specimens. Next, he removed his gloves, then those of his Indonesian partner. He stuffed them in a bag he had brought specifically for disposing of dirty material. He sealed it tight.

Then he turned back to the row of family members seated along the wall with another request. He told them he now wanted to briefly examine Dowes and asked for permission.

"All I want to do is listen," Uyeki appealed, showing them his stethoscope. "I just want to listen to him. No blood. I'm not going to do a full exam." They agreed.

Squatting next to the mattress, he listened to Dowes's chest. He listened with the patient sitting up and with the patient lying down. Uyeki heard an odd crackling sound, a kind of coarse gurgling each time Dowes inhaled and each time he exhaled. Uyeki had heard this in other patients, but never this loud. It was the sound of fluid, of a man drowning.

"He has pneumonia and he has it in both lungs," he reported to the relatives. "I strongly recommend he go to a hospital."

From beneath his straight black bangs, Uyeki's detective eyes watched them, trying to gauge their reaction. But he knew what the answer would be. The relatives refused. They insisted that Dowes had to complete two more days of traditional treatment. Uyeki had no choice but to defer to their wishes. He didn't want to make a scene. Nor did he have any way of forcing them to comply. Though he had come to the village with a security officer, the escort's charge was limited to keeping Uyeki safe should the locals turn unruly. So after only two hours in the village, the medical team was back on the road in their four-wheel-drive vehicle, skirting the volcano, bound for the district capital of Kabanjahe twenty-five miles away.

But before they left, Uyeki urged the relatives to get Dowes to a hospital if his condition worsened. Uyeki suspected it would. It was clear Dowes was going to die. Uyeki did not realize how quickly.

It had been nearly a year since Indonesia had confirmed its first human cases of bird flu. A government auditor in the suburbs of Jakarta and his two young daughters had died abruptly in July 2005. The outbreak was unnerving on several counts. Not only were disease specialists unable to determine the original source of infection despite weeks of investigation, but the sequence of the three cases strongly suggested that the virus had been passed from at least one victim to another. Perhaps most worrisome was that the virus was now striking people in Indonesia at all. The strain had already been exacting a toll elsewhere in Southeast Asia, but Indonesia, a country of vast size, sprawl, and poverty, was the most daunting front so far.

Then, less than a month later, the world got some surprisingly good news from a pair of international research teams. By applying cold, hard numbers and sophisticated computer models to the outbreak in Southeast Asia, the two teams had each determined that the world could nip a human flu epidemic in the bud given the right mix of antiviral drugs, quarantines, and social controls. This would be unprecedented. Mankind had never before been able to stem an emerging influenza pandemic but throughout history always had to ride out the storm, waiting for the plague to burn itself out. Now, armed with high-speed computers, advanced statistics, and modern medical science, humanity might be able to bend nature to its will.

At Imperial College London, computational biologist Neil M. Ferguson led a group of researchers in simulating an emerging epidemic in rural Thailand. They chose Thailand because detailed demographic data was available. But the researchers said the conclusions could equally apply to similar low-density communities in Southeast Asia. Starting with a single human infection, Ferguson and his colleagues found that the virus remained limited to the local area for about thirty days before shooting like buckshot across the rest of the country over the following two months. They said this created an opportunity to contain a nascent pandemic at its source if about 2 million courses of the antiviral drug oseltamivir were rushed to the scene and dispensed to those within three miles of the initial outbreak. Oseltamivir is marketed by the Swiss drug company F. Hoffmann–La Roche Ltd. under the brand name Tamiflu. Officials would also have to shutter schools and workplaces to limit human contact and would have to place the entire area under quarantine.

A separate team led by Ira M. Longini Jr., a biostatistician at Emory University in Atlanta, ran a slightly different model that also simulated an outbreak in a theoretical population based on data from northeastern Thailand. These researchers likewise examined the impact of antiviral medicine, limits on human contact, and quarantine and came essentially to the same encouraging conclusions. Their analysis was even more optimistic, finding that no more than 1 million courses of Tamiflu might be needed.

Together, these two studies, published almost simultaneously in the rival journals *Nature* and *Science,* offered the WHO a strategy for confronting a pandemic. It was perhaps the only hope, and WHO was quick to integrate the findings into its emergency planning. "No attempt has ever been made to alter the natural course of a pandemic at its start and the behavior of influenza viruses is hard to predict. Nevertheless, the rapid containment approach is considered one of the key elements of pandemic influenza preparedness," WHO wrote in describing its new strategy.

But the researchers made clear that success rested on some very perilous assumptions. The nature of the virus itself was a crucial variable. If the epidemic flu strain were good at reproducing itself, say for instance if each sick person on average infected at least two others, containment would be unlikely. If, however, the strain were less contagious, with each sick person infecting fewer than two others, chances of checking the disease would be better.

Success depended equally on the speed and effectiveness of the response. Ferguson's team stressed that health authorities would have to almost instantly recognize the initial case of an epidemic strain and detect most of the subsequent ones. They would also have to identify nearly everyone who may have had contact with the victims, including relatives, schoolmates, work colleagues, and anyone else who had ridden in the same bus, shopped together in same market, eaten in the same restaurant, or prayed in the same mosque, temple, or church, and then get Tamiflu into their hands. The drugs would have to reach an overwhelming majority, preferably more than 90 percent. Longini's group concluded that containment could only work if "the stockpile were deployed at the source of the emerging strain within two to three weeks of detection." Any longer and the highly infectious virus would breach the containment zone, confronting the world with an unstoppable epidemic.

WHO's emergency containment plan incorporated this narrow window, stressing that success depended on mass administration of antiviral drugs within three weeks of the "timely detection" of the first case of this new epidemic strain. A twenty-day supply of Tamiflu would have to reach 80 percent of the targeted population, defined as everyone within at least three to six miles of every detected case, or between ten thousand and fifty thousand people in each area at risk. This would require a massive global undertaking to tap international stockpiles of Tamiflu, airlift the drug into the country, ship it to the hinterlands, and distribute it among the locals. (If the outbreak occurred in a city, such efforts would be pointless. The population density would offer so much dry tinder to the virus that a global conflagration would be unavoidable.)

WHO would rely heavily on the government of the stricken country to quickly detect and honestly report any suspicious outbreaks, initiate the containment effort, and then efficiently transport antiviral drugs to the affected area, no matter how remote. Movement in and out of the containment zone would be sharply restricted. Cooperation from the community would be essential. People would have to stay put, stay home, and take their medicine as prescribed. The sick would be isolated, those exposed to them quarantined. Lab testing of their samples would be crucial for tracking the evolution of the outbreak and calibrating the response. To succeed, this highly disciplined operation would have to last a month or more. "It will require excellent surveillance and logistics mechanisms as well as an ability to ensure compliance with policy directives," WHO acknowledged.

The power of the computer simulations was awesome, and the chances they promised humanity to dodge a pandemic were previously unimaginable. But what would happen when they collided with the realities of Asia?

In the weeks before Dowes was tracked to his mountain hideaway in May 2006, he had watched sickness burn through his family. It had taken his older sister, his younger sister, and three of their children. His own ten-year-old son had held on longer than most. Dowes and his wife barely moved from the boy's bedside, but eventually, after suffering eleven days of fever and ever-greater difficulty breathing, he, too, succumbed.

The members of this Sumatran family had perished despite the efforts of the best hospitals in the island's main city of Medan, the third-largest in Indonesia. Though overseas laboratory analysis in Hong Kong had confirmed the cause was bird flu, Dowes did not believe it. When his relatives took antiviral drugs as prescribed for bird flu, they only got worse. When they gave blood to be tested for the virus, they only grew weaker. Nearly everyone who entered the hospital was fated to die, not to survive. So when Dowes fell sick two days after his son's death, he decided to make himself scarce.

The outbreak in his family had already attracted international attention because of the cluster's size. The WHO had sent Uyeki to Indonesia along with the agency's premier epidemiologist, Dr. Tom Grein. An intrepid German investigator, Grein had repeatedly responded to some of the world's most terrifying outbreaks, including Ebola and Marburg epidemics in Africa and mysterious clusters of respiratory disease in remote areas of northern Afghanistan and southern Sudan. He and Uyeki had worked together before and were familiar with each other's routines.

When Dowes fell ill, WHO's flu specialists surmised he was another link in an unprecedented chain of human transmission. Now his disappearance was setting off warning lights well beyond the islands of Indonesia. He was on the move and potentially contagious with what seemed to be the most transmissible strain of the virus yet. This was the way epidemics started. Indonesian authorities put out a bulletin to local health officers: Find Dowes. Uyeki and Grein had reached Sumatra the very day he went to ground. They joined the hunt.

Dowes and his family had rented a public minibus to smuggle him even deeper into the wilderness. They drove along a rough country road that cut through orange orchards, fields of high corn, and glistening terraces of rice. They passed the occasional brick church, onion-domed mosque, and villagers drying tobacco on wood frames set out in their yards. In places, tall brush brimming with wildflowers pressed up against the roadside. The green slopes of Mt. Sinabung beckoned in the distance, its upper reaches in shadow, the lower ones in crisp sunshine. After several hours they reached their destination.

Jandi Meriah is built on the opposing slopes of a slender valley bisected by the Loborus River, which courses through a deep, lush ravine. A bridge connects the two halves, and a dirt and gravel road ascends each side, meandering through the hamlet to the fields and orchards above. The villagers grow oranges, durian fruit, corn, chili peppers, and chocolate. But Jandi Meriah is particularly famed in North Sumatra province for its witch doctors.

When Dowes arrived at the small, clapboard dwelling three days after falling ill, he could still walk and talk. He was feverish and coughing badly, but he seemed relieved to be there. The reclusive witch doctor, or *dukun* in Indonesian, agreed to treat him and said there would be no charge. It is believed that demanding payment is the surest way for a sorcerer to lose his mystical gift, though gratuities like cigarettes are always welcome.

The witch doctor's formal name was Suherman Bangun. The villagers, with a mixture of affection and awe, called him Pak Dirman. Bangun had been born forty-three years earlier in Medan before retreating into the mountains to farm corn and exercise his mystical talents. Though Bangun was a Muslim, many of those who sought his help, including Dowes, were Christian. "His reputation is very famous," Bangun's wife, Rintang Boru Ginting, boasted to me when I later caught up with her in the village. "His patients come from many faraway places." Rintang herself had been his first patient, she recalled. Not long after the birth of their second child, he had removed a tumor from her neck, carving it out with a razor blade, then sealing the flesh by spitting betel nut juice on it and reciting a spell.

Dowes proved to be a more trying case. The nighttime treatments failed to dispel the affliction. Each new dawn found Dowes fighting harder for his breath. Late on his fourth night at the witch doctor's home, hours after Uyeki had made his futile appeal to the family, Dowes got up to use the bathroom. Since there was no toilet there, his uncle wrapped his arm around him and helped him walk next door. Dowes could hardly breathe. He was staggering, on the brink of collapse. His condition had taken an abrupt turn for the worse since the previous day. "He was almost unconscious," one relative later recalled.

Bangun advised the family there was nothing more he could do. So the uncle wrestled Dowes to the Suzuki jeep parked out front. Together with an aunt and two cousins, they drove out of the valley and began racing through the dusky daybreak back toward Kabanjahe and the district hospital. Before they made it, Dowes died.

In the highland village of Kubu Sembilang where Dowes and his family had lived, the Suzuki jeep was a rare luxury. The vehicle had originally belonged to Dowes's father, the patriarch of the Ginting clan, until he died in 2001. Though most everyone in Kubu Sembilang came from generations of hardscrabble peasants, Dowes's father, Ponten, had also been a gangster. In fact, in that corner of Sumatra, he had been the godfather, running protection and extortion rackets throughout the environs. It was said he had won his power by conjuring evil spirits. His criminal pursuits had earned the family a measure of respect and riches. But the Gintings remained essentially farmers, spending their days in orange and lime orchards and fields heavy with red chili peppers.

Ponten had five children, three boys and two girls, with Dowes in the middle. The eldest sister, thirty-seven-year-old Puji, peddled the family produce at a traditional market in the neighboring town. Villagers from the surrounding area of Karo district came to buy clothes, shoes, and other dry goods from stalls in the heart of the market. Puji's stand was near the edge with the other fruit, vegetable, and meat mongers, about twenty yards from a kiosk hawking live chickens. After she fell fatally ill, WHO investigators hypothesized she had contracted the flu virus at that nearby poultry stall, though samples later taken in the market all came back negative. Investigators also suggested she could have been infected by contaminated chicken droppings that she used as fertilizer in her garden.

Puji started feeling unwell on April 24, 2006, less than a week before the people of her village were to celebrate their annual harvest festival, Merdang Merdem. She began to cough and her temperature climbed. Her sister massaged Puji's throbbing muscles and tried unsuccessfully to nurse her to health. Her family decided to proceed with their thanksgiving anyway.

Tradition is an essential ingredient of daily life in the district. The Karo people are Bataks, a distinct ethnic group within Indonesia known for a stubborn streak and a fierce adherence to their own culture and beliefs. In the past, the Bataks were storied as ferocious warriors. Today they are leading officers in the Indonesian military and some of the most formidable lawyers in Jakarta. Inevitably the fearless bus drivers who career through the capital's tortured traffic are also Bataks. While the Javanese of Indonesia's main island are refined, aloof, and at times duplicitous, the Bataks can be coarse but also refreshingly candid.

The festival brought together all five Ginting siblings, their spouses, and their children. They gathered at the small house, facing the village soccer field, where Puji lived with her husband and three sons. It was a wooden dwelling with a sun-bleached blue facade and pitched corrugated metal roof crowned by a satellite dish. Chickens rooted around in the dirt out back. Directly next door, her sister lived with her husband, two daughters, and son in a larger, more robust, cream-colored home built from concrete and tile, also capped by a satellite dish. Just beyond that was where her brother Dowes resided with his wife and two sons. His home was the most modest of the three, a small, unpainted clapboard structure.

The celebration began just after sunrise and continued until nightfall, the family feasting on curried chicken and roasted pork served in a spicy sauce prepared from pig's blood, while the children played in the dirt road out front. In olden times, the harvest festival would last seven days, culminating with a communal feast and folk dancing that doubled as matchmaking for the girls and boys of the village. The holiday had in recent years been condensed to one day. But as they always had, relatives who'd moved away from the village came back to celebrate. So the youngest of the Ginting siblings, a twenty-five-year-old kid brother named Jones, who had left the village to start his own family in the district capital, returned with his wife and two sons.

Once the party finished, it was too late for Jones and his family to go home. They stayed over at Puji's house, sleeping in the same confined quarters with her and her children and two other relatives. It was a rough night, nine people crammed into one small space with Puji coughing uncontrollably.

Three days after the festival, the family brought Puji to the emergency room at the district hospital. She could hardly breathe, and her skin was turning blue from a lack of oxygen in the blood. The doctors gave her medicine and advised the family to take her to Santa Elisabeth Hospital in Medan, about fifty miles away along the coast, where she was admitted into the intensive care unit. Santa Elisabeth was one of the best medical facilities in Sumatra, a seventy-five-year-old Catholic hospital opened by Indonesia's former rulers, the Dutch. Inside the colonial structure, the black-and-white marble corridors were spotless. The interior courtyards were meticulously maintained with manicured lawns, sculpted shrubbery, and flowering bushes. Patients and their families waiting in the main lobby, with its high stained-glass windows, spoke only in hushed tones. By contrast with the crowds and chaos at Indonesia's public hospitals, everything about Santa Elisabeth was orderly and organized. But that wasn't enough. On May 4, after only twelve hours in intensive care, Puji died. The precise cause of death was never determined. No samples were taken, and no one suggested at the time it could be bird flu.

But by the time Puji was buried a day later, six other family members were already sick. Most had gone to their village's rudimentary health post complaining of fever and breathing difficulties, but found no relief. As they continued to grow weaker, even talking became a chore.

"The doctors were shocked," recalled the district health chief, Dr. Diana E. Ginting, who was no relation to Puji and her family. "The chest X-rays were all white." Their lungs had filled with fluid. The clinic doctor was the first to suspect they had caught a novel flu strain and sent them to Adam Malik General Hospital in Medan, which had been designated by the Indonesian government as the referral facility for all bird flu cases in the region.

As soon as they arrived, it was clear they desperately did not want to be there. "They didn't accept that it was avian influenza," recounted

Dr. Nur Rasyid Lubis, the hospital's deputy director and head of its outbreak response team. "They tried to refuse all treatment. We could treat them only because their condition made it impossible for them to resist. But if they had been healthy enough to walk, they all would have run away."

The family members balked at giving blood for testing. Several initially refused to go to the isolation ward set up for suspected bird flu cases. "None of us believed it was bird flu," Puji's sister-in-law later recounted. "We thought it was black magic. Everyone in the family was getting sick and no one else was. Someone had put a spell on our family."

That account was offered to me by the wife of Puji's younger brother Jones. I must have given her an inquisitive look because she added, "Black magic is very common in our place."

True to local practice, relatives and in-laws had streamed out of the mountains to keep vigil at the hospital. They crowded into the large room where their kin were lined up in two facing rows of metal cots and scoffed at the insistent advice of hospital staff that they put on masks and gloves to protect themselves.

Puji's oldest son died just one day after being admitted to the hospital. One of Puji's sisters succumbed next.

As Jones watched his kin perish in the beds around him, he confided to his wife, "If I stay here, I'll be dead." Jones's fever had topped 102 degrees and showed no sign of breaking. Fresh X-rays revealed that his left lung had entirely clouded over and his right lung was nearly as bad. He told his wife it was time to leave.

"I was losing hope," Jones later recounted in a raspy voice, almost a whisper. His stare was blank beneath thick eyebrows, and his cheeks were sunken. He had been a young tough. Now his arms and legs, covered with elaborate swirling tattoos of red and green, had grown emaciated. He had dropped twenty pounds during his struggle for survival, and his tank top hung off his shrunken frame. "I was afraid of the hospital," he continued. "I thought I was going to die. I can't remember much, but I remember running away."

At that moment in the retelling, Jones fell silent. His wife, a chatty twenty-three-year-old with long hair parted in the middle and tied in the back, picked up the thread. "We were watching our family die one by one," she recalled. "We thought the hospital could not cure him."

Jones had slipped out of the hospital with his wife and two young sons and hailed a taxi on the street. They told the driver to take them to a congested, working-class quarter of Medan jammed against the edge of the city airport. The houses were jumbled together along narrow alleys. Young men lingered on cramped front porches. Drainage ditches ran past the doorsteps and chickens pecked in the fetid black water. Every few minutes, an airplane would pass overhead with a tremendous roar. A year earlier, an Indonesian jetliner had crashed into the neighborhood moments after taking off, killing at least 149 people, including dozens on the ground, yet the incident was soon all but forgotten outside the quarter. This was the kind of place where Jones could disappear.

Jones also had relatives there to take him in. One, a distant relation, was a doctor with a practice on the main boulevard. Jones went to see him. The doctor examined Jones, prescribed some medicine, and offered an unwelcome diagnosis. "You have bird flu," the doctor said, urging the feverish fugitive to return to Adam Malik Hospital. Jones balked.

If Jones could not find the answer in the city, he reasoned, he should go home to the mountains. He and his family hired another taxi and fled Medan. They drove deeper and deeper into the jungle, following the steep winding road and treacherous switchbacks up and up and up to the highlands of Karo, where cool breezes offered relief and promise of a cure.

By the time health officials in Sumatra realized they had a problem, it had already been two weeks since the eldest Ginting sister, Puji, fell sick, and six others were already desperately ill.

District health and veterinary officials drove to Kubu Sembilang and summoned the villagers. "Bird flu is here, now, among you," an animal-health officer told the hundreds of people who gathered at the village hall. "To break the chain of infection, we have to depopulate your livestock."

The cooperative spirit lasted precisely one day. Trouble began when provincial veterinary officials drove up from Medan, pulled rank on their local counterparts, and announced that all the samples taken from local poultry were negative for bird flu and the mass cull, though prudent, had been unnecessary. The villagers were livid.

Kubu Sembilang was in revolt by the time a team of disease investigators from the provincial government arrived in the highlands later that week. Local officials pleaded with them to go home. "We're frightened something might happen to you," one local health officer warned them. "You might be attacked." The investigators retreated.

A day later, the provincial team returned reinforced by disease specialists who had flown in from the national health ministry in Jakarta. This time they managed to enter Kubu Sembilang. But the villagers refused to answer any questions: What was the possible source of the infection? How did the victims catch it? Was anyone else in the village sick? After several frustrating days, the team from the national health ministry withdrew to Jakarta and never came back. They were too scared. Villagers had threatened to stab them.

"We chased them away," Puji's brother-in-law admitted to me when I met him a few weeks later. The man had a hard face and bitter eyes. His tone grew more menacing the longer I peppered him with questions. "There is no such thing as bird flu. It's bullshit," he continued. "This is all simply black magic."

That last line was recited to me over and over as I explored Kubu Sembilang. There were no sick chickens in the village and only one sick family. What else could it be but an evil spell?

A neighbor laid it all out for me. "This is something supernatural," began Tempu Sembiring, a fifty-year-old orange grower with a balding pate and salt-and-pepper mustache. An unlit cigarette bobbed between his lips as he recounted the tale. "Their father used to be the chief gangster in this area. He took his power from a spirit called Begu Ganjang at a holy place in Simalungun. This is revenge for worshipping at that place."

According to Batak lore, Begu Ganjang can be conjured by offering him the liver, heart, and gizzard of a red rooster wrapped in four banana leaves along with seven pieces of chili pepper and chunks of banana. These are to be placed around the perimeter of a late-night ceremony, at the four points of the compass. The rooster's blood is drained into a coconut husk at the center and then the spirit summoned. He is believed strongest at a mystical site in the district of Simalungun on the shores of Lake Toba, in the heart of the Bataks' highland domain. Once called, Begu Ganjang will do his master's bidding, delivering wealth and protection. But the spirit will remain faithful only as long as periodic oblations are offered, and unpaid debts to Begu Ganjang can be collected from kin in blood.

"Ponten was quite powerful," Sembiring continued, lowering his voice to a whisper as he mentioned the patriarch's name. "When he died five years ago, his friends came all the way from Jakarta. The condolence flowers stretched all along the road. They stretched from there," he said, pointing at the Ginting homes in the distance, "all the way to here." He scanned his finger past the soccer field, the village church, and the dirt track leading up into the orchards. "The family had to be sacrificed. It is a payback for what happened in the past."

The village of Kubu Sembilang was in full rebellion as the investigation moved into its second week. WHO had received final lab confirmation that the outbreak was caused by bird flu. But Karo district residents objected that their poultry were not to blame, and, in mid-May, half a dozen farmers beheaded a chicken and drank its blood to demonstrate that their birds were healthy. A week later, with Indonesian officials designating Karo a bird flu–infected area, scores of poultry traders furious over the potential loss of revenue descended on the provincial capital, Medan, again slaughtering chickens and drinking their blood in protest.

The WHO team's initial effort to visit the village was "significantly delayed" by security concerns, according to an internal agency report. "It was later learned by WHO that quite strong hostility was expressed by some villagers, including a threat to bodily harm of laboratory staff," the report said.

When the WHO investigators, accompanied by Indonesian health officials, finally entered the village nearly three weeks after Puji fell sick, they recruited twenty local volunteers to monitor fellow residents for fever and set up a temporary health post on the soccer field, offering free medical care. The investigators bravely pieced together the chronology of the outbreak. They tried to collect samples from family members to see how widely the infection had spread. But, the WHO team reported, "Such requests were universally refused." The investigators traced those who had contact with the victims, providing them Tamiflu for protection. But many of those who were closest to the Gintings refused to take it.

Before I'd set off myself for Kubu Sembilang, I had worried I'd also get a hostile reception. A relative of the Gintings in Medan had given me the name of an uncle in case I needed help. When I got to the village, the man was away. But an aunt and cousin politely greeted me and invited me into the spartan eatery they ran, offering roasted pig and blackened dog. Then they served me up an earful.

"The doctors gave us Tamiflu but we didn't take it," recounted the aunt, Mamajus Boru Karo. She was thirty-eight years old with long filthy hair, parted in the middle. Her face scrunched up as she retrieved the unpleasant memory. "Why should we take it? We don't have bird flu."

"The doctors said we should take it as a precaution," offered Sanita Ginting, the twenty-nine-year-old cousin, her brown eyes opening wide.

"We're healthy," Mamajus retorted. "If we took it, we could die."

Sanita nodded in agreement and acknowledged she, too, had thrown aside the medicine. Seated at a scuffed table by the window, her gaze drifted toward the semipaved street outside. A few goats, chickens, and dogs scavenged among the potholes for discarded morsels of food. A highland breeze chased scraps of litter down the road. Sanita's attention returned to the cramped, concrete room with its chipped green paint. She, too, was getting agitated.

"We were scared of the health officials and of the WHO," Sanita admitted. The doctors had given her relatives medicine and shots, and yet they got worse and worse until they died. "It's possible they murdered them with injections."

"Why would they do that?" I asked.

"To keep them from infecting anybody else," she answered.

I looked at Mamajus. She was seated cross-legged on a wooden chair in the doorway. She concurred.

"Of course we're suspicious," Mamajus said. "They told us it was bird flu even before they had any lab results. They were just making it up."

"They think we are stupid people, uneducated," Sanita interjected. "We didn't go to advanced schools like them so they think they can say anything. But if no poultry are sick, how can it be bird flu? It's irrational. We can think even if we're stupid. Where's the proof?" Then came the refrain: "It was black magic."

When I had spoken earlier with provincial health officials, they told me they had tried to take blood samples from everyone who had contact with the victims.

"I didn't give a sample," Sanita responded.

"I didn't give a sample," Mamajus echoed. "I don't know anyone in my neighborhood who gave blood samples." She paused, glancing at Sanita for confirmation, and then continued, "We got angry. We said, 'Why do you want to take our blood? That's only for sick people.'"

Without specimens from the villagers, the world's picture of this unprecedented outbreak was, at best, incomplete. I decided to ask the health officials from Karo district itself. They were the ones closest to the scene and fluent in the local dialect. I finally caught up one evening with Dr. Diana Ginting, the health department director, for a talk literally in the shadows. She was still on duty despite the late hour, her jacket pulled tight against the chill, because a new bird flu outbreak in poultry had been confirmed hours earlier. As we sat on the dimly lit steps of the government building in Kabanjahe, the once-sleepy parking lot in front of us had been transformed into a war zone. The district was mobilizing for a mass cull, marshalling scores of uniformed security forces and civilian officials to carry out the late-night operation.

As Dr. Ginting distributed surgical masks to the gathering forces, she recalled what had earlier transpired in Kubu Sembilang. The doctors had tried to calm the people and coax them into giving blood. "It wasn't easy," Dr. Ginting said. "They kept wanting to know why."

When the medical staff had put on masks in preparation for drawing blood, many villagers took offense. Nearly everyone refused. Across the province, health officers were eventually able to collect thirty-two blood specimens, including many from doctors and nurses who had treated the victims. But in Kubu Sembilang itself, investigators ultimately took just two samples, from a neighbor and the local midwife. "The people here are very difficult," Dr. Ginting said. "Maybe we weren't good enough at convincing them."

Tortured by the diagnoses of big-city doctors, Jones, the youngest of the Ginting siblings, had bolted the hospital, fled the suffocating back alleys of Medan, and absconded for the highlands of his ancestors. But he arrived shadowed by death and local officials.

He initially sought out a witch doctor who lived in a hamlet amid the cornfields about ten miles from his home. This healer poured some water into a cup and recited an incantation. Then he gave it to Jones to drink. Next, applying the standard treatment for a fever, the medicine man prepared a paste of *beras kencur* from crushed rice and galangal (a root similar to ginger) and rubbed it on his patient's face and body. The remedy did not work. Jones was on fire.

His neighbors were not happy to see Jones when he returned to his home in the district capital, Kabanjahe. They were terrified of the curse. They had also heard that Jones had become the target of a major manhunt. The provincial health department, alarmed that a contagious patient was on the loose, had asked local officials to track him down and ship him back to the hospital in Medan. The word on the street was that police had also joined the hunt for Jones and his family.

"They'll arrest us if we don't go back," Jones's aunt warned him.

Jones objected, "If I go back there, I'll die for sure."

Jones and his wife debated what to do. The family told Jones they would defer to his wishes. But with his condition deteriorating, he was no longer in any shape to keep running.

After three days on the lam, Jones was readmitted to Adam Malik Hospital, a sprawling urban medical complex of white buildings with red tile roofs that, like many public hospitals in Indonesia, was in need of a little convalescence of its own. The corridors were clean, but their tile floors were chipped and the air was sour. Many of the fluorescent lights were out. While groundskeepers kept busy raking the lawns, the grass was perpetually overgrown, and many of the bushes were draped with drying laundry. The hospital was named for a former Indonesian vice president. But a letter had fallen off the sign over the main entrance so that it now read ADA MALIK, which is Indonesian for THERE'S MALIK!

Though Jones had surrendered to health authorities, he was not done fighting. He refused to take Tamiflu or accept intravenous medication and injections of antibiotics. As hard as it was to breathe, he would rip the oxygen mask off his face. The nurses were terrified of him. When they came close, he would wrestle them away, flexing his tattooed biceps, or lunge at them with a fork.

The doctors treating him in the special "red zone" set up for bird flu patients looked around and found a young resident physician whose family was from Karo district. He could speak to Jones in his local dialect. Soon the resident won a measure of cooperation. But only a measure. Jones began taking his medicine but still resisted giving the blood samples required to monitor his progress. "He thought it would make him weaker. For people like him, one drop of blood is worth a plate of rice," one doctor told me.

The red zone was demarcated with a red line painted on the hallway floor outside the two rooms where the most infectious patients were segregated. Hospital staff donned masks, helmets, rubber gloves, gowns, and boots before crossing the line. But as Jones slipped in and out of consciousness, now hooked up to a ventilator, more than a dozen unprotected family members pitched camp on the outdoor terrace just beside his room. They filtered in and out of the ward all day long, wearing neither masks nor gloves despite the timid protests of the medical staff. "We weren't afraid of catching the disease," his wife explained. "If we have to die, we'll die. No one else would look after him. Even the nurses were reluctant to look after him."

Dr. Luhur Soeroso, the silver-haired chief of the pulmonary department, took a deep interest in Jones. "This is a very rare case," he told me as he thumbed through the patient's file. When I met Soeroso, he had an air of composure and comportment that seemed out of place amid the bedlam beyond his office door. He wore a pink shirt with a pair of Mont Blanc pens in his breast pocket and a red tie fastened with a gold clip. On each hand he wore a gold ring, one inlaid with small diamonds, the other with a large black stone. He had just completed writing an article about Jones that he was hoping to publish in the *New England Journal of Medicine*. While all the other family members who caught bird flu had died within twelve days of falling sick, Soeroso noted that Jones was still alive and battling the virus six weeks later.

Jones was also unusual because he had developed brain abscesses after three weeks in the hospital and was suffering from severe headaches, mental distraction, and crippling muscle pain. Soeroso attributed this to a secondary infection caused by parasites that Jones had long ago picked up in his village. The intense battery of antibiotics and other medication had depressed his immunity, allowing the parasites to attack the central nervous system. He had tested positive for a parasitic infection called toxoplasmosis. A CT scan had revealed multiple pockets of pus in his brain.

The family's stubborn resistance to blood tests had made treatment difficult. But then the medical staff struck a deal with the family. If Jones would agree to give blood specimens, the family could invite a witch doctor to the hospital to see him.


Agenda Purba was laboring in his rice paddy when the Mitsubishi van drove up. He looked over to see who it was. Several passengers were strangers, but he recognized at least one man from his village. The visitors said they had a sick relative in Medan and needed Purba's help urgently.

"Can't you bring the patient to me? I have no way to get there," Purba demurred.

"He can't leave the hospital," one visitor responded. "We'll take you to him."

Purba was in no mood to make the long trek to the coast. It was already late in the afternoon. "I'll give you the right ingredients so you can do it yourself," he said.

"No, please, can't you do it the first time?" the visitor pressed.

Purba thought it over and agreed. "But on one condition," he added. "Bring me all the way home again."

Along with Suherman Bangun, Purba was one of the most prominent witch doctors in the village of Jandi Meriah. He was a skinny fifty-six-year-old with disheveled, graying hair and an oddly elongated face, carved with deep creases. A few wispy white beard hairs sprouted from his chin. His manner was agitated, almost manic, and when he spoke, he would look into space and squint, his heavy lids settling over melancholy, light brown eyes.

Even as a boy, Purba had realized he was special. He could always climb the tallest coconut trees in the village. But it was only when he turned twenty-one that he became a mystical healer. "I got the inheritance from my grandfather. He had this power. Before he died, he gave me the power by touching my left arm," Purba told me, rolling up his sleeve, extending his arm and tapping it with his right knuckles to illustrate the bequest. "I can still communicate with him through dreams."

Within years he became famous for his specialty: curing hernias. "If someone's colon is sticking out their anus, I chew up some betel nut and spit on his stomach," Purba explained with the cool detachment of a doctor describing an upcoming outpatient procedure. "I don't have to touch him. I just light a cigarette and hold it. When it burns down to my fingertips, that's the moment the colon goes back inside the body. But I can make it happen whenever I want. If I say you'll heal tomorrow, you'll heal tomorrow. If I say 'heal now,' you'll heal now." Over the years, Purba claimed his reputation had attracted patients from as far as Medan and even Java island. "I can do hernias, snakebites, and scorpion bites," he boasted. "But I don't do broken bones."

Before Purba had driven off to Medan, he'd rushed home to slip on his one clean shirt and say good-bye to his wife. She'd insisted he promise to stop along the way and eat something, reminding him he'd been under the weather and needed to keep up his strength. He promised. "I didn't want her to get angry at me," he recalled.

The van pulled out of Jandi Meriah about an hour before sunset and reached Adam Malik Hospital six hours later. It was midnight by the time Jones's relatives had hoisted the ailing young man from his bed and carried him out to the terrace for the ceremony, setting him down in a chair. Purba began to prepare the ingredients. He laid out twenty-one betel pepper leaves, each about the size of his calloused palm, arranging them in rows. Then, into each, he deposited palm blossoms called *mayang*, some pasty white lime called *kapur sirih*, reddish brown chunks of an astringent called *gambir* produced from local vegetation, and bits of the orange-colored betel nut from an areca palm called *buah pinang*.

Purba asked Jones his name. Jones told him, and the witch doctor began to chant over the leaves, praying for the young man's recovery. Then Purba lifted the first of the stuffed leaves to his lips. He chewed it up, puckered, and softly spit it onto Jones's forehead. Bending over the patient, he gently blew the slime over the flesh. The witch doctor lifted the second leaf, chewed it up, and spit it onto Jones's chest, repeating the same procedure. He continued until he had finished all the leaves, slathering the torso, arms, legs, hands, and feet, alternating between left side and right, making particularly sure to cover all the joints.

"Afterward, he looked better," Purba recounted. "I asked him how he felt. He said he felt better. He felt relieved." Purba told him to go back inside the ward and go to sleep. Then he passed a bag of ingredients to the family, instructing them to repeat the same spitting procedure four times a day: in the morning, noontime, afternoon, and night. Finally Purba asked to be taken home to his village. It was dawn when he got back.

Just three days later, the Mitsubishi van reappeared beside Purba's paddy. The family had used up all the ingredients. They wanted Purba to return to the hospital and personally repeat the ritual. He again acceded to the request.

None of the staff at Adam Malik hospital ever objected to either of the late-night rituals or cautioned Purba that he might be exposing himself to a killer disease. No one suggested he take Tamiflu. No one sampled his blood or monitored his health. But had they tried, Purba would have considered it silly. He knew for a fact the source of the illness striking Jones and his family, and it certainly was not bird flu.

"I know the truth because I speak to the spirits," the witch doctor told me as we chatted on a wood bench in a thatched shelter on the village outskirts. "I know that Jones's father made a deal with another, black-magic witch doctor. Both of them have died. But I can communicate with the spirits, so I know what happened." According to Purba's detailed account, Jones's father, Ponten, had petitioned a witch doctor in the Alas River valley to help him become rich and powerful by conjuring the spirit of Begu Ganjang. The witch doctor had obliged. But Ponten never rewarded the witch doctor as promised. "The father broke his vow," he continued. "Now the family has to pay. Now the whole family has to die."

But Purba added that he had been able to shift the course of fate. He alone had been able to stay the execution of the remaining family members. "I'm the one to save Jones," he said. He grinned, baring a mouthful of teeth stained black by decades of betel nut. "There will be no more casualties. Seven is enough. I drew a border around them at the hospital." Purba stood up to demonstrate and etched a line in the dirt with the tip of his flip-flop. "It stops here," he assured me, "because I protected the family with my magic line."


Jones was discharged from Adam Malik Hospital ten weeks after he had been admitted. It had been a long, torturous stay. Besides losing weight and developing brain abscesses, he suffered what doctors reported was permanent lung damage. He had also become exceedingly bored during his convalescence. As his strength returned, he had begun to wander, repeatedly slipping out of the infectious disease ward for a coffee in the hospital's small cafeteria, potentially putting other patients and staff at risk.

"I never expected to come home," Jones recalled when I met him less than a week after he was discharged. "It was too long in the hospital. I lost hope." He was still taking medicine, and the doctors had instructed him to come to the hospital for a checkup every week for the next six months. So he, his wife, and their two sons had returned to the crowded, working-class neighborhood near the airport and temporarily moved in with an aunt. Jones would spend much of his time sleeping on a thin woven mat on the floor, which was where I found him when I first came to visit. He gradually roused himself and sat up on the mat. But his gaze remained vacant. Only when I asked about his plans did he smile. "I want to go back to farming, back to my orchards and grow oranges," he said softly.

Jones was now the head of the Ginting family, the sole surviving son. But he did not know that. No one had yet had the heart to tell him that during his own long recovery, his older brother Dowes had died.

Throughout the Ginting family's ordeal, senior Indonesian officials dismissed any possibility that the virus had been passed from one person to another. They were afraid the world might conclude this was the start of an epidemic and isolate Indonesia, crippling tourism, staggering the economy, and inciting panic. Health Minister Siti Fadilah Supari later boasted about how she'd convinced her president, Susilo Bambang Yudhoyono, that CNN reports about human transmission of the virus in Sumatra were lies. "From the lessening of the tension in his face, I knew that he trusted me," she later wrote. Supari would continue to maintain over the coming years there had been no human spread, at times accusing those who disagreed of trying to sabotage Indonesia.

But WHO's flu specialists were quickly convinced that nearly all the stricken family members had actually caught the disease from one another. Most had fallen sick so long after the eldest sister, Puji, that it was highly improbable they had caught it from the same source as she. More likely, Puji had infected them.

Dowes, however, could not have caught the bug from Puji. He had taken ill too long after the others. If Dowes had the virus, he had caught it while caring for his son, who had caught it from Puji. That would be an ominous precedent, marking the first time the virus had been found to hop from one person to another and onto a third. But until samples from Dowes tested positive for the strain, this would only be speculation. There were no hospital specimens for him, unlike for most of his relatives. There were just the samples that Tim Uyeki and his colleagues collected when they located Dowes on the hilltop in Jandi Meriah.

Before leaving the man's bedside, Uyeki had taken the tubes and wrapped them with a cold pack in double plastic bags, which he then put in his backpack for the return trip to Kabanjahe. There, on a cement landing just outside the district office, he repackaged the samples. He made sure the vials were closed tightly and relabeled them so it was clear for the laboratory what they contained. He placed them, along with new cold packs, into a plastic bag, which he put inside another bag and then inside another. A provincial health officer drove this hazardous delivery out of the mountains, back to Medan, where a waiting WHO official carried them on to an evening flight to Jakarta. It was about 11:00 P.M. when this official arrived by taxi at Indonesia's national health laboratory. The samples were split up, half to be tested by Indonesians and half at a U.S. Navy lab in the capital. The scientists worked through the night.

That evening, a senior WHO official in Jakarta e-mailed the agency's regional headquarters in New Delhi reporting on Uyeki's success in finding the fugitive, Dowes, and obtaining the samples. The outcome of the tests would be crucial. "If he turns out to be positive, then we will have to consider going in for rapid containment measures," the official wrote. These would involve an intensified investigation, widespread distribution of Tamiflu, and possibly, for the first time, a mandatory quarantine of the affected villages.

His missive received an urgent rebuttal. It came not from the regional office but from the top, from the chief of the global influenza program at WHO's main headquarters in Geneva. "In response to the possibility of a rapid response raised in the e-mail, my sense from various communications is that the level of suspicion and hostility in the area is high. Trying to mount a rapid response, especially one involving many outsiders and oseltamivir coupled with quarantine and isolation, seems most likely to lead to a very bad outcome, especially in terms of more suspicion and hostility and rumors," wrote Dr. Keiji Fukuda, the global influenza chief. Only if the virus spread beyond the Ginting family should a rapid response be considered, he said. Fukuda was not optimistic about its prospects even then. "If intense social mobilization doesn't quickly reverse the population's suspicion," he added, "we would probably have to contemplate a heavy security force/military backed operation which is, I suspect, very unappealing to us all."

Uyeki had come down for breakfast at his modest hotel in the Sumatran highlands when, at 7:00 A.M., he got a call on his cell phone. Both labs had results. The specimens had indeed tested positive for the virus.

Fifteen minutes later came a second call. Dowes was dead. His body had just arrived at Kabanjahe Hospital, and people there were reportedly in panic. Uyeki was worried they'd blame him for the death since he'd worked on Dowes only hours earlier. "OK," Uyeki responded, "we're on our way."

When he arrived, he found the relatives were actually quite calm, even relieved. They'd been advised by the witch doctor, Purba, that seven people would succumb and Dowes was number seven. That meant the dying was over.

Tom Grein, as a seasoned veteran of Ebola and other hemorrhagic fever outbreaks in Africa, knew what to do next. He and Uyeki asked the local health staff for supplies: large-gauge needles, formaldehyde, rubber boots, chlorine bleach, and a backpack sprayer, the kind used for insecticide. They suited up in their gowns and other protective gear. They used the needles to take additional lung tissue specimens, blood, and other samples from the body. Then they donned the boots and filled the sprayer with bleach. They sprayed the corpse to disinfect it, setting aside the man's wristwatch to give to his family, placed the body in a black plastic bag, and hauled it over to a waiting coffin. They closed the coffin, sealed it, and disinfected it. They disinfected the hospital room. They disinfected each other. And when they were almost done, they disinfected the Suzuki jeep that had had transported Dowes to the hospital all too late.

In Jandi Meriah, Uyeki and his Indonesian colleagues had left behind boxes of Tamiflu with instructions that relatives, the witch doctor and his wife, and anyone else who'd been exposed take the tablets for protection. Health teams also distributed the drugs in three other villages, providing medicine to the Gintings' neighbors, their relatives, and the driver and conductor of the minivan that brought Dowes to Jandi Meriah. A total of fifty-four people who might have been infected by him were placed under voluntary quarantine and asked to remain in their homes, or at least in their villages, for two weeks. Few listened.

Heni Boru Bangun, a forty-seven-year-old in-law, was among those who had the closest contact with Dowes, helping to care for him when he was in hiding. She had stayed in the same house with him for four days. Less than a week after he died, Heni broke quarantine, traveling to the next province to visit her mother.

I asked Heni whether she had taken the Tamiflu she'd been given. She said she had finished only two of the ten tablets required for a full course. "I was afraid of the side effects," Heni explained. I looked at her adult son, who had been watching me suspiciously. "They gave me ten tablets but I only took three or four," he volunteered. His wife, who had also helped look after Dowes, had declined to take any.

Heni disappeared into the back room of her house and returned with the thin white box of Tamiflu. She opened it and dumped the contents on the wood floor. Out tumbled the eight remaining yellow and white tablets still in their wrapper for me to see—and, with them, some harsh truths about our hopes for bending nature to our will and forestalling a pandemic.

According to the computer models that underpin WHO's planning, the difference between success and cataclysm is measured in days, and the conditions for snuffing out an emerging pandemic are unforgiving. Yet in the dry run of North Sumatra, the virus had been spreading among the Gintings long before it was diagnosed, and in some cases, rather than seeking medical care, they ran. When public health officials finally responded, they were repeatedly chased off. The victims and their contacts refused to share information and samples vital for containing an outbreak, likewise rebuffing appeals to take antiviral medicine, obey quarantine, or even take rudimentary steps to avoid exposure to those stricken. It was as if the highlanders had done everything imaginable to accelerate the spread of the disease.

"If this were a strain with sustainable transmission from human to human, I can't imagine how many people would have died, how many lives would have been lost," said Dr. Surya Dharma, who was head of communicable disease control for North Sumatra.

If there was any encouraging news from North Sumatra, it was only this: the novel strain had not spread beyond the one family. In its current form, the virus was still tough to catch. But influenza viruses inexorably mutate. It had now, for the first time, demonstrated the capacity to jump not only from one person to another, but also onto a third.

Only nine years earlier, flu specialists had assumed that this strain couldn't infect anyone at all. When it had, alarms sounded around the world.

# CHAPTER TWO

# A Visitation from Outer Space

Keiji Fukuda had been anticipating this call his entire career. It came in August 1997, when he was busy caring for patients at San Francisco's Mount Zion Hospital, deep into a clinical rotation he would do a couple of weeks each year to keep his skills as a physician sharp. After he came off the wards, a hospital staffer mentioned that the CDC, where Fukuda worked most of the year as the country's top influenza investigator, had been trying to reach him.

Fukuda suspected something was wrong. He quickly returned the call, which had come from a laboratory in the CDC's influenza branch, and the lab director filled him in on some tests her scientists had just completed on a sample from Hong Kong. It looked as though someone had been infected by a new virus, a novel strain of flu called H5N1. The victim had died.

"It was a jolt," Fukuda recounted. "It was an unusual call. But it was the call you are kind of always waiting for in the field of influenza."

His mind instantly started to race. This was a strain that had never before infected humans, at least as far as scientists knew. That meant no one had immunity to the pathogen and everyone could be vulnerable. "How many other people have been infected?" Fukuda wondered to himself at the time, adrenaline pumping. "Are we missing anyone else? Right now, what's going on?"

It took a few hours to arrange a larger conference call. The CDC hooked in Fukuda along with Dr. Nancy Cox, the chief of the influenza branch, who had been tracked down while vacationing with her family at a horse ranch in Wyoming. Flu was their field, some would say their obsession, and they instantly understood what was at stake. "The idea of a pandemic coming on is one of the things you know is always possible," Fukuda said. "Perhaps this is the start of that pandemic."

Joining them on the call from Hong Kong was Dr. Margaret Chan, a Canadian-trained doctor who ran the city's health department. She had never spoken with Fukuda before. Nor did she know much about pandemic flu. Over the next decade, she would become intensely familiar with both, gaining a perspective on influenza shared by few on Earth. But in 1997, it was all new. Chan peppered the CDC specialists with questions. "Is this a big threat or not?" she asked. They admitted they weren't sure. They explained that the lab might have made a mistake. Even if the test result was accurate, it might reflect merely a single, isolated infection. But they cautioned her that the case could also be a harbinger of something larger. Chan was quick to appreciate the horrific implications. She told them she needed help. Hong Kong couldn't get to the bottom of this alone. Before the call was over, Fukuda knew he was bound for Asia.

Fukuda finished his rotation in San Francisco, then headed home to pack. He and his wife and two young daughters lived in a suburb of Atlanta, where he had worked for the CDC since 1990. In 1996, he had become chief of epidemiology in the influenza branch.

Fukuda had been bred for the job. His parents were both doctors. His mother was Japanese, his father Japanese American. Born in Tokyo, Fukuda became a New Englander at age three when his family moved to Vermont so his father could take up a medical post there. The future flu hunter earned his own medical degree at the University of Vermont and did his residency in San Francisco before studying public health at the University of California–Berkeley. Influenza became his calling at a time when few infectious-disease specialists paid it much mind. Starting in the 1980s, HIV-AIDS exploded on the American scene and monopolized much of the scientific attention and research money. Like others in the small influenza fraternity, Fukuda felt "his bug" was slighted. Even short of pandemic, seasonal flu was a proven scourge. "It certainly wasn't on the radar screen," he lamented. "People dying year in and year out in fairly large numbers in many countries and still it remains invisible." Among flu researchers, it was a common grievance.

Yet among them, Fukuda was rare. His fervor was a quiet one. When I met him nearly a decade after the first H5N1 case, his buzz cut was graying but his round face was still youthful. He was one of the few veteran flu specialists whose brow was clear of lines and furrows. I concluded this was due not to any deficit of passion but rather a balance of passions. (Fukuda, who claims he is only a mediocre musician, had decorated an entire office wall with pictures of antique cellos.) His manner is authoritative but reserved. He is brilliant in analysis but measured and soft-spoken in delivery. He is ever polite, even when his colleagues spy irritation smoldering behind his rectangular-framed glasses. Above all, he is calm and calming. So when he worries, those around him get scared.

Fukuda arrived in Hong Kong late one sultry August evening at the head of a small CDC team. By the next morning he was already cloistered at the city health department, mapping out a systematic investigation into the outbreak.

The victim had been a three-year-old boy named Lam Hoi-ka, who had died in the spring. He had been a healthy youngster, the second son of an affluent Hong Kong couple. The father owned a company that manufactured decorative candles at a factory in mainland China. The mother was in charge of the firm's marketing. They lived in a satellite town called Tseung Kwan O in the lower reaches of the New Territories, an expanse of former wilderness just beyond the congestion of old Kowloon that the Chinese had ceded to the British crown colony of Hong Kong a century earlier. Tseung Kwan O is really more city than town, a modern, highly planned community with scores of high-rise apartment towers, some as tall as fifty stories, amid American-style shopping malls and brightly painted schools and kindergartens. The town sits on a bay of the same name, which means General's Bay. But the British had called it Junk Bay after the multitude of traditional sailing vessels that once plied the waters. The community is no longer oriented to the sea but toward cavernous rail and bus depots. Each morning, the apartment towers disgorge thousands of office workers, headed for downtown less than a half-hour commute away.

Hong Kong health officers had already visited Hoi-ka's family. It had been a hard interview. The family was distraught, and it took a long time to explain to them how an entirely new virus could have claimed their son. Investigators had gathered what information they could from the parents and their housekeeper about the boy's recent history. None of it seemed exceptional. When Fukuda arrived in Hong Kong, he asked to see the family again, but they refused to talk anymore. He didn't press it.

Fukuda had other angles to explore. His first priority was to determine whether this was simply much ado about nothing. The positive test result for H5N1 was so startling that he suspected it could be the result of a contaminated sample. Perhaps the boy had never been infected at all. So Fukuda and his team traced the precise route of the specimen. They went to Queen Elizabeth Hospital, the primary referral hospital on Hong Kong's Kowloon peninsula, to see the intensive care unit where the boy had been treated. They checked to make sure that hospital staff had followed proper sampling procedures, and they examined the equipment. They asked whether any other patients had a similar respiratory disease and could have been the source of the virus. They asked about the health and habits of the staff. Perhaps some nurse had tracked the virus in. The team also visited the health department's lab at Queen Mary Hospital on Hong Kong island, reviewing how the tests were conducted. They explored whether the virus might have been inadvertently introduced by technicians working at a veterinary lab in the adjacent building. After nearly a week of investigating, there was still no evidence of contamination anywhere.

But the conclusive proof came in the lab. The team conducted an analysis using antibodies that would find the virus and cling to it. The antibodies were tagged with fluorescent stain so they could be identified under a microscope. When these were added to a sample from the dead boy's windpipe, they attached to the inside of the human cells in the specimen. The antibodies did not simply float on top. This meant the virus had truly infected the boy's cells and was not an accidental visitor.

The cruel progression of the boy's disease also seemed to confirm he'd been struck by a form of viral pneumonia arising from influenza. As Fukuda pored over the medical records, he learned that Hoi-ka was initially taken to a doctor after coming down with a high fever, sore throat, and dry cough. The doctor prescribed aspirin and antibiotics and sent him home. But the condition worsened. The boy didn't seem to be thinking right or acting right. He was oddly irritable. He wasn't his usual alert self. The doctor couldn't put his finger on what was wrong. So he admitted Hoi-ka to a small private hospital. A day later, he was moved to the intensive care unit at Queen Elizabeth Hospital and put on a respirator, but to no avail. The boy was losing consciousness. His lungs were failing. Five days later, he was dead.

Fukuda read the records and thought, "Wow. This is really bad. No matter what they could have done, they weren't going to prevent him from dying." Fukuda was struck by the relentless advance of the disease. "It was clear from the time the child got in. He started getting sicker and sicker and sicker and then he died. You don't expect to see that in young healthy children." It reminded Fukuda of the healthy young men he had seen in San Francisco in the 1980s struck down almost overnight by AIDS.

It also made him think of his daughters. He couldn't keep his mind off them. He wasn't frightened, not yet. There was just one case. But he felt so sad. How awful for parents to lose a child out of the blue to such a mysterious intruder. "This," he thought, "is just like a visitation from outer space."


Hoi-ka's young lungs had been ravaged. The final diagnosis gave the cause of death as pneumonia and acute respiratory distress syndrome, in other words a severe inflammation of the lungs that culminated in their wholesale destruction. But the virus didn't stop there. The boy suffered kidney failure, and his blood was poisoned, unable to clot normally. His liver and kidneys also showed evidence of Reye's syndrome, a separate, potentially fatal disease that can occur when aspirin is taken to treat viral disease.

This was indeed flu. But it wasn't the flu most people know. Sure, ordinary seasonal flu can put you on your back with a fever, a cough that keeps you up at night, and aches that rack your bones. But in the vast majority of cases, these symptoms pass in just a few days. This new strain meant far more than a runny nose and chills.

Often, viruses will adapt themselves to their human hosts over a period of time. They will soften their edges and temper their nastiest qualities. After all, the objective of any virus is to reproduce itself inside its host and then jump to another so it can continue to replicate. A death that comes too quickly does little to serve these ends.

But H5N1 is an interloper, an unrefined newcomer, all fury without the seasoning of age. In human cells it has discovered a fresh target and it pursues its prey deep into the body, penetrating much farther than ordinary flu. Seasonal flu usually strikes the upper respiratory tract, the nose, and nearby throat. The result is sniffles. This novel virus, by contrast, advances on the lungs themselves, attacking the branches of the bronchial tree and the myriad little buds on their tips called alveoli, where the life-sustaining task of exchanging carbon dioxide for oxygen occurs. The pathogen infects the coating of mucus that protects the membranes of the lungs. But unlike regular flu, which cannot reach much beyond surface cells, this newcomer penetrates into the tissue itself. It thrusts deeper. It spreads farther, often infecting both lungs at nearly the same time. As the pathogen relentlessly erodes the cells of the deep lung, you find yourself increasingly short of breath. Your cough is often bloody and you may bleed from your nose and even gums.

Once inside the cells, the virus begins reproducing madly. Even in the throat, which is at best a secondary target, the amount of virus can exceed that of normal flu. Down in the lungs, there are at least ten times more than in the throat, in some instances even a hundred times more. This is a virus in a hurry. And the human body, which has never encountered anything like it, has no ready arsenal of antibodies to choke off the process.

The body can still marshal its innate, all-purpose defenses. But in doing so, it mounts a counterattack so furious that some scientists believe it's more lethal than the virus itself. The besieged cells raise the alarm by dispatching messenger proteins called cytokines. These in turn prompt a counterattack by the body's defensive forces. Immune cells of different stripes, including killer T cells, macrophages, and other white blood cells, flood into the infected lungs, ferociously attacking the virus and the cells that have fallen into enemy hands.

This is the fateful battle on which life could turn. There is no restraint, nothing held back. The body throws everything it has at the intruder without regard to the tremendous collateral damage this causes the lungs themselves. The cytokines keep firing and firing in what scientists call a cytokine storm. (Researchers debate whether the massive response is due to the tremendous amount of virus in the system or the nature of the pathogen itself. Unlike regular flu viruses, which have a way of calibrating the immune response, this strain may not.) Ever more immune cells are summoned to the front and continue to blast away. The carnage mounts. The lung cavities fill with dead and damaged tissue, mutilated mucus cells, and other cellular wreckage. The lungs become rigid as the cells that make liquid to keep the lungs flexible are annihilated. The seal between the bloodstream and the air passages ruptures. Red blood cells and plasma leak into the lungs. The alveoli sacs are swamped with fluid and debris and are no longer able to exchange carbon dioxide for oxygen. If you listen closely, you can hear the liquid crackling. Your breathing accelerates. You desperately press all your chest muscles into helping you suck down precious oxygen. You're gasping for air. This is the acute respiratory distress syndrome, known as ARDS, that struck Hoi-ka. On a series of X-rays, it appears like a white ghost consuming your lungs. You're drowning.

But the virus is not content to remain a solely respiratory disease like regular flu. From the lungs, the pathogen sets upon other vital organs. It invades the digestive tract, perhaps when you swallow contaminated sputum, but more likely via the blood. The virus gets into the gut, often causing diarrhea and sometimes vomiting. It can assault the liver and kidneys, as it did with Hoi-ka. It can provoke heart failure. It can attack the eyes. It can even breach the brain and spine, on rare occasions causing encephalitis and related seizures. This virus has shown a singular facility to smuggle itself through the bloodstream and proliferate throughout the body.

Yet in the end, the lungs are where this microbe concentrates its energies and takes its heaviest toll, whether by killing directly or inviting a suicidal counterattack. The lungs are also the means by which it casts its net for further prey. In this one regard, it is much like its seasonal cousin. They both spread their sickness through contaminated droplets coughed or sneezed into the air, one of the most efficient forms of transmission known.

Influenza viruses, like all viruses, defy definition. Are they alive? On one hand, they lack the cell structure shared by all other living things from humans down to the amoebas that students study in science class. But at the heart of a virus is genetic material, the basic blueprint of all life, and viruses share the imperative of all living things to reproduce and pass on their genes before they perish.

The flu virus contains eight pieces of RNA, accounting for all of its genes. These determine everything about the virus: how it's structured, how it reproduces, how it spreads, who or what it infects, and how sick it makes them. In particular, the genes program the proteins that do all this heavy lifting.

The virus itself is a microscopic sphere studded with two types of these proteins. One is the hemagglutinin, a spiky protrusion that the virus uses to break into the cells of its host. The other is the mushroomlike neuraminidase, which the virus uses to break out again. The two are called H and N for short. Each can take slightly different forms—scientists have so far discovered sixteen H subtypes and nine N subtypes—and the subtle variations that define them can mean life or death to the host. Human antibodies can recognize certain of these surface proteins and disarm them, but they are powerless against those that are totally new.

The flu virus most often enters the human body by hitchhiking a ride on an inhaled droplet. As the virus is whisked through the nose and down the windpipe, it brushes along the cells lining the airway. These cells have special receptors. Some are a poor fit for the specific hemagglutinin subtype that has been inhaled, so the virus keeps going. But some are just right. When the hemagglutinin finds the right receptor, the spiky protein plugs in, allowing the virus to fasten to the outside of the cell like a pirate ship preparing to board another vessel. The human cell seeks to stem the attack by engulfing the virus, then walling it off while trying to digest it with acid. But all the cell has done is speed its own demise. The acid triggers a remarkable transformation of the flu virus. The hemagglutinin turns itself inside out, baring a hidden weapon often likened to a molecular grappling hook. The virus uses this to pull itself even tighter against the cell. The membrane of the virus dissolves, and the viral genes stream into the core of the cell like marauding buccaneers.

A virus cannot reproduce on its own. But once inside the human cell, it finds everything it needs to do so. The viral genes hijack the cell's own genetic machinery and deliver fresh commands. The innocent cell is soon churning out viral proteins, which are assembled into a brood of new viruses. Within hours, up to a million progeny are ready to explode forth and continue the mission.

But escaping the cell is not so easy. The same receptors on the cell's surface that first attracted the virus now try to ensnare its offspring. So the neuraminidase steps in to cut them loose, leaving behind the lifeless wreckage of a human cell.

Most flu viruses fall into one of two types, either influenza A or influenza B. While the former can infect a wide array of species besides humans and cause severe illness, the latter tends to be found only in people and is less virulent. Influenza A viruses are further categorized by the proteins on their surface. The strain that struck Hong Kong in 1997 and continues to menace the world today is identified, for instance, by its H5-type hemagglutinin and its N1-type neuraminidase, hence influenza A (H5N1). Like all flu viruses, H5N1 originated in waterfowl, and that's where most flu strains stay, circulating benignly among wild birds. But over the centuries, a few strains have succeeded in crossing the species barrier, either directly or via intermediate hosts like pigs, to infect people. Over time, these have evolved into the ordinary seasonal viruses that we usually associate with flu, losing much of their bite as humans develop immunity. The prosaic H1N1 strain that in recent years has kept millions of people in bed each winter, for example, is descended from a virus that first infected humans in the early twentieth century, sparking the catastrophic Spanish flu pandemic of 1918. The swine flu strain that erupted in early 2009 is also an H1N1 strain and it too can trace its lineage back to the Spanish flu. But swine flu, which scientists believe emerged only recently from pigs, has diverged so far from other H1 viruses over the years that vaccines against the seasonal H1N1 variant afford little protection against this new arrival.

So what distinguishes avian and human flu strains? Research suggests there may be several factors explaining why some viruses circulate primarily or exclusively in birds while others spread easily among people. Much of the suspicion centers on the receptors in the human respiratory tract. The hemagglutinin of human flu viruses fit better into one type of receptor common in the nose, sinuses, and upper reaches of the airway. Avian viruses, including H5N1, prefer a different type of receptor that is characteristic of birds but relatively rare in the upper airway of humans. Instead, this avian-type receptor is found deep in our lower respiratory system. Scientists surmise this is why H5N1 primarily strikes the lungs, as opposed to the throat and nasal cavity. This could also explain why the virus remains hard to catch, since these avian-type receptors are buried and relatively inaccessible. And when the deep lung does get infected, the virus has a long trip back up the windpipe before being coughed or sneezed into the air, making it hard to spread.

Yet all that separates the world today from an unprecedented calamity could be a slight retooling of the virus, some evolutionary tinkering with the hemagglutinin to make it a better fit for the receptors in the nose and throat, a change in another protein allowing the virus to better replicate in the temperatures of the upper airway, a few other genetic tweaks. Some researchers estimate it would take at most a dozen minor adjustments.

Scientists were aware of H5N1 even before it killed Hoi-ka in 1997. A version of the strain had initially been detected four decades earlier in chickens in Scotland. But no one thought it could jump the species barrier to people. It was strictly avian. That's why its sudden appearance in the lab sample from Hong Kong was so startling. By making the leap, the virus had satisfied two of three conditions for a pandemic. It was novel—no one had been exposed, so no one had immunity—and it had proven it could infect people. Now it just had to show it could get around.

One day before Hoi-ka died in May 1997, Hong Kong's public-health laboratory at Queen Mary Hospital had received a specimen of fluid from his windpipe. The sample was one of more than eighty collected every day from patients at the city's hospitals and clinics. They were all sent to the lab, a small but hectic facility on the seventh floor of the clinical pathology building, erected on a hillside overlooking the western approaches to Victoria Harbor. The waters below were crowded with freighters emerging out of the mists of the South China Sea. Inside, the staff busily tested the samples, sorting them for flu, hepatitis, HIV, and other viruses, categorizing them by subtype when appropriate.

The technicians suspected that Hoi-ka was suffering from influenza and placed his specimen in a cell culture designed to grow the virus. When they looked at it a little later under a microscope, sure enough, it was flu. But what sort? Using an antibody test, they determined it was a kind of influenza A. They assumed it was one of two run-of-the-mill subtypes, either H3N2 or H1N1. The lab had chemical reagents to identify each of these seasonal strains. Yet every time they ran the test, they came up empty. They were stumped.

Dr. Wilina Lim was the adept yet unassuming chief of the virology lab. She spoke in quick, clipped sentences and had a no-nonsense manner that won her the respect of colleagues. Lim was sure the boy's sample had to be seasonal flu. Since she never expected it to be something utterly new, she wasn't particularly worried. Lim concluded that one of the ordinary strains must have evolved ever so slightly, which would explain why her lab's reagents no longer picked it up. Still unaware that the boy had died, Lim decided to divvy up the specimen and ship these samples out for further analysis in specialized flu labs overseas, including the CDC in Atlanta and Mill Hill in London. She also sent a sample to a veteran Dutch virologist named Jan de Jong, who worked at an institute outside Utrecht in the Netherlands. Though they had never met, Lim and de Jong shared a fascination with odd and offbeat viruses and over the years had compared notes from time to time.

More than two months passed. Lim never heard back. Then, on a Friday in early August, she got a call from de Jong. He was coming to Hong Kong, arriving in two days, and would like to see her. He didn't say why. Lim assumed he was just passing through. She said she'd be pleased to finally meet him. She reserved him a room at the Ramada Hotel in Tsim Sha Tsui, a teeming quarter of narrow streets jammed with shops hawking clothes, shoes, and electronics, where the air was pungent with the smells of tropical cooking and the night skies blazed with neon.

First thing Monday morning, she picked him up. They set off for Queen Mary Hospital. De Jong wanted to see her lab. Lim was behind the wheel of her Nissan, seated on the right like all drivers in Hong Kong, de Jong to her left.

After five minutes, as they approached the harbor tunnel, de Jong looked over and asked, "Do you have any idea what virus you sent me?"

"No, I don't know," she responded. She was betting it was some idiosyncratic version of the common H3 strain.

"It's an H5."

"What?" she asked. "H5?"

Lim was bewildered. She had never come across an H5 strain. She wondered to herself, "Where did this H5 come from?"

De Jong didn't say much more about the test results during the ride, but privately he had a suspicion. It could be contamination or some confusion in classifying samples. That was why he wanted to inspect her lab, why he had come all the way to Hong Kong. But once they arrived at the hospital, he quickly saw her operation was well run.

A few hours later, Lim called Margaret Chan, the health department director. The scientists at the CDC in Atlanta, which had independently reached the same results, had yet to inform Hong Kong of their findings. So Lim's news came as a shock.

"Are you sure?" Chan pressed. "H5N1? I have never heard of H5N1 infecting people."

"That's why I'm calling you," Lim explained.

"Please educate me," Chan told her.

Chan's expertise was not infectious diseases. Her early medical interest had been pediatrics, followed later by women's and family health issues. In reality, she had never intended to be a doctor at all. Growing up in Hong Kong, Chan had trained to be a teacher and for a year taught English, math, and home economics to elementary school students. But when her sweetheart, David, left for Canada to attend college in 1969, she followed him to the far side of the world and enrolled in a Catholic women's college in Ontario. Before long, they were married. When David then decided to brave the rigors of medical school, she concluded the only way to assure his continuing attention was to become a doctor, too. Chan had little background in college science. But she did have the late-night tutoring of her new husband. After receiving their medical degrees from the University of Western Ontario, she and David returned to Hong Kong. Chan joined the government in 1978, rising quickly through the ranks of the health department. Sixteen years later, she was running it.

Chan has a charm that makes her seem taller than her modest height. Her manner is eminently self-assured yet empathetic. Her black hair is coiffed and her preferred lipstick bright red. Her brown eyes radiate warmth from behind large, round lenses. She speaks with authority, whether lecturing on health-care politics or the therapeutic qualities of her mother's recipe for pork tenderloin soup. (Chan swears it's good for stamina. Years later, after leaving Hong Kong and rising to the top post at WHO, she would get the Chinese herbal ingredients shipped to her in Geneva.) But in a realm of outsize egos, she is quick to admit what she doesn't know.

When Chan heard about the test results, she asked Lim to bring de Jong over to see her at the health department. The next day, Chan took the visiting researcher out for lunch at the Hopewell Centre, a circular, sixty-story skyscraper that for a time had been Hong Kong's tallest building. Over a lunch of dim sum, she continued to press him for an explanation.

"What is the implication of this?" Chan asked.

"Dr. Chan," de Jong answered, "if this is true, we are heading for something really serious."

By the time the CDC team arrived in Hong Kong, Chan's health officers were already hunting for clues, trying to pinpoint the source of the boy's infection. But even a seasoned sleuth like Fukuda would find them elusive. Hoi-ka had died three months earlier, and the trail was getting cold.

The investigators got their first break when they learned the virus had already been implicated in a previous outbreak. Earlier that year, a baffling plague had raced through three farms in the rural northwest of Hong Kong. Every single bird at one farm had fallen over dead. So did most of those at the other two. About five thousand chickens had crashed out. Researchers from Hong Kong University, who studied influenza at a veterinary lab in Queen Mary Hospital, were stumped. They forwarded samples to a high-security lab run by the U.S. Department of Agriculture in Ames, Iowa. "It was one of the most highly virulent influenza viruses we'd ever worked with," recalled Dennis Senne, a USDA microbiologist. "We'd never seen anything like it before. We'd never seen anything that killed so quickly."

Senne inoculated ten chickens with the virus to test its pathogenicity. Some died within a day, the rest a day later. At first Senne couldn't believe the virus was responsible. "We thought they'd somehow suffocated in the cages," he said. But when researchers examined the dead birds, they found the lungs had been devastated. Genetic analysis revealed the culprit was an H5N1 flu virus. The discovery surprised the researchers back in Hong Kong, since this stripe of flu had never before troubled the city's poultry. Yet they didn't bother to stroll over to the adjacent building where Lim had her public-health lab and share the chilling news. Why should they? This virus had never been known to infect people. It was considered strictly avian.

When Fukuda learned in August there'd been an outbreak among birds, he was convinced it was somehow related to his case. Yet he wasn't sure how. The infected farms were clear across Hong Kong in Yuen Long, more than fifteen miles from the boy's home, and Hoi-ka had been nowhere near them. The investigators searched his apartment and found bird droppings in the air-conditioning ducts. Was this the missing link? Or was it the live poultry market they discovered close by? The evidence was at best circumstantial.

The trail soon led to the boy's nursery school. Several weeks before he fell sick, the staff had brought in three baby chicks and two ducklings to help the children get closer to nature. The cages were placed on the ground and the youngsters encouraged to peer inside, even to touch the birds. That was in the spring. By the time the investigators arrived in August, both ducklings had died, as had two of the chicks. The final chick had been removed, and no one knew where it was. There were no samples to take, no evidence that the birds were the source of infection. But suspicions ran strong.

On the remote chance they'd detect some vestige of virus, health officers scraped up dirt and dust from the school grounds for testing. The lab analysis found no trace. The investigators asked after the health of other children at the school and the staff. But there had been no unexplained absences. Nor had there been any other unusual illnesses.

In fact, there seemed to be practically none anywhere in Hong Kong. Fukuda and his colleagues examined medical records from hospitals and clinics from around the city, searching for atypical respiratory cases that might signal a swelling wave of infections. They visited intensive-care doctors and medical directors, urging them to report patients who were ending up on respirators with undiagnosed ailments. The phones started ringing, but they were all false alerts. "We didn't see a big upsurge in respiratory illnesses going through the city," he recalled. "We weren't hearing about kids dying of mysterious illnesses."

The team had even gathered two thousand blood samples, including several dozen from Hoi-ka's schoolmates and teachers, to check for antibodies indicating exposure to the virus. These would ultimately all test negative except for a few from otherwise healthy poultry workers. The virus had gone silent. Fukuda's anxiety about a nascent pandemic began to ebb. "This is perhaps an odd infection that we don't quite understand where this one child became infected," Fukuda thought. "Perhaps it's a one-off, a freak event."

Yet even as he prepared to wrap up the probe and head home, he couldn't put his mind to rest. Was this the end or the beginning? Could this virus still spark a global epidemic? He didn't know how to answer the question.

Before Fukuda left Hong Kong, he met health department officials one last time. "We don't fully understand what happened," he admitted to them. "Look, we don't know how this virus works. We don't know what's going to happen." He urged them to step up surveillance for additional cases in city hospitals. He privately wondered whether he'd be back.

There was no doubt the threat would ultimately return. Flu pandemics are inevitable. Divining precisely when one will strike may be no easier than predicting the timing of earthquakes and hurricanes, but global epidemics are just as certain.

Three times in the twentieth century, novel flu strains crossed the species barrier from animals to humans, then circled the globe. The Spanish flu in 1918 claimed about 50 million lives, according to official estimates. But because many deaths occurred in far-flung corners of the world, beyond the range of medical chroniclers based in the United States and Europe, scientists and historians now believe the true toll could have reached 100 million. This plague killed about 675,000 Americans, more than the American death toll in all the wars of the twentieth century. Until recently, Spanish flu was considered the worst-case scenario. Bird flu has made the experts reconsider.

A better-case scenario is illustrated by the two subsequent pandemics, the Asian flu in 1957 and Hong Kong flu in 1968. Together, they claimed an estimated 3 million people worldwide. What distinguished the Spanish flu from its successors was how sick it made people, not whether it made them sick in the first place. In rough terms, all three pandemics were equally contagious, infecting a quarter to a third of the world's population. The coming pandemic will likely do the same. Even in the mildest scenario, hospitals and other medical care will buckle.

It is not just precedent that makes a flu pandemic inevitable. There is a dynamic to the virus that accounts for the historical pattern of recurring pandemics, and that dynamic continues to hold today. Influenza is among the most capricious and mutable of viruses, and it is this very unpredictability that makes a pandemic a sure thing.

All viruses, influenza and otherwise, can be grouped by the nature of the genetic material at their core. Some are built around DNA, or deoxyribonucleic acid. These viruses contain an inherent self-correcting mechanism that discourages the microbe from mutating. When the virus reproduces, this genetic spell-check detects inadvertent changes in the code and fixes them. But other viruses, including all influenza, are built around a second form of genetic material, RNA, or ribonucleic acid, and lack this proofreading mechanism. As a result, many of the copies contain subtle and not-so-subtle errors.

Among viruses, flu is exceedingly sloppy, constantly spinning off mutant progeny. Most of these copies are defective, with mutations that impede their ability to spread and reproduce. But a different kind of accidental change, instead of undermining survival, can take the virus in a new, more perilous direction.

The flu virus reproduces so vigorously that there are enough viable copies to propel it forward. The other copies, those with damaging mutations, are cast aside, as the virus is swept along a path of ever-shifting forms and threats. Ultimately, the virus hits upon precisely the set of mutations needed to infect people and spread like a common cold, the recipe for pandemic. It's just a matter of chance. Each time the virus replicates, it's a roll of the dice. Each new bird that's infected, each person who's exposed, each one who's sickened is another toss. Throw the dice enough times and they're sure to come up snake eyes.

As if the threat of mutation wasn't enough, the virus can also take a shortcut through what's called genetic reassortment. Flu viruses are notoriously promiscuous because of a rare gift for swapping genes with other flu viruses. Though most other viruses can't do this, flu can go out and acquire entirely new attributes. This is because the genetic material in a flu virus—unlike nearly all other RNA viruses—is composed of separate segments that can each be individually replaced. If two different flu strains infect a person or even a pig at the same time, a new hybrid strain could emerge that is both lethal and has the tools to spread with ease. The poster child for gene swapping is swine flu. It was produced by the recent, seemingly improbable encounter of two different flu viruses: one known to circulate among pigs in the eastern hemisphere and another among pigs in the western hemisphere. The latter strain was a so-called triple reassortant, born from even earlier flu strains originating in humans, birds, and swine.

With the wholesale genetic changes that reassortment allows, it doesn't require too many rolls of the dice to splice together a pandemic this way. "You can move a whole lot of characteristics in one go," explained Robert Webster, a veteran virologist at St. Jude Children's Research Hospital in Memphis, Tennessee, and the dean of avian flu researchers. "Flu is an RNA virus and it's also a segmented RNA virus. That gives it a double whammy."

Among flu strains, none unnerves disease specialists as much as H5N1 bird flu. In the decade after it surfaced, the virus spread over a swath of Earth unprecedented for a highly lethal avian virus. It extended its reach among animals, even infecting mammals like tigers and leopards. It grew more tenacious. The disease persisted longer in birds and spread more easily among them than only a few years earlier. The dice were being rolled faster and faster.

Researchers have concluded that the continuing outbreaks "appear out of control and represent a serious risk for animal and public health worldwide." No matter how many times governments claim they've expunged the virus, it returns. In some countries, like Indonesia and China, the disease has become deeply entrenched in poultry, posing a permanent threat of contagion to their neighbors. Global eradication, according to senior animal-health experts at the UN Food and Agriculture Organization, "remains a distant and unlikely prospect."

Yet this strain is not the only avian virus menacing humanity. A little-noticed but equally novel avian strain called H9N2 has also proven it can infect people, including several in Hong Kong and mainland China since 1999. This pathogen, like its better-known cousin, has quietly spread across the birds of Asia and the Middle East and on to Europe and Africa. Studies have suggested that human cases of H9N2 are more common than generally acknowledged, and human transmission may have already occurred. Most worrisome, scientists say the H9N2 virus is actually a better fit for receptors in the human airway, giving it perhaps an edge in the pandemic sweepstakes. "The establishment and prevalence of H9N2 viruses in poultry pose a significant threat for humans," an international team of researchers reported.

A separate family of novel strains, the H7s, has meantime been circulating in both North America and Europe. Several of these pathogens have also shown an increased affinity for receptors in the human airway. Researchers have urged "continued surveillance and study of these viruses as they continue to resemble viruses with pandemic potential."

But the H7s, like H9N2, so far remain fairly benign, far less lethal than H5N1. The latter, with a recorded human mortality rate of about 60 percent, is so savage that most flu specialists agree it is the one to be most feared.

Some medical scholars dissent. Although another flu pandemic is inescapable, they doubt that H5N1 will be the source. They note that years have passed, tens of millions of birds have been infected, and countless people exposed without the virus crossing the pandemic threshold. "If it was going to happen, it would have happened already," said Dr. Peter Palese, chairman of microbiology at the Mount Sinai School of Medicine in New York. Moreover, he suggested that H5N1 wasn't nearly as virulent as many of his colleagues claim. "I feel the virus is awful for chickens. But this is not a virus that has been shown to really cause disease in humans except in unusual circumstances when the dosage has been extraordinarily high," he told me, adding that a person has had to practically sleep with a sick chicken to catch a bad case. Perhaps there is some hidden, immutable attribute of the virus that precludes it from ever spreading easily among people. Maybe the dice are loaded, never to come up snake eyes no matter how many times they're tossed.

This line of reasoning is comforting but, unfortunately, unconvincing to many other virologists. "Such complacency is akin to living on a geological fault line and failing to take precautions against earthquakes and tsunamis," wrote a leading team of flu specialists. How much time does a virus need to become a pandemic strain? Scientists don't know. There's scant information about the virological events that preceded previous pandemics. Had the 1918 strain been smoldering in animals for many years before it crossed to people? Had the 1957 and 1968 strains been circulating for a long time, bouncing between birds and people, but gone unnoticed because these pathogens did not cause mass poultry die-offs like H5N1? Is a decade a long time for a virus to evolve into an epidemic strain? With severe acute respiratory syndrome (SARS), for example, it wasn't too long. Malik Peiris, the Hong Kong microbiologist who identified the virus behind the 2003 outbreak of SARS, cites evidence that people were exposed to that microbe for quite a number of years before it finally acquired the ability to be transmitted among humans. Once it did, it spread like fire. Flu could do the same.

"The virus has evolved in alarming ways in domestic poultry, migratory birds, and humans in just the last four years," Margaret Chan told a conference of American business leaders in early 2007. "Global spread is inevitable."

Chan's remarks came a month after she'd become director general of WHO. She herself had traveled far since the scare of 1997. In 2003, after staring down outbreaks of both bird flu and SARS, she had left Hong Kong for Geneva. Before long, she was the agency's assistant director for communicable diseases and its special envoy for pandemic influenza, which she identified as the most serious health threat facing humanity. By the time she ascended to the world's top health post in January 2007, she was well schooled in flu and convinced that a pandemic was coming.

"If you put a burglar in front of a locked door with a sack of keys and give him enough time, he will get in," she later warned at a summit of international health policy makers in Seattle. "Influenza viruses have a sack of keys and a bag full of tricks. They are constantly mutating, constantly delivering surprises." She cautioned that a pandemic strain would be unstoppable once it became fully transmissible. No corner of the world would be spared. So no country could count on outside relief as with earthquakes or tsunamis. "This will almost certainly be the greatest health crisis experienced for almost a century," she said.


But back in the fall of 1997, as the mystery of Hoi-ka's death had faded with Hong Kong's steamy summer, flu had all but vanished from Margaret Chan's mind. She was facing a new crisis. A public health clinic in Hong Kong had been mistakenly dispensing toxic mouthwash to sick babies instead of syrup for their fever. Many of the children had developed diarrhea and vomiting. The public was clamoring for an explanation. The scandal captured the city's grim mood as an historic year approached its end. Months earlier, with the world watching on television, Britain had ended more than 150 years of colonial rule by relinquishing sovereignty over Hong Kong to China. But the sheen quickly came off the handover. The Asian financial crisis that autumn rocked Hong Kong. The stock market crashed. The property market tanked. Tourism dried up. Even the weather was rainier than usual.

In late November, Wilina Lim's lab received a sample from a two-year-old boy who had been briefly hospitalized in another building at Queen Mary with a fever, cough, and sore throat. The lab staff tested the specimen for seasonal flu. They drew a blank. But now they had the chemical reagents required to check for H5N1. When they ran this test, it came back positive.

Lim called over to the health department headquarters. Dr. Thomas Tsang, a senior medical officer responsible for infectious diseases, got the news. "Not again," he said to himself, thinking of all the work this would mean.

Tsang led a team of Hong Kong investigators to the boy's home. It was located in Kennedy Town, an older urban quarter of aging apartment buildings and street-level shops at the westernmost end of Hong Kong island, far from the site of the original case. The neighborhood is tucked between the island's steep green slopes and the sea, the cranes of Kowloon's port just visible across the channel. The boy was

Vietnamese, the son of a migrant construction worker. Though the youngster suffered from a congenital heart condition, he had succeeded in recovering from the attack of flu and was back in the family's cramped apartment when Tsang came calling.

The interview was difficult because the parents spoke little other than Vietnamese. It took a lot of patience, and the investigators often resorted to hand gestures to convey the intent of their questions. Tsang asked about the boy's recent history, in particular whether he'd had contact with ducks, chickens, or other birds. The parents insisted he hadn't. But when the health officers produced a calendar and went over it day by day with the couple, they noted a Vietnamese festival about a week before the boy got sick. "What did you do for the holiday?" Tsang asked. The mother remembered she had bought a live duck or goose at the market. She had slaughtered it at home, littering the apartment with feathers and feces. To Tsang, the source of infection seemed clear.

Lim had also e-mailed the CDC in Atlanta with the results of this most recent test. When Keiji Fukuda was notified of this second case, he had no doubts this time that the results were correct. "OK, are we off to the races?" he thought darkly. The initial case was no longer a freak occurrence. His mind sped through possibilities. "Is this the first of many cases? Are there more cases going on? Is this the tip of the iceberg?" Fukuda packed his bags again and on Friday, December 5, headed with a fellow CDC investigator for the airport. The probe would be under the auspices of WHO, but CDC staff would carry it out.

The flight from Atlanta to Hong Kong takes about twenty-four hours. As Fukuda sat in his economy-class seat, he had a long time to reflect. He reviewed everything he would need to know. He thought about what his years on the trail of influenza had taught him and about the findings of the earlier investigation in Hong Kong. His mind groped for what was crucial, sorting and filtering the information. He sketched out the scientific surveys he'd want to conduct this time. He plotted out what he'd do as soon as he arrived. He didn't want to waste a second. "Question one, two, and three," he said to himself. "Are there other people infected? Are they passing it to each other? Is there an animal source?" So far, thankfully, the answer to the first question seemed to be no.

When the plane landed, Tsang met him at the airport. While officers were clearing Fukuda's passport through immigration and customs, Tsang whisked him into a side room to begin briefing him on the outbreak.

"I've got good news and bad news," Tsang quipped.

Fukuda waited for the rest.

The good news, Tsang said, is that he'd be taking Fukuda to a nice dinner. The bad news was that while Fukuda had been in the air, Hong Kong had confirmed two more cases.

Tsang brought Fukuda to his hotel in the downtown Wan Chai neighborhood, the former red-light district of *The World of Suzie Wong,* which in recent years had gone through a commercial renaissance. Then they went to work. They reviewed the lab data and the findings from the preliminary investigation, mapping out the next steps. It would be the first of many late nights spent together.

Fukuda and a growing CDC team was set up with an office in Wu Chung House, the imposing thirty-eight-story tower on Queen's Road East that was home to the health department and assorted other government agencies and private enterprises. From the windows of their corner room on the seventeenth floor, the Americans looked out at one of Asia's great skylines and each night watched the lights in the opposing buildings go dark. The team consistently worked past midnight, recapping the progress of their probe and debating its mystifying results. Cases continued to surface all over the city with no apparent geographic pattern. The ultimate source of infection remained elusive. "I've never been in an investigation where the stakes were both so high and information was so little about what was going on," Fukuda said. "I don't think I have ever slept less over a sustained period of time."

Early each morning, the questions would rouse him, at five o'clock, four o'clock, even three thirty, and he would resume his self-interrogation. "What are we missing?" he pondered. "What are we not asking? Is there anything obvious going on?"

He and his growing team of CDC investigators would meet for breakfast and compare notes on the day. By eight o'clock, they were back at the health department. They had dragged the desks to the center of the room to form a single large rectangular table and covered it with computer printers and laptops they had brought from home. At the far end of this command center was a whiteboard, where they recorded each suspected case, jotting down the age, gender, location, and crucial dates of the illness. The ultimate outcome was marked at the end of each listing. A downward arrow meant death.

The scrutiny from politicians, foreign officials, and particularly the press grew intense. The earlier investigation in August had received little media attention. But when Fukuda returned to the health department in December, the hallways were already crammed with reporters. Walking to the bathroom, he recounted, was an "exercise in photography." Press conferences became high-pressure events where the subtleties of epidemiology were often lost in the journalistic scrum, buried beneath shouted questions and the forest of microphones. One day after he arrived, Fukuda faced the media. "There's a possibility these cases are the only cases that appear and the virus completely vanishes. Another possibility," he added with foreboding, "is that these viruses may increase." By the middle of December, the number of cases had in fact reached double digits.

Hong Kong residents with little more than sniffles streamed to hospitals and clinics, fearing the worst. In normal times, these health centers took only a few samples each day from suspected flu cases. But now the samples were being wheeled into Wilina Lim's lab by the trolleyful, at least five hundred specimens every day. And health officials were demanding that the lab analyze them immediately for H5N1, rather than screening them first to see if they were flu at all. The workload overwhelmed the small influenza lab and its staff of two technicians. Lim yanked staff from the hepatitis and HIV labs and put them to work on flu, more than twelve hours a day, seven days a week.

Most samples came back negative. But the few that were positive made a terrifying impression. To Lim, this virus behaved like an alien. Her lab, adhering to international norms, normally tested for flu by culturing the virus in special cells harvested from the kidney tissue of cocker spaniels. Scientists consider these canine cells to be especially good for growing flu viruses, far better than most other types of cells. So it was little surprise when Lim's technicians placed the virus into the canine cells and it grew. But when they inserted it into a variety of other cells to see what would happen, they were stunned by what they saw under the microscope. "This one, it grows well in all kinds of cells," Lim recalled. "You can see the cell change very quickly. It seems that whatever cell you put it in, it just grows."

The long hours were also taking their toll on Margaret Chan, who was often up until the middle of the night speaking with anxious WHO officials in Geneva. But as the face of the health department, she tried to remain cool and upbeat. Once she went too far. Accosted by reporters after delivering a luncheon speech, she was asked whether she still ate chicken despite the flu scare. "Yes, I eat chicken every day. Don't worry," Chan told them reassuringly. Her response was scientifically sound, since the virus cannot spread in cooked meat. But to many in panicky Hong Kong, her answer seemed frivolous and out of touch. It sparked a brief furor in the local media, and she later admitted she'd fumbled the public relations.

In the privacy of her office, by contrast, Chan and her senior lieutenants were growing extremely worried. "Has the situation got out of hand?" Tsang recalled them wondering at one meeting. The disease seemed to be spreading exponentially. They suspected it might now be jumping from one person to another. They feared that hundreds of thousands in the city could soon fall sick. They still didn't understand where the virus was coming from, so they didn't know how to stop it.

Hong Kong's senior health officials looked to Fukuda for reassurance. But now he, too, was frightened, and his anxiety showed. Unlike in August, he was in the middle of a storm with no end in sight. Moreover, time was running out. In just a few more weeks, Hong Kong would enter its regular winter flu season, and Fukuda feared that the convergence of prosaic and novel infections would overwhelm the city's hospitals since there was no easy way to tell them apart. Even worse, if both strains were circulating at the same time, the opportunity for them to swap genes and spawn an epidemic strain would be tremendous. "All of that was very pressing," he said. "We were racing against time."

"It was striking that this was not regular influenza we were looking at, whatever it was," Fukuda later told me. Ordinary flu preyed on the weak: infants, the elderly, and the infirm. But this strain had demonstrated that even the young and healthy—especially the young and healthy—were its targets. "What does it look like?" he asked. "It looks like young people dying from something new. So it really brought us back to 1918."

The parallels were eerie. Ordinary flu has what scholars describe as a U-shaped mortality curve, with deaths concentrated among the very young and very old and a far lower proportion among those in between. The milder pandemics of 1957 and 1968 adhered to the same pattern. But during the Spanish flu of 1918, more than half the deaths were among those between eighteen and forty. This gave the disease a W-shaped mortality curve, reflecting the heavy toll in the middle of life as well as at the beginning and end. The avian flu outbreak in Hong Kong was much the same. And after the virus resurfaced in 2003, spreading its reach across much of East Asia, the deaths continued to follow this disquieting pattern. The largest toll was among those between ages ten and nineteen, followed by those in their twenties. The overall case-fatality rate was highest among those between ages ten and thirty-nine.

"Most of the time in public health and in medicine," Fukuda continued, "there's a fair amount of uncertainty, but you rarely come across issues where there's a really high degree of uncertainty and what you're sitting on may be something like a 1918. You feel like, 'I don't know what is going to happen. I don't know what is going on. But what is going on is not good, and what it reminds me of is the worst not-good of the century.'"

Researchers have yet to account fully for why the Spanish flu and avian flu, alone among contemporary flu outbreaks, manifest this W-shaped curve. "Explaining the extraordinary excess mortality in persons 20–40 years of age in 1918 is perhaps the most important unsolved mystery of the pandemic," wrote researchers at the U.S. National Institutes of Health. The answer could lie with another uncanny similarity between the two viruses. Historical accounts from 1918 and experiments on a version of the Spanish flu strain resurrected in the lab reveal that it also provoked tremendous cytokine storms, those withering counterattacks by the body's immune system. Scientists speculate that the young and healthy may be most vulnerable because, ironically, they have the most robust immune systems, thus the ones that launch the most vicious and ultimately suicidal responses. These victims may be undone by their own strength.

Scientists' understanding of this novel bird flu strain is still evolving, and the more they learn, the more they worry. Some now suspect that bird flu is moving down the same path as the virus responsible for the deadliest epidemic in human history. "This is a kissing cousin of the 1918 virus," warns Michael Osterholm, director of the Center for Infectious Disease Research and Policy and a frequent commentator on the pandemic threat.

Spanish flu, like bird flu, is thought by scientists to have been a wholly avian virus that developed solely through a series of internal mutations, as opposed to the genetic reassortment that spawned the 1957 and 1968 strains in addition to the swine flu of 2009. Some of the mutations discovered in the 1918 virus look familiar. "Notably, a number of the same changes have been found in recently circulating, highly pathogenic H5N1 viruses that have caused illness and death in humans and are feared to be the precursors of a new influenza pandemic," wrote a team of researchers led by Jeffrey Taubenberger, the American scientist who first fully analyzed the genes of the 1918 virus. Just since 1997, bird flu has become more like the Spanish flu strain. A series of studies shows bird flu has grown more virulent and less susceptible to antiviral drugs. "The H5N1 avian flu viruses are in a process of rapid evolution," said researcher Elena A. Govorkova in 2005. "We were surprised at the tenacity of this new variant." A later lab study suggested that bird flu may have already become more ferocious than the 1918 virus, laying even greater waste to the respiratory system and, fiendishly, targeting those lung cells specifically involved with repairing damaged tissue. In September 2006, WHO brought the world's premier flu specialists to Geneva to scrub the evidence. Malik Peiris, the renowned microbiologist from Hong Kong, told the three dozen participants at this private session something that took their breath away. If the virus continued to develop along the same path, ultimately emerging as a pandemic strain through internal mutation rather than genetic reassortment, its high lethality could persist. He concluded there was no scientific reason for expecting a decrease in the fatality rate, currently at 60 percent of recorded cases. His comments, though later reported by WHO, were largely overlooked by the media. Their import was horrifying. Once the virus evolved into a form easily passed among people, it would be expected to infect a quarter of humanity. So even if the actual fatality rate was only 50 percent after accounting for overlooked mild cases, this could mean the deaths of nearly a billion people.

That figure is so big as to be incomprehensible. Researchers would rather dwell on scenarios more akin to 1918. That strain claimed fewer than five percent of those it infected. If the coming pandemic follows suit, the global death toll would only be 62 million, according to one extrapolation.

The World Bank originally projected that a severe pandemic could cost the world economy $800 billion in the initial twelve months. By late 2008, the World Bank had nearly quadrupled this estimate, concluding that an epidemic would cost about $3.13 trillion during the first year. Even a mild pandemic, like the 1968 Hong Kong flu, would cost $450 billion, and a moderate one like the 1957 Asian flu would reach $1.3 trillion.

The gloom was suffocating in those final weeks of 1997, like the cold, foul mist wrapping Hong Kong's steep slopes and settling on its myriad islands. A small, tongue-shaped islet called Ap Lei Chau had become the latest focus of the city's collective anxiety.

Five-year-old Chan Man-kei had been playing with friends at her kindergarten, a brightly decorated school on the ground floor of a public housing project in Ap Lei Chau, when she started throwing up. Her parents had been called. A doctor had referred her to nearby Queen Mary Hospital, where her lab samples tested positive for bird flu. About a week later, on Tuesday, December 16, Hong Kong health officials announced that two of her younger cousins had also been hospitalized in Queen Mary's isolation ward. They, too, might have the virus.

Hours after that disclosure, Fukuda and Chan addressed the press about the heightened prospect of human transmission. "It's a possibility in this case and one of the things we are concerned about," Fukuda acknowledged at the evening news conference. Chan agreed that Man-kei might have infected her cousins. "They live together at Grandma's and play together," she said. The health department was trying to crack the case, she told reporters, assuring them, "We are working at breakneck pace."

Ap Lei Chau was connected to the southern shore of Hong Kong island by a bridge. A decade earlier, the government had erected several dozen brown-and-gray high-rise apartment buildings on the green hillsides of Ap Lei Chau, and about ninety thousand people now lived there, making it one of the most densely populated islands on Earth. Man-kei's two younger cousins and a third sibling stayed with their parents and grandparents in an apartment barely three hundred square feet in size. This was often where Man-kei spent her days.

When health investigators arrived at the apartment, they were rebuffed. The family patriarch, a sixty-four-year-old watchman, refused to speak with them or provide blood samples to see if he'd been exposed to the virus. Other family members were also reluctant to talk. They were afraid of being stigmatized or shunned by neighbors spooked by this new, mysterious plague. Fukuda arranged to meet one relative secretly at a café, the first of several clandestine interviews health officers conducted over the course of the outbreak. Eventually, they were able to tease out a history of the children's recent activities, noting what they had done together and what they'd done separately. (Only one of the younger cousins ultimately tested positive for the virus. All three children survived.) But the source of their infection remained elusive.

Chan was frustrated. She couldn't visit Ap Lei Chau herself because of the media frenzy this would cause. Nor was she getting a good sense of the family and its surroundings from the investigators. She kept sending them back to scare up more details, instructing them to observe the neighborhood at different times of day and different days of the week.

"I want to know exactly what is going on," she insisted. "What do the children do?"

"They play in the car park," came the response. "That's their playground."

Chan wanted to be able to picture the parking lot. She told them to take photographs and draw her a map. They did. Nothing seemed amiss.

But when the team returned to Ap Lei Chau the following Sunday, they spied cages of geese in the parking lot at the base of the apartment tower. There were several stalls near the entrance. Perhaps they'd been there on previous occasions and the children had passed too close. The investigators snapped some photographs and later presented them to Chan. They were all thinking the same thing.

Samples were taken from the grubby cages and tested. "Bingo," Tsang recalled. "We found a positive swab in one of the stalls."

They had identified a likely source for the infection. But could it account for both Man-kei and her cousin? Or had Man-kei caught the virus and then passed it to her cousin while they were frolicking on their grandmother's carpet? Could the virus already be spreading from one person to another? Neither Chan nor Fukuda nor anyone else could ever rule that out.


Hong Kong at Christmas was a city under siege. People were flooding emergency rooms to be tested for the virus. Even medical personnel were calling in sick, fearful they'd been infected. Drug prices had spiked, and panicked calls were overwhelming the health department's hotlines. Though only a handful of sick chickens had been discovered in the city's live markets, poultry sales had plummeted, and restaurants were banishing popular Cantonese chicken delicacies from their menus. On Christmas Eve alone, three more suspected human cases had been announced. Now the first of the seasonal flu cases were also surfacing.

There was no holiday break that year for the six members of the

CDC team. Fukuda had never before missed a Christmas with his wife and daughters. So the health department arranged a Christmas Day lunch cruise for their CDC guests. But they couldn't escape the oppressive mood. Word came that four more suspected cases had been identified, the highest one-day total so far.

"I don't know whether this disease will stop or spread," Fukuda once again told reporters two days later. He was appearing at a press conference on Saturday, December 27, with Chan and Hong Kong's agriculture chief, Lessie Wei. There were now about twenty confirmed or suspected cases. The press was demanding to know what more the government would do to stem the crisis. Since poultry were thought to be responsible for most of the outbreaks, would officials have them killed? Chicken hawkers had already been ordered to clean and disinfect their cages. On Christmas Eve, Hong Kong had barred all imports of chicken from mainland China, the primary source of the city's poultry and a possible origin of the infection. "I feel at this point in time," Chan responded, "the measures are sufficient."

The call that changed everything came that night at two in the morning on Sunday, December 28. Chan was at home in bed. It was the agriculture department. There had been a die-off of chickens at Cheung Sha Wan, the city's main wholesale poultry market. About fifty birds from a batch of three hundred in a single stall had abruptly fallen over dead with swollen chests and necks, internal bleeding, and other symptoms of avian flu. And there was more bad news. A week earlier, a similar outbreak had occurred at a farm in the New Territories. The disease had moved gradually along a row of cages, claiming its victims. The test results had just come back positive for bird flu.

Chan couldn't fall back asleep. "What is going on?" she wondered. Only days before, city inspectors had toured Hong Kong's farms and reported that there were no outbreaks. "Margaret," she told herself, "the size of the problem is bigger than what it appears."

She didn't have proof of a widespread poultry outbreak. But if there was one, it could finally explain the unrelenting series of human cases. She thought about the holding pens, nicknamed "chicken hotels," where Hong Kong's birds were kept overnight. As demand for poultry had dropped, retailers in the city's live markets found they couldn't sell many of their chickens. At the end of each day, birds from various stalls and markets were gathered together before being distributed again for sale in the morning. Chan's technical advisors had told her this system was perfect for disseminating disease throughout the markets. "Enough is enough," she thought.

After a long wait for sunrise, she contacted researchers at Hong Kong University. They had begun sampling poultry in the city's markets just before Christmas. Most of the test results had yet to come back. Chan urged them to take as many specimens as possible and to hurry up. She was about to recommend drastic action but didn't yet have the scientific evidence to support it. "When you are ahead of the curve in dealing with new and emerging infections, science is always lagging behind," she later explained. But despite the uncertainty, she wouldn't wait. "Don't be afraid to make major decisions," she told herself. "Don't be afraid to be wrong."

Chan spoke that morning with her boss, Chief Secretary Anson Chan, who ran the Hong Kong administration. They scheduled an emergency meeting for later in the day. As health director, Chan would tell her that all the chickens in Hong Kong had to go. But there was no guarantee the city's political leaders would sanction such a costly measure. "I am prepared, if they don't accept that, to resign," she thought. "I will resign, because if the environment does not allow me to do my job to protect the people, then that is the proper action to take."

Chan's driver ferried her up to the Peak, the highest point on Hong Kong island and home to a white colonial villa called Victoria House. Once a taipan's summer retreat, it was now the official residence of the chief secretary.

Chan and agriculture director Lessie Wei briefed the chief secretary on the latest poultry outbreaks. Hong Kong had already closed its borders to imports from China, yet the poultry infections persisted. If the disease continued to circulate among birds, Chan explained, the public health threat would mount, especially if the virus mutated or reassorted. That's why all 1.2 million chickens in Hong Kong had to be culled. About three hundred thousand ducks and geese that were kept in close contact with the chickens would also have to go.

The chief secretary cautioned that the economic implications were huge. A mass slaughter could severely harm the livelihoods of countless chicken farmers and traders.

"People will not like it," Chan admitted, "especially when it affects their vested interests."

The chief secretary continued to press. The poultry sector would demand the government pay compensation. The bill could be tremendous.

"Yes, it's going to cost money," Chan agreed.

"What if we don't solve the problem after killing all of the chickens?" the chief secretary asked.

"Then we need to go after all the ducks and geese," Chan said.

"And what if we still don't solve the problem again?"

"I'll be accountable," Chan answered. If she needlessly put Hong Kong through the trauma of a mass slaughter, she'd accept the consequences. "I'll deliver my head on a platter. I'll resign."

Finally, the chief secretary accepted Chan's recommendation. She agreed to take it to Hong Kong's leader, Chief Executive Tung Chee-hwa.

Within hours, Chan announced that Hong Kong would kill every last chicken. The slaughter would start the following afternoon. They were all to be gone within a day.

From the start, it was clear the slaughter would not go as planned. The government pressed 2,200 public employees into the operation, giving them masks, canisters of carbon dioxide, and orders to gas the birds to death. But these were dogcatchers, park rangers, and other civil servants with no experience in killing birds and unsure even where to find all of Hong Kong's farms. Day laborers were hired as reinforcements, and even some market traders joined the effort, slitting the throats and snapping the necks of their birds. The result was literally bloody chaos. The chickens clawed and scratched and scampered for safety. Flies swarmed. Farmers resisted. By the end of the first day, less than a fifth of the chickens targeted had been slaughtered, and many of those remained unburied. Television showed plastic garbage bags of carcasses heaped high while stray animals and vermin scavenged through the carnage, spreading fear of contagion. Nearly a week into the slaughter, fugitive chickens still roamed the streets.

Many traders and laborers in the city's nearly one thousand poultry markets were incensed. On the third day of the slaughter, New Year's Eve, Chan accompanied Chief Executive Tung to the Cheung Sha Wan wholesale market, an expanse of weathered stalls with rusting corrugated metal roofs in an old industrial quarter of Kowloon. They walked through the parking lot, crowded most other mornings with small trucks heavy with poultry, and toured sheds crammed with wood and bamboo cages, now empty but still caked with droppings. The chickens had gone silent. Instead about two hundred market employees confronted the officials, shouting objections and waving placards with slogans scrawled in red paint.

The *South China Morning Post*, Hong Kong's leading English newspaper, captured the prevailing public skepticism. In a front-page editorial, the paper asked whether the mass slaughter would eradicate the bird flu virus and allow poultry sales to recover. "There must be serious doubts whether either of these aims will be quickly achieved, if only because the central question surrounding the spread of bird flu has still not been answered: where does it come from? Until we know the answer, the killing of a million birds cannot hope to quell the public's understandable fears. And, more importantly, nor can it be certain to stop any more cases of bird-to-human transmission of the deadly H5N1 virus." A week later, this skepticism turned to harsh criticism over what the *Post* labeled the "botched" operation. "What is amazing is that the Government should have embarked on this task without any accurate estimate of its scale or duration," the newspaper editorialized. "Equally astonishing is that it should have bungled matters so badly as to raise the possibility that other animals may have become infected."

Yet even as the *Post* published those words, the final chickens were meeting their fate, and Chan was starting to find her vindication. The researchers at Hong Kong University had finished sampling nearly 350 chickens at markets around the city. They discovered the disease was even more widespread than expected. One of every five chickens had been infected.

Chan had accurately identified the source of the crisis. But had she found the solution?

Reporters asked Chan shortly after she announced the slaughter how she would know if it was a success. Chan told them it would have to put a complete stop to human cases. The normal incubation period for flu was a week. To be confident, Hong Kong would wait twice as long. If there were no more cases by the end of the second week of January, she would declare victory.

Just two days into the waiting period, Chan received a report of a new case, a three-year-old boy. "Son of a gun," she thought. "What is going on?" Chan was scared. But when she reviewed the boy's history with her staff, they concluded he had likely been infected just before the mass cull began. For two weeks the newspapers counted down the days and the population held its breath. On the final day, two new cases were announced. But these victims, too, had fallen ill before the slaughter.

There were to be no more. The final count stood at eighteen. Of those, six had died. Hong Kong would be the source of no new human infections, not that year and not for the next decade. The city had banished a killer.

Researchers later concluded a pandemic had been averted. This took an unprecedented effort that marshaled some of the world's leading disease specialists and courageous investigators. It required exhausting lab work in Hong Kong and Atlanta, tapping some of the most sophisticated techniques then available to medical science.

Yet that alone was not enough. Hong Kong had aggressively pursued the pathogen from the instant city health officials learned of it. They took radical action to eradicate the virus though it was a gamble that carried a huge economic cost. The government acted with openness, even when hammered with criticism. As a result, humanity's first brush with this novel strain would be its most successful. Over the following years, no other government would match this achievement, and the standard set by Chan and her colleagues would too often be honored in the breach.

Hong Kong's success also lay partly in its nature, rare in Asia. As a small, mostly urban outpost, there were few agricultural interests to contest the imperatives of public health. Nor did most residents of Hong Kong live among livestock. Their main exposure to the virus, researchers later concluded, was at live markets. And though an age-old preference for fresh meat made these markets an integral part of Cantonese culture, the government overcame the inertia of tradition by restructuring them. When they reopened, Hong Kong barred the sale of live ducks and geese, believed to be the original source of the infection in chickens. Live chicken sales resumed, but all imported birds were screened for the virus before entering the markets.

But the virus would prove implacable. Even in Hong Kong, it would resurface in 2001, killing chickens in market after market. As a pre-emptive strike, the government ordered a second mass slaughter of all poultry in the markets and imposed a mandatory rest day each month when they would be emptied, unsold poultry killed, and the stalls cleaned before restocking with new flocks. Undeterred, the virus struck yet again the following year. But over that period, it never again jumped to people.

Fukuda eventually would go on to become WHO's global influenza chief. Yet he always remembered the Hong Kong investigation as the most rewarding in his life. He lauded Chan's leadership as heroic. He had no way of knowing that nearly a decade later he would be reunited with her in fighting the virus as it spread to much more difficult terrain far beyond the horizon.

# CHAPTER THREE

# The Elephant and the Lotus Leaf

The operation, as Dr. Prasert Thongcharoen called his lonely campaign against a killer, started on a hungry stomach. After seventy years of life, he had developed an abiding fondness for farm-fresh eggs and looked forward to those occasions every few weeks when a pair of young friends would present him with dozens of the finest from their family homestead outside of Bangkok. So when they returned to the Thai capital empty-handed one warm December day in 2003, Prasert was disappointed. He also suspected something was terribly amiss.

A distinguished man with silver-and-black hair combed back off a high forehead and eyes keen behind thick aviator glasses, Prasert had accumulated countless honors and accolades during his pioneering career. He had been dean of the medical school at Bangkok's Mahidol University, chief editor of the Thailand Medical Association's journal, and a fellow of the Thai Royal Institute. He had also earned a reputation as something of a gourmet. He often cooked for the staff at Siriraj Hospital, where he continued to do research as an emeritus professor, preparing chicken-leg curry and other favorites in the small departmental kitchen. He experimented at home with new recipes, trying them out on his children. He was so passionate about fresh food that he once flew home from Hong Kong with a newly butchered goose in his hand luggage.

The doctor was partial to eggs sunny-side up for his breakfast and had been expecting a fresh batch as usual. Two of his longtime friends, the chief reporter at one of Thailand's leading newspapers and his wife, had taken advantage of a national holiday marking the birthday of Thailand's revered monarch, King Bhumibol Adulyadej, to escape Bangkok's suffocating traffic and hectic pace. They had decamped to their farm about fifty miles to the east amid the mango orchards of Chachoengsao province. After a brief vacation, the couple returned to the capital and came to Prasert with apologies. "What happened?" he asked. The response astonished him: "There are no chickens, so there are no eggs. The farm is usually full of chickens. But the chickens all died. We don't know why."

The couple traditionally visited their farm twice a month and for years had been furnishing Prasert with six or seven dozen eggs on their return. This time they brought him only a riddle. It was a grim and disquieting puzzle that would become the focus of a personal crusade. Almost alone, he would press his kingdom's leaders to admit that a plague was raging in the Thai countryside and threatening to leap beyond the borders. "I had no eggs to eat," he later told me. "And so the operation began."

Six years had passed since the novel strain of flu had crossed to people in Hong Kong in 1997, and the virus had been nearly invisible since then. In the interim, the only confirmed human cases in the world had come several months earlier in a Chinese family. At this moment in late 2003, though Asia was on the brink of an unprecedented outbreak in both birds and people, flu was on few minds. The virus, it seemed, had vanished, lulling much of the world into a naive confidence that the pandemic threat had passed.

This would become the pattern. Over the next few years, the H5N1 strain would repeatedly raise its head, surfacing in ever more countries and provoking grave public health warnings anew. Yet each time, the virus would stop short of epidemic and then retreat. Politicians would boast they'd cornered the bug. Media attention would fade. Much of the public would forget.

In 1997, H5N1 had been but one hurdle shy of pandemic, yet that recognition vanished in the silence that followed the Hong Kong outbreak. The virus was a subtype against which no one was immune. It had demonstrated it could infect people. Only the last of the three conditions for pandemic had remained unfulfilled: that the disease could be relayed along a human chain.

It would fall to Prasert to prove that once again, a novel strain was circulating across the landscape of East Asia. Soon after, he would also reveal that the virus had begun to strike his countrymen. And eventually, as the world watched anxiously to learn whether the virus could be passed from one person to another, Prasert would play a central role in confirming it could. Although Prasert held the title of WHO consultant, he waged this campaign mostly by himself. For while WHO's flu specialists back in Geneva also suspected something was not right, they had to defer, as they all too often must, to the assurances of a sovereign state that all was well.

When I first met Prasert in 2005, the bags beneath his eyes had grown heavy and the lids were sagging. The years had generously etched his forehead with fine lines. But his curiosity and intellect remained acute. He was still Thailand's most eminent virologist and the man who literally wrote the book on flu in his country. His 1998 monograph on the disease begins, "Influenza has been an epidemic illness since ancient times but the flu virus never remains the way it was. It never stops changing."

When the Asian flu pandemic of 1957 swept through Thailand, Prasert was just one month out of medical school and beginning his hospital residency. As pandemics go, it was comparatively mild. But for the three months that it throttled Thailand, the flu flooded Prasert's hospital with the sick and dying. Several dozen of his fellow residents fell victim. The young physician was one of only two healthy enough to attend to patients. Eleven years later, when pandemic revisited Thailand in the form of the Hong Kong flu, Prasert was already emerging as a leading researcher. He devoted his efforts to deciphering the virus's behavior by scrutinizing the pattern of mass absenteeism at Thai schools and industries.

Prasert received me in his sixth-floor office at Mahidol University's microbiology department, where he had established the virology program decades earlier. Siriraj Hospital, home to the department, is a teeming complex of white-and-cream towers rising on the west bank of the Chao Phraya River, which winds through the heart of the Thai capital. The dark waters below bustle with low-slung ferries shuttling commuters and shoppers between the shores, carnival-colored long-tail boats popular with sightseers, and the occasional tugboat dragging a sand barge slowly downriver. On the opposing bank shimmer the fairy-tale pagodas and gold steeples of the Grand Palace, the Temple of the Emerald Buddha, and the other signature sites of royal Bangkok. Prasert's office was cramped but well air-conditioned, a relief from the hospital's steamy corridors. Several bouquets of silk-and-plastic flowers brightened the room. As Prasert's assistants brought us coffee, he slipped on the white lab coat of a man still very much involved in scientific inquiry and sank into the cushions of a purple sofa. Then, after offering a few observations about the merits of Thai coffee beans, he began to recount how he had forced his government to come clean.

In November 2003, Thai poultry farmers had begun to complain about a disease devastating their flocks. Their chickens at first seemed depressed and lost their appetites. The hens stopped laying eggs. The birds would soon start to wheeze and cough as the sickness ravaged their lungs. Some would develop diarrhea. Many would stagger around as their brains began to hemorrhage, melting into a bloody cranial mush. They would die suddenly, often overnight, and by the thousands. This disease was killing with a speed almost never seen in Thailand or anywhere on Earth.

Thai agriculture officials investigated and, suppressing the truth, reassured farmers it was nothing serious, just a minor outbreak of fowl cholera complicated by bronchitis. The country's deputy agriculture minister and livestock chief would both later blame the spiraling death toll on the weather, saying that an abrupt seasonal change had left chickens in a weakened state that made them vulnerable to catching these two infections in tandem. There was no reason to panic. There was certainly no reason, the agriculture officials said disingenuously, for anyone to doubt the quality of poultry raised in Thailand, which at the time was the world's fourth-largest chicken exporter, annually shipping $1.3 billion worth overseas.

Prasert heard the initial reports and had no reason to disbelieve them. But he was intrigued. Though he had spent years studying infectious diseases, from dengue fever to HIV-AIDS, fowl cholera was entirely new for him. "I just wanted to see about it," he explained. So in the predawn hours of a December morning, Prasert headed for the wet market that sprawls across more than a dozen acres beside the Bangkok Noi railway station. There, across from the creaky gray train carriages ferrying farmers from the countryside, Thais continue to shop as they have for generations. Prasert ventured into the warren of crowded aisles, slightly stooped, stepping carefully along floors slick with water and mud, past mounds of pineapples, mangos, hairy rambutan, and lavender dragon fruit, past chili peppers and curry powder heaped high in plastic baskets, past pig heads lined up along white tile countertops and catfish slithering in their trays. Even before he reached the chicken mongers, he could hear the repeated thump of their cleavers as they hacked the birds into wings and legs, feet and ankles, liver and intestines. He asked around until he located several peddlers whose families raised poultry in Nakhon Sawan province, about a hundred miles north. Prasert had heard it was at the center of the fowl cholera outbreak.

"They said they didn't believe it," he recalled. "They said it wasn't like fowl cholera. If they have some chickens that are sick with that, they give them tetracycline, and they get better. But these chickens, by the next morning, they're all dead. So they thought there was something more than that."

Prasert was unsure what to make of this skepticism and filed away their dissent. "I didn't pay much attention to that then," he said. But within two weeks, when his friends came to him with apologies instead of eggs, he remembered what the farmers of Nakhon Sawan had said.

His ears now attuned to reports of ailing chickens, Prasert was starting to hear about a bird flu outbreak in southern China. Though Chinese agriculture officials would not disclose the spreading infection for nearly two more months, in late January 2004, the virus had in fact been circulating there for years. This was no secret to Chinese farmers, veterinarians, and researchers, and the rumor soon reached Prasert. He surmised that the flu virus afflicting Chinese poultry was now making chickens sick in Thailand. He suspected that migratory birds had spread the disease from China. But he still needed proof.

With Thai livestock officials continuing to hide evidence of bird flu, Prasert turned to two colleagues at the Ministry of Public Health. Together they plotted to smuggle poultry samples from the countryside into Bangkok for testing. Since agriculture officials had barred them from visiting chicken farms, Prasert and his colleagues called a farmer in Chachoengsao province, home to the county's largest concentration of chickens, and recruited him as an accomplice. They arranged for him to wrap two dead chickens in newspaper and then leave them at the edge of his property like innocent rubbish. The pair of health ministry officials would drive out to the province to collect the contraband carcasses.

When the officials arrived back in the capital, they called Prasert. "We've got them," they reported. "What should we do now?"

"Bring them here," he instructed.

As a medical facility, Siriraj Hospital was not authorized to test animal samples, just those from humans. But Prasert was sure that veterinary labs would refuse to run tests of the dead chickens, fearing government reprisal, or would decline to tell him the results. "We had to do it ourselves. It was a must," he said. Prasert immediately turned the samples over to lab researchers at his hospital, who began secretly analyzing them for both influenza and fowl cholera.

The proof he was seeking came later that evening, "December 19, twenty-hundred hours," Prasert recounted, flashing a broad smile. The researchers brought him lab results demonstrating that the birds were infected with influenza. The tests were not precise enough to determine whether the virus was specifically H5N1, which had surfaced in Hong Kong in 1997 and had been circulating in China ever since. Siriraj Hospital did not have the right chemical reagents to tease this information out of animal samples. But Prasert was now sure that something more than fowl cholera and bronchitis was troubling Thailand's chickens. And if it was bird flu, he was convinced it had to be

H5N1. That meant that the people of Thailand and beyond were in peril. They had no natural defense against this novel strain of flu.

Prasert placed an urgent call to senior officials at the health ministry. "I have evidence that this is avian flu," he notified them. He told them they had to be vigilant. He urged them to prepare for a medical emergency. But he was rebuffed. "For agricultural production, we have nothing to do with that," one senior health officer responded curtly.

Prasert was aghast. "This is a public health concern," he scolded, "and I will not close my mouth. I will talk even louder. I'm retired already and no one employs me. It's not like I'm a government official."

Prasert was not alone in suspecting the worst. Inside Thailand's health ministry itself, infectious-disease specialists were already speculating that the strange sickness decimating the country's chickens was flu. Officials in the ministry's disease-control department later told me they had heard rumors as early as November 2003 about flu outbreaks among birds in China and Vietnam. So the ministry sent out letters across Thailand alerting doctors in the provinces to be on guard. The public health surveillance system, which Thailand put in place months earlier in response to the SARS epidemic, was urgently revamped to monitor for bird flu. But health officials remained mute in public. They issued no advice on how to avoid the virus, no warning against contact with sick birds. There was no public guidance about how to recognize the disease if you caught it and what to do if you did. No effort was made to import and stockpile Tamiflu, which was considered the main remedy for the virus.

Nor was Prasert the only one with hard lab evidence of a flu epidemic out on the farm. Across town at Bangkok's elite Chulalongkorn University, researchers had confirmed it was bird flu back in November and even determined that the virus was the H5N1 strain. Associate Professor Roongroje Thanawongnuwech, head of the university's veterinary diagnostic laboratory, would later disclose that the virus had been found in samples taken from Suphan Buri province, about sixty miles northwest of the capital. His lab used a test called the RT-polymerase chain reaction to reveal the virus's makeup. Researchers compared their results with GenBank, a public database run by the U.S. National Institutes of Health containing tens of thousands of known genetic sequences, and discovered that the Thai virus was very similar to those previously isolated in Hong Kong and Vietnam. Roongroje later reported that these findings had been turned over to Thailand's national livestock department. But again the public was not told.

Thai health officials and researchers were cowed. They were under intense pressure to keep mum. At stake were the country's lucrative poultry business and its export markets, primarily in Europe and Japan. Hundreds of thousands of Thais were estimated to work on thirty thousand poultry farms and in related industries. While some infectious-disease specialists in the health ministry were anxious to sound the alarm, they were silenced by a powerful agriculture ministry whose primary aim was promoting farm interests.

"We were fighting with each other and not the virus," admitted Dr. Kumnuan Ungchusak, a senior Thai health official, in a remarkably frank lecture delivered in May 2006 at a medical conference in Singapore. As the director of epidemiology in the ministry's disease-control department, Kumnuan had been uniquely positioned to investigate the threat to human health that began emerging with the mass poultry deaths in November 2003. But he continued, "Nobody dared to speak that this was H5N1." He urged his listeners from countries facing the epidemic—by that time the virus had spread to birds in more than fifty countries—not to repeat Thailand's mistake in withholding information. "We should have declared this a couple of months before so we can save a lot of poultry and we can save a lot of lives."

I was in the audience as Kumnuan recounted his experience and was surprised by his public candor. After he finished, I asked him privately to expand. Why wouldn't anyone in the health ministry reveal the truth? "At the time," he explained, "there would be a large impact on exports. There was a lot of reluctance to inform the public about it. This is a very bitter story." But specifically, I pressed him, who gave the order to keep quiet? "I don't want to pinpoint," he demurred.

What Kumnuan remained reluctant to say was that the poultry sector's pull had extended beyond the ranks of the agriculture ministry to the very top of the Thai government. The industry had influential advocates in the tight inner circle of Prime Minister Thaksin Shinawatra. A telecommunications tycoon turned politician, Thaksin had come to office in a landslide in 2001 vowing to turn around the struggling Thai economy by running it like a corporation. He had indeed succeeded in resuscitating the economy by 2003, achieving some of the highest growth rates in Asia and promising even faster growth in the year ahead. He had also demonstrated his impatience with those who got in the way of his economic juggernaut. Thaksin labeled himself the CEO–prime minister. His critics called him autocratic and vengeful.

In mid-January 2004, nearly a month after Prasert produced proof of a flu outbreak among Thai chickens, Deputy Agriculture Minister Newin Chidchob was still vowing it wasn't so. "Irresponsible media and some groups of people are trying to spread this rumor," he told reporters. "There is no bird flu here." His assurances were endorsed three days later, on January 19, by David Byrne, the European Union's health commissioner, who was visiting Bangkok on a previously scheduled trip. Part of his job was to protect European consumers, and, after a briefing from agriculture officials, he pronounced Thai chickens to be safe. "There's absolutely no evidence of the existence of bird flu in Thailand," Byrne said in remarks he would soon furiously retract.

The next day, as television cameras rolled, Prime Minister Thaksin and his cabinet ministers tucked into a luncheon feast of spicy chicken soup, minced chicken salad, chicken biryani, chicken teriyaki, and grilled, boiled, curried, and fried chicken, capped by a healthy serving of the Thai leader's trademark bravado. "Come and join us. Are you scared?" the prime minister taunted reporters. With anxiety mounting among consumers, Thaksin's response was to order up the repast and claim all was well. "It's the best chicken in the world, Thai chicken," he offered between bites. "It's very good. It's safe."

But WHO already knew better. A week earlier, on January 14, the agency's office in Bangkok had received a confidential tip from a government epidemiologist reporting that bird flu had been detected in a recent poultry outbreak and tests had determined it was an H5 virus. Though more analysis was needed, the scientific implication was that the strain was a novel H5N1. This troubling disclosure was forwarded to WHO headquarters in Geneva, where infectious-disease specialists asked the Bangkok office to get more specifics from the Thai government.

These inquiries were repeatedly spurned. WHO's chief representative in Thailand and his staff were informed by their government counterparts that only the health minister could release details, and she remained tight-lipped. So over the coming days, the WHO team in Bangkok grew increasingly exasperated, venting frustration in e-mails and conversations with Geneva.

WHO had been on edge practically since the new year. Poultry outbreaks of bird flu had already been detected in South Korea, Vietnam, and Japan. Just a week earlier, Vietnam had confirmed human cases, stoking public fear of a flu pandemic for the first time since Hong Kong's outbreak six years earlier. Could Thailand be the next front? It was crucial to know. Yet there was little the agency could do. It couldn't force the Thai government to come clean. Nor could WHO itself go public unless the government cooperated.

As a United Nations agency, WHO was established by its member countries and, like other UN bodies, respects their national sovereignty. Though WHO has been growing increasingly assertive in recent years, pressuring governments on occasion to hunt for infectious diseases on their turf and disclose them when discovered, the agency still largely defers to local politics. Even when an outbreak becomes apparent, WHO cannot dispatch investigators to a country without a formal invitation. And even after they get on the ground, these teams are barred from the field until authorized by the country to proceed.

The agency's critics fault it for becoming a prisoner of its politics. They accuse it of too often bowing before the dictates and deceits of its member countries, of placing a higher price on diplomatic nicety than on truth. But senior WHO officials, including some who have personally braved the world's most horrible pathogens, scoff at the contention that they're weak or cowardly. They counter that WHO is a creation of international politics and thus, by definition, a creature of one. Otherwise, they say, WHO and its mission could not exist at all.

Nor is WHO some kind of global health department with an army of doctors, nurses, ambulance drivers, and inspectors. It has no labs or hospitals of its own. The oldest disease-control program at WHO, even older than the agency itself, is its global monitoring effort for flu. The perils of pandemic combined with the economic and health impact of seasonal flu made this initiative an early priority. Yet even this program depends entirely on a network of outside labs—at latest count more than a hundred in eighty-plus countries—to track the evolution of flu viruses and help develop suitable tests, drugs, and vaccines.

Mostly, WHO supplements the efforts of individual governments, offering specialized expertise and scarce materiel like stockpiled vaccines for meningitis and yellow fever. To accomplish this, the agency relies on an extensive network of consultants from around the world, both public and private, to help investigate outbreaks, treat the sick, test samples, train local health staff, and deliver medicine, vaccines, and equipment.

These outside allies are people like Prasert, whose career was devoted to forging the institutions and disciplines of modern medical learning in Thailand that now make his country among the most advanced in the region. Yet there was always time for WHO. On a curriculum vitae stretching for several pages of publications and affiliations, Prasert prominently highlights his position as consultant to the World Health Organization. For over three decades, he served on various advisory committees for the agency, most notably the expert panel on viral diseases. He has run a WHO collaborating laboratory for AIDS research and edited an agency monograph on dengue fever.

In Geneva, senior agency officials describe their role in coordinating all this outside expertise by using words like *secretariat, catalyst,* and *platform.* What they mean is that they're like the salaried fire chief of a vast volunteer brigade.

———

WHO was born of the optimism that followed the Second World War, when international cooperation in the shape of the freshly minted United Nations and its agencies promised a new chapter in human history. Founded in 1948, WHO set its objective as nothing less than the "attainment by all peoples of the highest possible level of health." This was an ambitious goal. Yet advances in medical science at the time seemed to be bringing down the curtain on epidemic diseases that long plagued mankind, notably polio and smallpox. By the 1960s, however, WHO suffered a colossal setback with the failure of global efforts to eradicate malaria. It was emblematic of a broader resurgence of infectious disease as microbes mutated, outsmarting new medicines and vaccines, exploiting environmental degradation, poverty, population growth, and humanity's lapses in vigilance.

As a young American physician, Dr. David Heymann had played a starring role in the eradication of smallpox. He and his WHO team had tracked it to its final havens in India. But soon after, as a new recruit to the CDC, he confronted a pair of entirely new threats. In the summer of 1976, he was dispatched to help investigate a mystery pneumonia spreading through an American Legion convention in Philadelphia. The outbreak, which sickened more than two hundred people and killed nearly three dozen, was ultimately blamed on a previously unknown illness dubbed Legionnaires' disease. By the end of that same year, Heymann was in Zaire, responding to the first recognized outbreak of a horrible hemorrhagic fever called Ebola. He would end up spending thirteen years in Africa and, during that time, track the Ebola virus deep into the rain forests of Cameroon.

Heymann would later point to 1976—with its outbreaks of Legionnaires' disease, Ebola, and also swine flu in the United States—as an inflection point in public health history. Man's conceit was that modern medicine and potent drugs had given him mastery over emerging diseases. But the events of 1976 started to rekindle the world's concern about these threats, Heymann told me, and the appearance of the AIDS pandemic dashed any remaining illusion of invincibility.

"HIV-AIDS really caught the world off guard," he said. "This really changed the thinking. The world realized the vulnerabilities."

In 1995, WHO tapped Heymann to establish a program on emerging and communicable diseases. Storm clouds were gathering at all points of the compass: pneumonic plague in India, cholera in Latin America, resurging tuberculosis in Russia and Ukraine, Ebola in central Africa, meningitis across the whole of that continent, and an unprecedented epidemic of dengue fever in nearly sixty countries. Under Heymann, the agency overhauled its intelligence gathering, integrating a system developed by the Canadian health department that mines the Internet for reports and rumors of disease outbreaks. Next Heymann and his colleague Guenael Rodier set up what they called a global strike force, tapping disease investigators from more than a hundred universities, hospitals, and ministries who could get their boots on the ground within two days of any reported outbreak.

Then came SARS. In a matter of weeks in 2003, this novel respiratory disease spread to four continents, striking the economic heart of Asia, putting global air travel in jeopardy, and raising the specter of a worldwide epidemic. WHO's rapid response contained the epidemic before it became entrenched. This success consolidated the agency's role in managing outbreaks around the world. That largely explains why WHO, and not the CDC, took the lead in responding to the human cases of bird flu when they erupted in 2004.

SARS was a close call. It underscored the need to rewrite the global code of conduct called the International Health Regulations. The new rules, which took effect in the middle of 2007, require countries to notify WHO within twenty-four hours of any outbreak posing a global threat. Previously, the requirement applied only to yellow fever, plague, and cholera, a legacy of the nineteenth century, when European governments sought to forestall pestilence from the East. Now it was flu, again rising from the East, which posed the greatest menace.

The adoption of the regulations emboldened WHO. "When we come to an assessment that our assistance is needed, we have to push our agenda," said Dr. Michael Ryan, the burly Irishman who runs the agency's alert and response operations. But WHO is still ultimately constrained. Governments like the one in Bangkok can continue to tell it to buzz off. "At the end of the day, you are dealing with sovereign states," Ryan added. "That has to be respected."


One day before WHO was tipped off to the spreading epidemic in Thailand, a six-year-old boy with symptoms of pneumonia was rushed to Prasert's hospital. He had a fever of 104 degrees and was desperately short of breath. Within twelve hours, his breathing had grown so labored that the doctors placed him on a ventilator. It seemed at first to do little good, so they kept cranking up the pressure on the device until they could finally achieve an adequate flow of oxygen. An X-ray revealed that the boy's lower right lung had gone cloudy white, indicating that fluid was flooding the airspaces. The cloud spread a day later to the upper right lung. The next day, it progressed to the left one. The boy, Captan Boonmanut, had been brought to Siriraj Hospital from his home province of Kanchanaburi, located eighty miles from Bangkok near the western border with Burma. Outside Thailand, Kanchanaburi is best known for the Death Railway, built during World War II by Japanese occupying forces to supply its front lines, using Allied prisoners of war and Asian forced labor. At least sixteen thousand POWs perished from disease, hunger, and exhaustion, as did many more of the locals. This brutal chapter was captured in the Oscar-winning film *The Bridge on the River Kwai*, and the infamous steel-and-concrete bridge still stands, very much in use. But inside Thailand, Kanchanaburi today means rice paddies and chicken sheds.

Captan was a healthy youngster who had a country boy's love of farm animals. He would often play with the chickens that roamed his backyard. So when he had been handed a rooster during a fateful visit to his uncle's nearby farm, the boy had hugged it tight. Like many in rural Thailand, the uncle had raised fighting cocks and at first had high hopes for this particular rooster. But when it got sick, the uncle decided to do what most Southeast Asian farmers do with an ailing bird: eat it. Captan's parents told me how the boy had cradled the bird in his arms and kissed it during the final moments before it was slaughtered and converted to curry.

Captan fell ill within days. A nearby clinic diagnosed the illness as a common cold. When it got worse, his father brought him to the local hospital, where he was given injections of antibiotics. Then, as his fever climbed and his breathing began to race, he was rushed by ambulance to Siriraj Hospital, eventually admitted into the pediatric intensive care unit. His white-blood-cell count was plummeting. So was the level of platelets in his bloodstream. Doctors prescribed broad-spectrum antibiotics on the assumption that his pneumonia was caused by a bacterial infection—but to no effect. The disease was unrelenting. So the doctors shifted their diagnosis to a possible viral infection and began treating Captan with antiviral drugs. They notified Prasert, the hospital's most respected virologist.

The doctors had learned from Captan's parents about his history of close contact with poultry. His father had related the tale of the rooster. Family members further reported that all three hundred chickens on the uncle's farm had eventually died or been culled and that all but one of the chickens at Captan's home had also succumbed.

Prasert was afraid he knew what this meant, that he was seeing his worst fears materialize in his own hospital. But without definitive test results, he was reluctant to go public. "We had suspicions already but couldn't say anything. At that time, nobody could reveal information to anyone. The information the government was releasing was that we didn't have any avian flu," he said with narrowing eyes and an ironic smile. For all his credentials and earlier bluster, Prasert was wary of tangling with Thaksin and his ministers, at least for now. That very week, the agriculture ministry had threatened to sue another research institute and the media for allegedly damaging Thai national interests by exaggerating the number of chickens that had died nationwide. "What could I do?" Prasert asked. "I'm only a small, old man. Who would believe me?"

Subsequent study would reveal the viciousness with which the virus was assaulting the little boy's body. The disease was decimating his respiratory system, destroying the air sacs and capillaries in his lungs and inundating them with blood. The virus also invaded his intestines, where it established a beachhead and began to reproduce further. The pressure on the ventilator helping him breathe had to be turned up so high that even this was starting to take a toll.

Shortly after he arrived at Siriraj Hospital, initial tests confirmed that Captan had influenza A. A week later, on Thursday, January 22, another set of results came back and showed conclusively that it was the novel strain. Prasert now had proof that the second condition for a pandemic had been met. The virus was again infecting people.

Time was up. Prasert placed three calls in the following hours to officials at the public health ministry, including the minister and the director general of the Thai center for disease control. He rebuked them: "Bird flu has reached humans already." He also went public with his laboratory evidence of a flu outbreak in chickens, telling reporters that the H5N1 strain was widespread and the "cover-up" had to stop. His efforts were seconded by a top Thai lawmaker, a physician-turned-politician named Nirun Phitakwatchara. Nirun, a member of the Thai Senate, announced he'd learned from health officials about a second boy, this one from Suphan Buri province, who had also tested positive for bird flu. He accused the government of hushing up the outbreak for the sake of poultry exports. "I think it's very late but very late is better than not telling the truth," Nirun told me at the time.

The next morning, Thailand's Public Health Minister, Sudarat Keyuraphun, hastily summoned the Bangkok press corps. "There are two cases of bird flu, in a seven-year-old boy from Suphan Buri and a six-year-old boy from Kanchanaburi," she announced, adding that they were in stable condition. She said that everyone who had contact with the boys would be quarantined for ten days. She blamed the delay in disclosing the cases on the time required to finish testing samples.

The agriculture ministry followed right behind by issuing a statement confirming that chickens on a farm in Suphan Buri province had tested positive for the H5N1 strain. Samples from elsewhere in the country were still being analyzed. Newin, the deputy agriculture minister, announced that a mass slaughter of birds in central Thailand was already under way and that Thailand's poultry exports were to be suspended.

"It's not a big deal," Thaksin reassured the Thai public. "If it's bird flu, it's bird flu. We can handle it."

Tamiflu was urgently flown into the country and immediately administered to the sick boys.

Three days later, Thailand confirmed its first fatality from bird flu. In the early hours of Sunday, January 26, after taking an abrupt turn for the worse, Captan died.


Krisana Hoonsin could not sleep the night he paid eight laborers to slaughter all his chickens. He took a pill to help. When he awoke, he discovered that the silence blanketing the flat, lush province of Suphan Buri had enveloped his farm. Morning broke without the cackling and cooing he had known since he was a teenager. "It reminds me I'm not a chicken farmer anymore," he told me plaintively. "In a week, all the chickens in our district will be gone."

One day after Thai officials publicly confessed that bird flu had struck, Krisana sat heartbroken in a small, clapboard kiosk erected inches above a fishpond in front of his farmhouse. He wore a loose, black-checked work shirt and had a slight scar on his left cheek. His eyes were bloodshot, his dark brow deeply furrowed, like some of the nearby plots. Between the fingers of his rough right hand, the thirty-eight-year-old farmer clutched a lit cigarette, but he barely puffed. It would burn to a stub. Then, noticing just in time, he would rub it out and light another. This was still January, one of the coolest months in Thailand, but the midday sun was intense, so Krisana had taken refuge beneath the pitched, corrugated metal roof of the simple shelter. It was here that he had often come at dusk, when his chores were finished, and fondly gaze at one of his poultry sheds on the other side of the narrow country road. "My chickens would recognize me," he recounted. "They would stick their heads up and see me. Now it's empty." His voice cracked. "I still think of all my chickens."

His birds started getting sick two weeks earlier. He reported it to local livestock officers. Though they assured him it was only a minor case of fowl cholera, he was ordered to take draconian measures and put all seven thousand to death. Too upset to execute the sentence himself, Krisana hired a few locals. They marched down the tight aisles of the poultry sheds, wrestled the birds from the raised metal cages, and stuffed them alive into plastic feed and fertilizer sacks. The chickens were left to suffocate, then buried in a pit coated with lime at the edge of his property. "How can I express the feeling to see all our chickens die that way?" he asked. He let his sandals slip from his feet and rubbed his soles against the rough wood planks. "When you do chicken farming, it's like you're taking care of your own children. You love them. They love you back."

I had come to Suphan Buri early that morning with an energetic Thai journalist, Somporn Panyastianpong, who often worked as my translator. Before we left Bangkok, she had stopped to buy us surgical masks and rubber gloves, though the two health officials I had consulted were unsure whether these would adequately protect us. The disease was still so new, its precise lines of attack still uncharted. We drove north along the modern divided highway that connects the sprawling suburbs of the capital with Thailand's central wetlands. After an hour, shimmering green rice paddies opened up before us, many fringed with coconut palms. Storks, herons, egrets, and cormorants swooped and scavenged amid the neon fields. Peasants in straw hats meandered along the earthen dikes, hoes slung over their shoulders. A few ragged duck herders, barely teenagers, squatted at the edge of flooded paddies while their flocks waded into the murky waters, shaking their dark brown tail feathers and rooting around in the muck, hunting snails.

As we turned off the main road, we began see the scores of metal-roofed chicken sheds that Suphan Buri's farmers had raised in making their region one of Thailand's most prodigious poultry producers. These were long, open-sided structures on wooden stilts that all seemed to jut out over ponds and reservoirs. Under an ingenious system, chicken droppings are not cleared away but allowed to fall between the wooden floorboards into the water, which are stocked with carp, tilapia, and barb. The droppings serve as nourishment and save on fish food. The fish themselves often command better prices at market than the birds. But when the chickens die, the fish go hungry.

Now, as we ventured deeper down the rural roads, we drove past one eerily vacant shed after another. A legion of cullers had swept across the countryside ahead of us, killing an estimated 7 million birds over previous days in Suphan Buri and two other provinces. More than five hundred workers from the agriculture ministry were again fanning out across Suphan Buri to continue the mass slaughter. Teams clad in masks, rubber gloves, and high boots were storming through the sheds, cramming squawking birds into sacks and spraying disinfectant from tanks. Hundreds of Thai soldiers and several dozen prisoners were also being pressed into service, many with brightly colored shower caps to protect their heads. Though the government was targeting four hundred more farms on this day, Newin had warned that teams were running short of sacks and burial space. He told reporters that the cullers were now being forced to use the grounds of Buddhist temples.

Krisana's family had been raising hens in Suphan Buri for twenty-two years in a village called Baanmai. At first, their aim was to produce just enough eggs for income between rice harvests. But when Krisana took over the farm from his father in 1994, the ambitious young man decided paddy was the past and poultry the future. He immediately quadrupled the number of hens to four thousand, adding more in the following years. It proved a lucrative business. He built an airy, two-story house with a solid brick facade. An upstairs veranda with a cheerful blue-and-purple balustrade looked out over the emerald fields. He bought a new Toyota pickup, parked it out front, and hired a farmhand. He never imagined that one day his livelihood would be buried along with his birds in a hole in the side yard.

Over the years, he had grown accustomed to a few chickens dying suddenly and mysteriously. But he had never witnessed the kind of epidemic that had been stalking his province for the last two months. "It got bigger and bigger and spread from one farm to another before it reached our farm," he recalled. "Every night, three or four chickens would die." He consulted a veterinarian, who prescribed antibiotics. They had no effect. Local livestock officers could offer no explanation. The provincial livestock chief reprised the official line that the affliction was fowl cholera. But Krisana was starting to suspect something else. When Vietnam confirmed its poultry had been infected by influenza, Thai television carried reports with footage showing the symptoms. The birds suffered from stiff muscles, reddening skin, and chills, then died. "I saw it on the news and saw the same symptoms here, and I was sure it was bird flu," he said.

Krisana and his neighbors had alerted officials to their suspicions but were ignored. Now the farmers were livid. They were convinced their flocks could have been salvaged by a swifter government effort to quarantine contaminated farms. "They kept denying and denying and denying it was bird flu. If the government had admitted it earlier, they could have contained it," he said. A thin smile passed across his sullen face. "Instead, farmers kept transporting chickens and eggs from one place to another."

Slipping on his sandals, Krisana roused himself from his bench inside the kiosk. He led the way around the side of the house to give me a closer look at one of his chicken sheds, empty and deadly quiet. He shuffled along the wooden planks that served as a short causeway. Though the shed had been disinfected, I was still wary of following. I had the surgical mask and rubber gloves with me but Krisana had neither. I didn't want to be rude. So with some trepidation, I left them in my bag and poked my head ever so briefly inside the entrance. I tried not to breathe.

When we returned to the kiosk, Krisana's father came out of the house to join us. At seventy-six, Sompao Hoonsin was still vigorous, with thinning gray hair, and age spots on his broad face. He wore a jolly T-shirt with pictures of Winnie the Pooh and Tigger, but his manner was decidedly downcast. In his hand, he carried a small, handwritten note listing the family's liabilities in blue ink. They totaled nearly $40,000. The family, he explained, netted about $2,500 a month from selling eggs to small-time retailers in Bangkok, and this had long been enough for Krisana to support his wife, three children, and his father, who had raised him alone since he was a boy. But the debts mounted over the previous two years as the price for eggs weakened. They had to put up the house and land as collateral. "Finally," Sompao said, "things began to brighten." Prices had picked up since summer, and the weather had at long last turned favorable for laying: not too warm, with a gentle breeze. "We thought we'd be able to get out of debt and buy all our chicken feed without borrowing money," the elderly farmer continued. "Then suddenly, we had to bury all our chickens. We can't even earn one penny. All the children in the family, before they go to school, they're asking, 'Will we have enough to eat today?'"

Often overlooked in discussions about bird flu, amid all the anxiety over a possible human pandemic, are the staggering economic costs already incurred by Asia's farmers. For the poorest peasants, their few chickens were an insurance policy against hunger in bad times. For those who have proven more successful, like Krisana, the flu plague has jeopardized the investments and aspirations of a generation. Some of the farmers in Baanmai village, seduced by the riches that poultry promised, had gone so far as to ask their children to quit or forgo good jobs in Bangkok to help with the business back home. Now they faced bankruptcy.

A day before I stopped by the farm, Sompao had visited the bank to talk about the family's debts. He had hoped to defer the interest payments of about $250 a month. Nothing was resolved. Government officials had floated the idea that they might pay some degree of compensation for culled flocks. But even if they did, Krisana vowed he was through with chicken farming. "Once they announced the results of the lab tests, I got worried about my health and the health of my kids. We don't know if this flu would come back," he said.

His voice trailed off. A white government van pulled up at the edge of the front yard. Krisana watched as officers from the animal disease control department got out to examine the pit where his birds were buried. Satisfied that they were properly interred, the inspectors got back in and drove off.

Krisana resumed his thought. Perhaps it was time to go back to rice farming. He certainly had abandoned his ambition to keep expanding. Maybe self-sufficiency was the answer, he suggested dejectedly. "You can't imagine how it feels to sit and stay quiet when you see all you have suddenly disappear," he went on mournfully. "Everyone around here is in shock. We've lost hope in life."

Earlier in the day, Krisana had heard Thaksin might visit the province with his agriculture minister to inspect the culling operation and reassure local farmers. Krisana hoped they would. He said he had something for them—a lotus leaf. "If an elephant dies, you can't cover it with a lotus leaf," he quipped, reciting a Thai proverb. Don't try to hide a large mistake once it's in the open. "If a million chickens die, you can't cover it with a lotus leaf," he continued, embroidering on the original. He paused and reflected for a moment. Then he added, "I'd give them the leaf. But I'm not sure they would understand."

Hours after Captan Boonmanut died on January 26 in a nearby ward, Siriraj Hospital convened a seminar to discuss the gathering storm. Prasert was to brief his medical colleagues and review what it would take for H5N1 to spark a human epidemic. The first two conditions had been met. He wanted to discuss the third and fateful one.

Many of those at the seminar were relative newcomers to flu. Prasert reminded them there were two ways a bird flu virus could become transmissible among people. The virus could gradually undergo a series of discrete mutations making it progressively better suited to the human body. This first process was called antigenic *drift*. The other way, Prasert continued, was antigenic *shift*, in which the bird flu virus experiences genetic reassortment, swapping genes with an existing human influenza virus and creating an entirely new strain that is both highly lethal and as easy to catch as an ordinary flu bug. This latter transformation could happen overnight, he warned.

It had taken a whole lot of pushing and prodding to get the government to acknowledge that Thailand's birds were spreading the disease. But now, Prasert told his audience, an equally acute threat could be posed by the country's pigs. That was because swine could be what researchers called the mixing vessel, in which two flu strains exchange genetic material. Sick pigs in Asia, Europe, and Africa had repeatedly been found infected with a human strain of influenza. Prasert said he was also hearing reports, later confirmed by Chinese researchers, that pigs in China had come down with bird flu. If a pig caught both strains at the same time, the results could be catastrophic.

Over the coming months, as the scourge spread to a third of Thailand's provinces and across a half-dozen Southeast Asian countries, flu hunters would grow haunted by the prospect that the strain had cracked the code for human transmission. Yet even as the world was reawakening to this threat, Prasert was already probing how the virus might cross this final hurdle.

The next morning, his admonitions made headlines in Thai newspapers. But the reaction was not as he'd hoped. Swine farmers were enraged, fearing for their sales, and some threatened him.

The prime minister was asked by reporters about Prasert's warnings and brushed them off as the ranting of a mad old man, calling them overly imaginative and without basis in science. "Are the doctor and the media going take any responsibility if the virus does not spread to pigs?" Thaksin asked pointedly.

Dismissing the elderly virologist with a disparaging Thai word, *ai*, that can best be translated imprecisely as "goddamn," the prime minister accused the elderly virologist of going too far this time. "It was that goddamn doctor," he snapped, "saying it all by himself."

# CHAPTER FOUR

# Into the Volcano

The flu hit America early that year and it hit hard. It was shaping up to be the worst season in twenty-five years. That fall, in October 2003, Tim Uyeki had been summoned back to Atlanta while at an international influenza conference on the Japanese island of Okinawa. His CDC colleagues had urgently notified him about an unusual spike in severe flu cases among children. Texas and Colorado were being struck particularly hard. The culprit was a traditional strain of human flu but a new, unexpected subtype. As autumn turned to winter, the epidemic spread eastward until outbreaks were being reported in most states. The disease was taking an unusual turn in some children, resulting in neurological complications. Even worse, scores of children were dying. Uyeki, as both a pediatrician and influenza specialist, was tapped by the CDC to help run a national effort to identify and detail these fatal cases.

Now, in the waning days of December 2003, Uyeki found himself at his desk. The hallways of the CDC were depressingly empty except for a few other souls on flu duty.

He had been forced to cancel his Christmas vacation. He could have used the downtime. It had been a grueling year, much of it spent on the road. But the mounting pile of pediatric files beckoned. He had to sift them, study them, and try to divine why children were falling victim while the flu's typical casualties, the elderly, had this time been spared.

When he logged on to his computer on Monday, December 29, he came across an e-mail from Vietnam titled "Urgent." It was a copy of a request sent to one of Uyeki's colleagues by a virologist at Hanoi's National Institute of Hygiene and Epidemiology (NIHE). The Vietnamese scientist, Dr. Le Thi Quynh Mai, reported that Hanoi Hospital was treating a number of children with respiratory symptoms and doctors there were stumped. "We need to know what's causative of it," she appealed.

This entreaty was the first hint outside East Asia of a nascent outbreak that would soon transfix the world's flu specialists. In nearby Thailand, Prasert Thongcharoen had already concluded earlier in the month that bird flu was sweeping his country's poultry flocks. But he still had no inkling that it had spread to people. That would come three weeks later.

By then, global flu hunters would be streaming into Vietnam on the trail of the novel strain. Their pursuit would widen over the coming months to ever more provinces of Vietnam and then Thailand, the two countries to confirm human infections in 2004. (More countries, including Indonesia, would begin to report them in 2005.) And with each case, investigators would confront that terrible question: Had the virus been passed from one person to another? As the cases persisted and the deaths in Vietnam and Thailand mounted, it became increasingly clear there were indeed likely instances of human transmission. Yet the region's leaders and the senior brass of WHO itself remained loath to acknowledge publicly that the virus was flirting with the third and final condition for a pandemic.

As Uyeki reviewed the e-mail, he immediately thought of two possibilities, one worrisome and the other worse. "Could the situation be similar to what we are experiencing in the U.S.?" he wondered, thinking about the unusual uptick in seasonal flu. It would hardly be unprecedented for such a strain of human flu to circle the world. "Or," he pondered, "could these be highly pathogenic H5N1 virus infections?"

He had reason to suspect the latter. Though there was yet no public report of unusual poultry deaths in Southeast Asia, South Korea had officially disclosed a die-off two weeks earlier on a chicken farm outside the capital, Seoul. But Uyeki had little other information to go on. So he replied to the e-mail best as he could, laying out possible diagnoses, suggesting more than a half-dozen different viral infections. Topping the list were influenza of some stripe and an ailment called respiratory syncytial virus infection, or RSV, common among infants. He urged his Vietnamese counterpart to collect samples from the patients and test for those two possibilities.

Uyeki was already acquainted with the Vietnamese doctor and her colleagues in Hanoi. He had first gone to Vietnam three years earlier to collaborate with them on a study looking for evidence of bird flu in live poultry markets. He had stayed in touch, cultivating the relationship as he had with scientists across much of Asia. When SARS broke out in Hanoi in early March 2003, Uyeki returned to help contain the epidemic. He arrived just days after Vietnam's first case was identified and stayed for a month. Later in the year, he was back yet again, advising the Vietnamese on how to monitor for flu.

When Uyeki first joined the CDC's influenza branch in 1998, Keiji Fukuda had been on board for two years and had already helped run the investigation into Hong Kong's H5N1 outbreak. Now the two of them would return together to Asia yet again, trying to decipher whether the new threat was also a passing scare or a harbinger of something far worse.

Uyeki talks about his colleague as he would about an older brother. Both are Japanese American, graduates of Oberlin College in Ohio, and dedicated to a virus that others in Atlanta call "their bug." They are both among the best at what they do. But while Fukuda is reserved, precise, and methodical, Uyeki is exuberant. Fukuda speaks in carefully crafted arguments, commanding attention with an economy of words. Uyeki's discourses cascade from topic to topic, detouring through colorful details and intriguing distractions. Fukuda is the kind of man who organizes his day so he can drive his daughter to evening soccer practice. When I last met Uyeki, he was still regretfully a bachelor. His work habits are legend at the CDC. It is not unusual to find him at his desk until one in the morning or later.

So it was no surprise that Uyeki was still in the office when, shortly before midnight on December 29, Dr. Mai's e-mail reply arrived.

"Dear Tim," she wrote, "these childrens . . . have fever, cough, difficult breath." But other symptoms didn't look at all like flu, she reported. Some of the patients developed diarrhea a few days after the onset of illness and, she added, "died quickly."

"Died?" Uyeki thought. She hadn't mentioned that earlier.


Exactly a week later, Dr. Peter Horby got an urgent call on his cell phone. He had driven out of Hanoi that Monday morning to train Vietnamese medical personnel in a nearby province. Horby, a British epidemiologist, had joined WHO's Hanoi office only a few months earlier after working for several years at the Public Health Laboratory Systems in London, specializing in communicable diseases. He was still learning his way around his new home.

The call was from the director of Vietnam's National Pediatric Hospital back in Hanoi. He had a mystifying outbreak of respiratory cases. A week earlier, he'd brought these to the attention of senior officials at Vietnam's health ministry, but they'd brushed him off. So now he was turning to Horby. There was something about the call that told Horby he shouldn't wait. He broke away from the training session and directed his WHO driver to take him back to Hanoi, directly to the hospital.

The car pulled past the gate and onto the campus of the pediatric hospital, an oasis of soothing greenery and tropical decay in the middle of one of the capital's most crowded quarters. The institution's sun-bleached buildings with their ancient wooden doors and paint-chipped balconies were arrayed amid overgrown lawns. Stands of bamboo rose here and there. The grounds were still but for the chirping of birds in the generous shade trees and the occasional sound of a wailing infant. Uniformed nurses walked briskly along the scarred tile walkways. Orderlies in traditional conical hats shuffled past.

Professor Nguyen Thanh Liem, the hospital director, met Horby on his arrival. Three other doctors, including the heads of intensive care and infectious disease, were asked to join them. The doctors told Horby they feared that SARS might again be breeding within their walls. During the previous three months, they had admitted eleven children with unusual respiratory ailments, and seven had died. The other four remained hospitalized. Yet another child, the sibling of one of their cases, had succumbed a week earlier from a similar illness in a provincial hospital.

Horby inquired about the background of the children. There didn't seem to be an obvious pattern. They ranged in age from nine months to twelve years. They came from a variety of places outside Hanoi, mostly from the countryside but in a pair of cases from town. There were no reported outbreaks in their communities or in their schools.

The doctors escorted Horby to the intensive care unit to see the four surviving children. They had all been healthy just weeks earlier. At first it was just a runny nose, dry cough, and fever. Then the infection grew violent and spread to their lungs. When Horby reviewed the chest X-rays, they were desperately clouded. Their white blood count was low, suggesting the infection was not bacterial but viral. It was likely, Horby concluded, they would die.

"How unusual is this?" he asked the doctors.

"It's unusual," Liem replied. "They're not responding to treatment and we don't have a diagnosis."

Horby quickly surmised it wasn't SARS. That disease had largely bypassed children. It was more likely a pathogen called adenovirus, or perhaps flu. They agreed to conduct more tests on the patients and to press the health ministry about similar cases at other hospitals. WHO would supply masks, gloves, goggles, and face shields to the staff at the pediatric hospital.

A day later, Horby took a call from a journalist asking whether he'd heard reports about a massive die-off of chickens outside Hanoi. He hadn't. No one at WHO had. "That instantly started ringing alarm bells," he recalled. "The first day, we thought it was influenza. The second day, we were talking about possible avian flu."

Dr. Mai at NIHE came to the same conclusion later that week. She had finally succeeded in discovering her causative agent in a sample from one of the children. It was H5N1 bird flu. But WHO still wanted confirmation from an overseas lab with proven experience and turned to Wilina Lim in Hong Kong, the laboratory chief who had worked on the initial human cases in 1997. It took several days to get her the samples. Vietnamese Airlines had balked at transporting them, so another airline had to be found.

The results finally came back on Sunday, January 11. Two children had tested positive for the virus. And so had a third person, the mother of one of the dead youngsters.

WHO put out a worldwide call for reinforcements.

WHO was coming off a monumental victory six months earlier. The containment of SARS, a previously unknown killer that had spread to four continents before it was checked, marked one of the agency's greatest successes in a half century of history. But it came at a great price.

Agency personnel and their allies from dozens of countries had hustled day and night for months on end, often far from home, to uncover the extent of the SARS outbreak, crack its genetic secrets, and ultimately run it to ground. It was a sprint pace at marathon length. And as the death toll had mounted, so had the pressure. Individual governments made relentless demands on the agency. The global media's appetite for information was insatiable. The prospect of failure was chilling. After the final two countries, China and Taiwan, were declared SARS free in July 2003, the troops were utterly spent. In Geneva, where crisis had built camaraderie, the agency descended into internal bickering as all the disputes and grievances that had been repressed now bubbled up.

"People were just strung out," recounted Michael Ryan, who directed WHO's alert and response program. "Our systems survived. But I use the word survived because it's like surviving a nuclear explosion. We were still breathing. We were still feeling our limbs to see, were they all there."

When the threat of pandemic rose anew in January 2004, the agency was still reeling. "We were thinking, 'We don't want to do that again,'" Ryan said.

That was especially true for WHO in Vietnam, which had been

among the first countries struck by SARS and among the first to contain it. Pascale Brudon, the auburn-haired Frenchwoman who headed the agency's Hanoi office, likened the SARS experience to *The Plague,* by Albert Camus. She said the tension and sense of personal jeopardy had been profound, especially after the loss of Dr. Carlo Urbani. Urbani, an Italian infectious-disease expert assigned to Vietnam, had investigated the country's initial SARS case, and his early insights into the pathogen ultimately helped the world defeat it. But not before he, too, succumbed. He had been a popular figure, a hanggliding, motorcycle-riding musician of a man who, while previously working for Doctors Without Borders, had received the 1999 Nobel Peace Prize on that organization's behalf. His death was staggering.

When Horby notified Brudon in early January 2004 about what he'd learned at the National Pediatric Hospital, her reaction was, "Oh no, not again."

In Geneva, Dr. Klaus Stohr was the head of WHO's global influenza program. He had never doubted that bird flu would resurface, and he was waiting for the moment. "To prevent an earthquake or an eruption of a volcano, you always prepare for it," he recalled. "But when it happens, you're still surprised, still shocked."

Stohr wanted to get his flu hunters on the ground fast. But it was proving difficult to assemble a team. "There were some people, all international experts, who said 'Why should I go? Why should I jump into the frying pan?'" he recalled. They were thinking about their families. They were thinking about Carlo Urbani. "It's too hot for us to go right in the middle of a possible volcano." They demurred.

Horby was already on the ground and he, too, was thinking about Urbani. He had assumed some of Urbani's duties, and like his predecessor, was back in the hospitals, seeing desperately sick patients infected with an uncertain yet catastrophic agent. "It was a very worrying time," Horby later acknowledged.

But Uyeki, biding his time in Atlanta since he'd first learned of the outbreak two weeks earlier, couldn't get there soon enough. "You want to help and you want to find out answers," he told me. "Yeah, I was ready to go right away. Keiji and I, we're ready to go."

"What do I need?" Uyeki thought. He stocked up on antiviral drugs to dose himself. He collected his protective gear. As a matter of course, he had already been custom-fitted for N95 respirators, what most people call masks, and he replenished his supply. Then he and Fukuda started turning over their command responsibilities in fighting the seasonal flu still raging at home.

The flu outbreak that began that fall had jolted the American health-care system. It was only seasonal flu, but hospitals and doctors' offices were flooded with the infirm. Emergency rooms from coast to coast were reporting record numbers of patients, in some cases a hundred a day, and many waiting rooms were standing room only. Some hospitals made other patients give up their beds. Local government officials activated disaster plans. Just a week after Thanksgiving, flu shots already had run out.

But as nasty as that flu season was, again, it was only seasonal flu. In a pandemic, the health-care system could crumble. Just the initial rumblings of a pandemic, the first weeks of the swine flu outbreak in spring 2009, overwhelmed many American hospitals and clinics as patients with little more than common colds, or no symptoms at all, clamored to be checked out. A mild pandemic with a relatively low death rate would still sicken at least a quarter of the population, sending millions of petrified, sniffling Americans to the hospital. In a more severe epidemic, our broader society as we know it could be in jeopardy. That's the lesson of Philadelphia.

As a young reporter, I worked there for eight years—it was my first big city—and I got to know its streets well. I never realized I was sharing the ghostly geography of the worst calamity ever to befall the United States.

It was September 11, 1918 when the Spanish flu made its first recorded appearance in Philadelphia, striking the Naval Yard at the foot of South Broad Street. The virus had come ashore with scores of sailors transferred days earlier from Boston, a city already under siege. But Philadelphia's flu epidemic would evince its full fury only later in the month, after the city had experienced perhaps the greatest orgy ever of human-to-human transmission. Soon the city would be the hardest hit in the country, gripped not only by illness but by terror and social breakdown on a scale unprecedented in American history.

As autumn broke in 1918, the eyes of Philadelphians, like those of most Americans, were on the war in Europe. Two days after U.S. forces and their allies launched a decisive offensive in the battle of Argonne Forest, attention shifted to the home front with the city's Fourth Annual Liberty Loan parade. Billed as the largest in Philadelphia's history, this procession on September 28 would kick off the city's campaign to raise money for the war effort. As I study an old photograph of that Saturday afternoon in 1918, I can almost see death marching through my neighborhood, retracing the steps I walked daily. Five uniformed sailors, rifles on their shoulders, escort a festooned float bearing a navy patrol boat past the intersection of Broad and Chestnut. Hundreds of spectators are crammed beneath the classical columns of a building that decades later would become my local bank branch. At least two hundred thousand others pack the route along twenty-three blocks of Broad Street, cheering on the passing pageant of marines, sailors, and yeowomen, steelworkers, shipworkers, and makers of "shot and shell," with horse-drawn eight-inch howitzers, Boy Scouts, women of charity and relief, and Main Line debutantes riding farm equipment. Never would a flu virus more clearly demonstrate what it means to fully satisfy the third and final condition of a pandemic.

Philadelphians had barely boarded the streetcars for their Monday morning commute when the epidemic exploded. By Tuesday every hospital bed in town was taken. Thirty-one hospitals, and they were all turning people away. In the historic Society Hill neighborhood, the sick rushed to Pennsylvania Hospital, cofounded by Benjamin Franklin. "When they got there, there were lines and no doctors available and no medicine available. So they went home, those that were strong enough," a neighbor recalled. Five days after the parade, a doctor at Women's Medical College of Pennsylvania reported that students had begun filling in for hospital staff who were themselves laid low. "The experiences through which we are passing remind one of the historic records of the plague," wrote Dr. Ellen C. Potter, a medical professor at the college, in a letter to an academic colleague.

Just a week after the parade, on Saturday, October 5, doctors in Philadelphia reported 254 deaths in a single day. Five days later, the daily toll was 759, almost precisely triple. Hundreds of thousands were sick.

Philadelphia General Hospital, in West Philadelphia, was among the first to appeal for help. "Two-thirds of the nursing force were prostrate by the disease with none to replace them in the wards," reported sisters from the Roman Catholic archdiocese, who time and again answered the call. Almost half the doctors and nurses had themselves been hospitalized. Others had collapsed from overwork. Patients, many violently delirious, were getting minimal care. "Some of the poor sick had had no attention for over 18 hours and some had not been bathed for over a week," the sisters reported.

Isaac Starr was a third-year student at the University of Pennsylvania's School of Medicine. After a single lecture on influenza, he was dispatched to staff an emergency hospital opened in a partly demolished building at Eighteenth and Cherry streets. Starr and his classmates hauled twenty-five beds onto each of five floors. These filled right up with victims. "After gasping for several hours, they became delirious and incontinent, and many died struggling to clear their airways of blood-tinged froth that sometimes gushed from their nose and mouth," he later wrote. Many died without seeing a doctor. Corpses were "tossed" onto trucks, which hauled them away when filled. "The rumor got around that the 'black death' had returned," he wrote.

More emergency hospitals were opening every day in garages, parish houses, gyms, armories, nursery schools, and college frats, but often there was barely anyone to staff them. The city established one of the first at the poorhouse in the Holmesburg section. Its five hundred beds were filled in a day. In the second week of October, when a contingent of nuns came in relief, they discovered only twelve nurses caring for the patients. "One can imagine the distress, neglect and misery of these poor creatures. Some did not have their faces washed for days; their bed clothing had not been changed for a like period of time," one of the sisters recounted. Patients were moaning, coughing, delirious, some rising from their beds and frantically wandering the wards like specters. With only a single orderly for the whole hospital, the dead could lie unattended for hours until volunteers came to haul them out. "The first day we saw 13 bodies carried out to the dead-house within four hours," the nun continued. "The odor from this dead-house was something awful."

Nor was it just the city that was in the crosshairs. The smaller towns in its orbit were also succumbing. In Pottsville, the residence of a wealthy family was converted into a medical facility. "What sights and sounds met us when we entered that room where 84 patients were moaning and crying for help!" one nun wrote. "There were about forty babies in one room, all crying and perfectly helpless, their ages ranging from six days to two and a half years." All night, the stricken begged for water, ice, or a comforting presence in their final hours. The nun was horrified. "Some," she said, "were so far gone that worms were crawling out of their mouths."

On the streets of Philadelphia, cars bearing medical insignia were mobbed. College classes for pharmacy students were suspended so they could help fill prescriptions until drugstore shelves ran bare. Public services broke down. Nearly 500 police officers stayed off the job. About 1,800 telephone employees failed to show up for work, forcing Bell Telephone Company of Pennsylvania to take out newspaper ads warning it could handle "no other than absolutely necessary calls compelled by the epidemic or by war necessity."

Most people stayed cooped up in their homes, often low on food, at times dying there unattended. What volunteers from Holy Name Parish discovered in one Fishtown home was not uncommon. "In the parlor were the dead bodies of the married son and his wife who had died a few days previously," a nun wrote. "A daughter was dying in the adjoining room, alone, while her mother was seriously ill upstairs. The only attendant they had was the father who was too sick to realize what he was doing."

During the second week of October, 2,600 people died of flu in Philadelphia. Another 4,500 died a week later. There was no longer anywhere to put their bodies. At the city morgue, abandoned corpses were stacked three and four high in the corridors and spilling out onto

Wood Street. Bodies were piling up on the porches of row houses, in closets and garages, uncollected for days. "The smell would just knock you," Elizabeth Struchesky remembered decades later.

Police wagons, mortuary trucks, and even horse-drawn carts plied the street, and people were called to bring out their dead. "They were taking people out left and right. And the undertaker would pile them up and put them in the patrol wagons and take them away," recalled Louise Apuchase, who said her family was the only one in her neighborhood spared by the flu. "Directly across the street from us, a boy about seven, eight years old died, and they used to just pick you up and wrap you up in a sheet and put you in a patrol wagon. So the mother and father [were] screaming, 'Let me get a macaroni box.'" There were no more coffins. "'Please, please, let me put him in the macaroni box. Let me put him in the box. Don't take him away like that.'"

Nor were there enough embalmers. Nor gravediggers. "They had so many died that they keep putting them in garages," recounted Anne Van Dyke, whose mother had volunteered to shave the corpses.

The highways department finally dispatched a steam shovel to dig mass graves in a field at Second and Luzerne streets. Prisoners were pressed into service to bury decomposing bodies that others refused to touch. The few available caskets were priceless, and people were stealing them. A fresh supply had to be shipped in by rail under armed guard.

By the time the plague had finished claiming 12,897 Philadelphians in late November, the compassion and common decency that bound society together had been shredded. The nuns found babies without milk and adults without water. They even happened across children newly orphaned and abandoned in their homes. One nun later reflected, "It was the fear and dread of the scourge on the part of kindred and neighbors, who ordinarily would have cared for friends."

Much of the world still knows what it is to live with death. Not to take old age for granted. To see, in fact expect, that children will die. Most Americans, by contrast, have forgotten 1918.

Yet the American health-care system, with its promise of the highest quality care for those who can afford it, is intensive, expensive, and particularly vulnerable to the extraordinary demand for medical care that would accompany even a mild flu pandemic. "It's a more brittle system," Fukuda told me. "The ability to meet an upsurge in patients is not one of the virtues of that kind of system. Whereas in a lot of the developing countries, where you have more flexibility in terms of the health-care system, ironically it may be those systems that are able to cope."

In the United States, the health-care system has been under tremendous financial pressure to operate on the margin. Hospitals have been closing around the country, with the number offering critical care tumbling 14 percent between 1985 and 2000. By 2005, vacant ICU beds were rare. Some of these beds have been removed because of a severe nursing shortage. So, too, intensive care doctors have also been running short. Emergency rooms are being shuttered, about 10 percent of the national total between 1995 and 2005, and a survey of American emergency physicians revealed that almost 90 percent said their departments were routinely overcrowded. Ambulances are commonly diverted from one ER to another—on average, somewhere in the country, of once every single minute.

When researchers from the U.S. Government Accountability Office explored in 2008 whether hospitals were preparing for a mass casualty event like a pandemic, they learned that hospital executives were too preoccupied with day-to-day financial problems. The same researchers reported that federal funding for hospital emergency preparedness had decreased 18 percent from 2004 to 2007.

"Medical economics is really pushing toward downsizing of hospitals, reducing the number of staff, reducing the number of unoccupied beds," Fukuda said. "When you look at pandemic influenza, which is a one-period-of-time occurrence, that absolute increase in cases cannot be handled so easily. You cannot handle it without having a lot of staff. You cannot handle severe cases without having hospital beds."

Medicine would run out. Oxygen, crucial for treating those with lung disease, could be gone within days. The producers of medical oxygen are few, and the fleet of tanker trucks required to haul fresh supplies is far too small. There would be a tremendous shortage of ventilators. Most of this equipment is already being used in the everyday treatment of critical-care patients. In a severe pandemic, about 740,000 people would require ventilation, according to the U.S. Department of Health and Human Services, while studies put the existing stock at between 53,000 and 105,000.

Infectious-disease experts now debate whether the industrialized world might actually be more vulnerable today than it was in 1918. After nearly a century of medical progress, how could this be? No doubt there have been some astounding advances in hospital care. The development of antibiotics alone might save millions who would have died in 1918. Many, perhaps most, of the Spanish flu's victims succumbed not to the virus itself but to secondary bacterial infections that are now treatable.

But consider this: In the United States, 80 percent of all prescription drugs are now produced overseas. They are delivered to drugstores just hours before they're dispensed. In a pandemic, international shipping could come to a halt as countries impose travel bans and quarantines, companies suspend operations, and employees fall sick or stay home. Once again, pharmacy shelves would run bare. Nor would it just be flu medicine and antibiotics. Within days, medication for heart disease, high blood pressure, and depression would vanish, and insulin for diabetics would disappear. Many hospitals now maintain minimal inventories, receiving three rounds of medicine and other equipment each day. These supplies, too, could evaporate, and with them many forms of critical care.

Since the novel strain reemerged in 2003, Dr. Michael T. Osterholm, the director of the Center for Infectious Disease Research and Policy at the University of Minnesota, has been warning of the perils inherent in modern commerce. The economies of countries like the United States now more than ever depend on just-in-time supply chains and offshore sourcing of essential goods and services. "The interconnectedness of the global economy today could make the next influenza pandemic more devastating than the ones before it," he wrote. "Even the slightest disruption in the availability of workers, electricity, water, petroleum-based products and other products or parts could bring many aspects of contemporary life to a halt." With little surge capacity of their own, Osterholm projects that countries facing a major pandemic would run short on everything from soap and lightbulbs to gasoline and spare parts for municipal water pumps, and, of course, food.

Four days after lab tests had come back positive for bird flu in Vietnam, Klaus Stohr in Geneva convened an unpublicized conference call on January 15, 2004, with a half dozen of the world's leading influenza specialists. He wanted to know what he was up against.

"A critical situation, unprecedented," said Dr. John Wood, senior virologist at Britain's national biological institute. "We have to behave as if it could go to pandemic."

"Very, very serious," said Dr. Masato Tashiro, head of virology at Japan's national infectious-disease institute. The likelihood of human transmission is rising, and this, he said, "would be devastating. [Something we] have not yet experienced before."

Just two months earlier, Dr. Robert Webster from St. Jude Children's Hospital had coauthored an article for *Science* magazine warning that the world was unprepared for a flu pandemic. Now the dreaded scenario seemed to be unfolding. But the extent of the outbreaks caught even him by surprise.

"A very unusual event," Webster said.

These superlatives reinforced Stohr's concern. Stohr himself was not formally schooled in flu. Trained as a veterinarian in East Germany before the Berlin Wall came down, he had established himself as a national authority on rabies. His work caught WHO's eye, and he was recruited to the agency, where he was later tasked with restructuring an influenza operation then considered a backwater. He joined the fraternity of flu hunters.

"Klaus was very excited," a colleague said, recalling those uncertain days in early 2004. "It's one of the things they wait for their whole life. Pandemic, it's the big one."

Reared on the grinding shortages of the Eastern Bloc, Stohr was skilled at marshaling scarce resources. Now he cobbled together a global response from an agency strapped and weary. He massaged the bureaucracy, spinning out long lists of urgent tasks as he walked the halls, assigning them with dispatch. He stoked the enthusiasm and anxiety of his staff with talk of pandemic and helped position the agency to ensure it got a piece of this action.

The CDC had been eager to send in its own team and had already won a nod from senior Vietnamese officials. But WHO, flush with its triumph over SARS, didn't want to cede control of an emerging pandemic to the big boys from Atlanta. So WHO hurriedly dispatched Dr. Hitoshi Oshitani, its senior East Asian expert on communicable diseases. He was an astute, hard-driving Japanese doctor, a former Africa hand fascinated by diseases of the developing world. He had a humble respect for flu and little patience for politics. But he'd have to tend to both. Oshitani set out to assess the extent of the outbreak and negotiate with the Vietnamese government over permission for a larger team. He insisted that his investigators be allowed into the field. But the Vietnamese health ministry, wary of outside meddling, was reluctant to oblige.

"Avian influenza could be much worse than SARS," he admonished the skeptical officials. "If this avian influenza becomes a pandemic, it could infect two billion people. Millions of people would die."

It wasn't just a line. He personally thought he could be seeing the start of a global outbreak. Oshitani had already helped the world dodge one epidemic by steering Asia's response to SARS. "Hitoshi suddenly came alive again," an associate in the regional headquarters recounted. "For people like him, this is what life is about: crises. He immediately understood the implications." When Oshitani warned that avian flu could dwarf SARS, his exhortation sent shudders through his WHO colleagues. But it also had the desired effect on the Hanoi government.

With the door cracked open, Tom Grein was urgently detoured to Hanoi from a WHO mission in southern China. Grein, who would team up two years later with Uyeki to investigate the Ginting family cluster in Indonesia's North Sumatra province, became in essence the agency's player-coach in Vietnam. His was an all-star roster of epidemiologists, virologists, lab technicians, clinical specialists, and veterinary, logistics, and public-affairs experts that would eventually total nearly a hundred personnel. Finally, as part of this international effort,

Uyeki and Fukuda were bound for Vietnam along with five CDC colleagues.

Uyeki quickly peeled off into the field. Fukuda remained in the capital, helping set up a command center in a conference room just off the entrance to WHO's office in downtown Hanoi, and from there he helped direct the response. "We're not sure what's going on," Fukuda recalled, "and we have to sift through this pretty quickly." He pressed Vietnamese health and agriculture officials to cooperate, to share their intelligence about the outbreaks and ramp up efforts to contain them. "Don't be lulled into a sense of false confidence about small numbers," he urged in meeting after meeting. Though only a handful of human cases had been detected so far, the country could be at jeopardy.

Fukuda's counsel carried weight. He had unique credentials as a veteran of the Hong Kong outbreak in 1997. Yet these dynamics were different. The Vietnamese were not open to the kind of close partnership he'd established with Hong Kong's health director, Margaret Chan, which had been central to success. The international team itself was also different, larger and more unwieldy than the exclusively American one he'd led six years earlier. But the outbreak itself looked very similar. Again, it was mostly birds infecting humans. In urban Hong Kong, the source had been markets. Here in rural Vietnam, it was mainly farms. Still, for Fukuda, the killer was no longer a stranger but a known assailant.

The cases continued to come, the pace quickening. Healthy, mainly young victims kept turning up with breathing problems, rising fevers, and tumbling white blood counts. Many had diarrhea. Most died. By the third week of January, the virus had opened a second front in the south of Vietnam with initial cases in a young girl and teenage boy. Poultry outbreaks were also accelerating, proliferating faster than Vietnam could slaughter its afflicted birds and extending throughout the region. Under pressure from Prasert Thongcharoen, Thailand finally stopped its dissembling and confirmed both human and poultry outbreaks. By the first of week of February 2004, four more Asian countries had reported infected flocks: Cambodia, China, Laos, and Indonesia. Most of these countries had never been struck by any strain of avian flu before. Never had a highly lethal bird flu strain sparked as many outbreaks at once.

Each evening Pascale Brudon, the WHO's chief representative in Vietnam, gathered team members in the command center to compare notes and briefly unwind. The workload was tremendous, the days long, and the nights late. Yet it didn't seem to be enough for Geneva. "We had all of this pressure," Brudon said, "a lot from headquarters from Stohr, saying that this was going to be a terrible epidemic and it was going to be like Spanish flu."


The boy's mother was stumped. So was his grandmother. The thirteen-year-old had been among the very first in southern Vietnam to succumb to this new disease. Yet there had been nothing noteworthy about his habits. The women told Uyeki they didn't keep any poultry in their home. In fact there weren't even farms in this suburban enclave of Ho Chi Minh City, the former southern capital still known by most of its denizens as Saigon.

Uyeki had followed the outbreak south, remaining just a week in Hanoi before decamping to Ho Chi Minh. He had tracked the virus to the boy's home, determined to find the source of infection. "They had no idea what this kid did," Uyeki said later. "Because, like a normal kid, aside from school, they're playing. It's not unusual for a parent or grandmother to have no idea what this kid does after school." Uyeki could find no evidence of exposure to birds, much less sick ones. Nor was anyone else in the family ill. Could the virus now be lurking somewhere else? Perhaps in a new, undiscovered lair?

Sometimes, on disease investigations, fortune breaks your way. As Uyeki and his colleagues filed out of the family's home, they were an odd sight, a band of strangers in masks. Quickly, curious neighbors started congregating. They showered the investigators with questions. Then Uyeki started doing the asking.

"Oh, yeah," one neighbor piped up. "That kid, every day after school, he participates in cockfighting."

Uyeki's ears perked up. "Where?" he asked.

The neighbor led them down a narrow, paved street. After a couple of hundred yards, they came upon three or four woven baskets on the roadside, each containing a rooster. A little farther they found several more roosters and then a few more beyond that. This was the staging ground for the fights.

"We found out that this kid, along with other boys, every day after school he'd actually hold the cocks," Uyeki later reported. "So that's pretty good contact. If you're holding the rooster, the rear end or the feathers are pretty close to your face. That is what we believed was his risk." With this discovery, Uyeki had become the first to identify the threat posed by cockfighting, a popular pastime that over the coming years would be repeatedly implicated in bird flu deaths across Southeast Asia.

Other times fortune breaks against you. Days later, Uyeki and his Vietnamese colleagues pulled into a rural village in Tay Ninh province, about fifty miles northwest of Ho Chi Minh City, to investigate the case of a seventeen-year-old girl who had died of bird flu in a local hospital. They parked their van down the dirt road from her home. Chickens foraged in the dust. Pigs grazed on the roadside. Then, dressed in full protective gear, including baby blue disposable aprons and pants, masks, goggles, shower caps, gloves, and rubber shoe coverings, the team got out of the van and slowly approached the house.

But as they did, they discovered they weren't the only strangers in town. A Vietnamese television crew had also turned up to interview the family. A crowd was forming.

The girl's mother and grandmother were sitting in the doorway of their home, little more than a shack of bamboo and metal. They spotted the TV crew. They spotted the group of what could only have looked like space aliens. And the grandmother began to shriek. She was afraid. She was resentful of the intrusion at a time when the family was grieving.

But Uyeki wondered whether there was something more. He asked his translator what the old lady was saying. She was shouting over and over that her granddaughter had died of natural causes, insisting there was nothing untoward about the death. Uyeki suddenly realized the histrionics were meant for her neighbors. "She was trying to reassure them and also trying to deflect any potential stigma," he concluded. The family was terrified of being ostracized if fellow villagers believed they were infectious or, even worse, cursed. The investigators retreated, deciding the family should be left alone.


As long as the cases were coming one at a time, WHO felt confident that a pandemic virus had not yet broken loose. Each isolated human case was a dead end for the pathogen. Then investigators began to hear about the wedding party.

In late December 2003, a family in the northern province of Thai Binh, a verdant, rice-growing region in the Red River delta, had gathered to prepare for a marriage. On January 3 the couple wed. Four days later, the thirty-one-year-old groom started having trouble breathing. He died in the intensive care unit barely a week after his nuptials. Next his twenty-eight-year-old bride and two younger sisters came down with a cough and fever. Though the wife soon recovered, the sisters were taken to Hanoi, where they were admitted to the tropical disease institute, a six-story cream-colored building on the bustling urban campus of Bach Mai Hospital. Both ultimately tested positive for bird flu.

Bach Mai was already associated with tragedy. In the waning days of 1972, U.S. B-52s had leveled the hospital during a withering aerial campaign against North Vietnam that came to be known as the Christmas bombing. At least thirty people inside Bach Mai were killed in an attack emblematic of the war's excesses. After the conflict subsided, the hospital was slowly rebuilt, an inch at a time because of unexploded ordnance. Now, a generation later, Bach Mai again commanded the world's uneasy attention.

Peter Horby went to visit the sisters the day after they were admitted. "They were well," he recalled. They were walking around and taking the antiviral drug Tamiflu. But soon they started to deteriorate and within nine days were dead.

Horby set out for Thai Binh to learn more about the cluster of cases. "This was the first time we could investigate something like this," he said. "Everyone was wondering whether it was human-to-human." He interviewed the mother of the three victims. It was a tough conversation. She was distraught. He questioned neighbors. He tried to reconstruct the chain of events and scrutinized the timing of the cases to see whether everyone could have conceivably contracted the disease from the same, single source.

He learned that one sister had handled a duck while cooking. But she had not personally butchered the bird, considered a practice with a high risk of infection, and in any case the duck had seemed healthy. The bride and the other sister had no such exposure at all. There were no birds around their house or, for that matter, in their immediate neighborhood. They lived in town, not on a farm. But the sisters had cared for their ailing brother who, though never tested, almost certainly died from the disease. Horby deduced that the sisters had likely caught the virus directly from him.

Horby drove back to Hanoi and presented his report. The findings were explosive, especially because WHO had been reassuring the world's media there was no reason to panic as long as there was no human-to-human transmission.

"We really have to be sure," Brudon told the team members. "We must be really, absolutely sure."

For several long evenings, they cloistered in the Hanoi command center scrubbing the evidence. The conclusion was clear. "The team was quite convinced after the careful epidemiological investigations done by some members of the team," Brudon later wrote, "that we were in front of cases of H to H transmission." She agreed to release the findings.

Almost instantly, Horby said, "the shit hit the fan." The agency's upper echelons had not seen the disclosure coming. The next day WHO's director general, Dr. Lee Jong Wook, called Brudon. He was furious. He castigated her for endorsing what he thought an ill-considered statement. He said she had rattled the world for no reason.

Team members in Vietnam were stunned by Lee's response. They were sure their hard work had uncovered a fateful twist in the evolution of a killer virus, and this was their reward? They concluded that Lee had buckled under pressure from some of WHO's member countries, notably the Thais. With the virus also circulating in Thailand, that country's government was sensitive about any suggestion that the strain could become epidemic, fearful of the toll this could take on the economy. Thai Prime Minister Thaksin publicly attacked WHO's statement about human transmission, calling it bad science. "Normally, the ethics of researchers is such that if there is only a slight possibility of something happening, then they will discuss it among themselves. They will not say anything to the public to raise concern," Thaksin told reporters, assuring them that "the possibility of human-to-human transmission is 0.00001 percent."

In subsequent statements WHO adopted a far more reassuring tone, minimizing the significance of what the team in Vietnam had discovered. Three leading researchers based in Vietnam, including the director of its National Institute of Hygiene and Epidemiology, later cited the episode, recounting that "temperatures were running high, and any mention of person-to-person transmission of H5N1 was thought by some to be reckless." They added, "An air of tension . . . surrounds this disease, particularly in the corridors of power within the international health and political communities."

WHO would later clarify that "limited" human transmission was not a threat. The concern was "sustained" and "efficient" transmission. But the novel strain had now crossed another barrier. In 1997, it had demonstrated it could jump from animals to people. Back then, investigators in Hong Kong had also suspected that the virus could hop from person to person but never had conclusive proof. Now, nearly seven years later, it had shown convincingly that it could.

As reports of outbreaks among both people and birds continued to pile up, Vietnam resolved to take a terrible gamble. It would seek to exterminate the sickened flocks, the course recommended by international health specialists. But this would mean dispatching thousands of potentially susceptible peasants against an inscrutable enemy. They would fight the virus literally with their bare hands.

Inside the dimly lit coop of one farm outside Hanoi, workers chased the frenzied chickens, trying to pummel them with wooden rods. A bird bolted out the door into the sunlight and a woman lunged, snagging the errant fowl with her unprotected hands. She pounded it senseless, then stuffed its lifeless body into a sack with other casualties. Dozens of survivors scattered, dodging blows. Feathers flew. Droppings kicked up underfoot.

It was early February 2004, and I had arrived in Vietnam days earlier, anxious to see how authorities were tackling the budding epidemic. Poultry outbreaks had been reported across Ha Tay province, home to much of northern Vietnam's chicken industry. As I approached the village of Phu Cat just south of the capital, an agriculture official in a blue uniform motioned for me to stop my car. He circled the vehicle, spraying the tires with disinfectant to ensure I was not tracking death into his community. Then, for my protection, another official handed me a 3M mask, surgical cap, rubber boots, and a white Kimberly-Clark jumpsuit and instructed me to put them on. He then joined me in the car and guided us into the village.

We pulled up at a two-story shed. Thousands of chickens on this farm had been marked for death because they might carry the virus. A pair of veterinary officials huddled in the driveway. They were clad in broad masks, caps, and thick goggles, as well as protective suits, gloves, and boots. The local government had dipped into its budget to finance protective outfits for officials but could not afford to buy them for the cullers—the very people on the front line. Upstairs on the second floor was the large, low-ceiling coop where farmworkers were running down their prey, stuffing the battered chicken bodies into sacks, and then heaving them out a window into the bed of a truck below. Most of those called to battle wore only cheap rain slickers and flimsy masks.

One young culler emerged from the shed, dirt and bloodstains speckling his sandaled feet. "I didn't wear boots or gloves. I didn't buy any," he told me. Other farmworkers reported the same. "It's very dangerous for the people, but we can't buy everything we need to wear," added a neighbor. "I'm afraid I'll get infected from the chickens, but I have to do it because I can't ask anyone else to do it for me."

WHO officials were becoming alarmed. The workers in this nationwide slaughter were risking their lives. Even more ominously, they were giving the disease a prime opportunity to remake itself inside their bodies, potentially hatching a new strain easily passed among people.

Then, as springtime approached, the virus seemed to vanish abruptly from all of Asia. The outbreaks ceased like a fever breaking overnight. Most of the flu hunters headed home, still puzzled, but few believed they were gone for good.

"We need to think of it like a war," urged Dutch scientist Dr. Albert Osterhaus in a conference call with WHO.

"We may have no choice than to live with the virus," added Dr. Les Sims, who had been Hong Kong's chief vet at the time of the 1997 outbreak. "I don't believe we are going to get rid of this virus from the region even in the long term."

Though flu specialists had anticipated the return of the novel strain, they were startled by its sudden reappearance in late June 2004. The virus had always surfaced in the cooler weather, and here they were still in the summer. Despite the unprecedented culling of more than 100 million birds in East Asia, the scourge reappeared on farms across Vietnam, eventually returning to Thailand, China, Cambodia, and Indonesia and spreading, for the first time, to Malaysia. New human cases popped up in Vietnam and Thailand. Researchers were reporting that the virus was widespread in ducks, moving into wild birds, and growing ever more lethal in lab animals. Another study concluded that the strain had gained a permanent foothold in Asian poultry.

In August 2004, Vietnam's *Tuoi Tre* newspaper reported that a brother and sister in the southern province of Hau Giang had died under suspicious circumstances. The man, a nineteen-year old high-school student, had just taken his university entrance exam when he fell sick on July 23. Four days later he was admitted to the hospital with a fever, headache, and bloody cough and died after three more days. The local doctors diagnosed the case, apparently incorrectly, as septicemia or blood poisoning and never notified higher government authorities.

A day after he died, his twenty-five-year-old sister, a local teacher, began to complain of headaches, muscle pain, and difficulty breathing. She was dead within a week. A journalist for *Tuoi Tre* got wind of the cases and publicized them.

The hospital had already discarded samples from the brother, but lab technicians tested those from his sister and discovered she had bird flu. Health officials also reported that a third family member, a cousin, had also died recently but had never been tested.

Alerted, investigators came to the densely populated town deep in the luscious Mekong Delta, but they struggled to find the source of infection. Though the family raised ducks, chickens, and geese, the birds were all healthy. Investigators reported that "no link could be established with deceased or dying poultry." But they did learn that the sister had initially cared for her ailing brother, possibly accounting for her infection. This revelation, coupled with the sequence of the cases, persuaded some at WHO that for at least the second time, the virus had hopped from one person to another.

"My personal feeling is that this was almost certainly H2H transmission," a WHO epidemiologist would write in an internal memo later that year.

This time WHO didn't announce the conclusion at all. Senior WHO figures and Asian political leaders remained unwilling to acknowledge how far this fatal strain had come.

Scott Dowell was America's sentinel, watching from his post in Bangkok for threats on the horizon, when he got a call in September 2004 from Thailand's chief epidemiologist.

"We've got a weird situation," Kumnuan Ungchusak began.

He told Dowell that a local hospital had been routinely watching for cases of bird flu when a woman with severe pneumonia came in. She'd been around chickens, so the staff suspected the virus. Investigators from the health ministry were called. It ended up being a false alarm. But as the officials were preparing to leave, a nurse pulled one of them aside and asked about another woman, who had just died. She, too, had had severe pneumonia. No one had suspected bird flu in this case, Kumnuan said, because the woman hadn't been around any poultry. But then they learned the woman's daughter had also died about a week earlier in the countryside.

Kumnuan was going to drive up to the province in the morning to check it out. Did Dowell want to come?

Dowell was an American who ran the CDC's International Emerging Infections Program, headquartered in a sprawling office park that houses the Thai health ministry. Like the listening posts established by U.S. intelligence agencies during the cold war to monitor developments behind the Iron Curtain, Dowell's operation was on the front lines of a new struggle, watching for novel diseases that could threaten Americans and their national security.

As he and Kumnuan drove north to Kamphaeng Phet province on that Friday morning, the Thai doctor recounted more of the story. An eleven-year-old girl named Sakuntala Premphasri had lived with her aunt and uncle in a remote village about twenty miles off the main road. Their home was set back in the trees, a traditional, one-room house on wooden stilts with a sloping roof. Like everyone else in the village, the family kept chickens, and they ranged freely in the shady space beneath the house, where the girl often played with her friends and sometimes slept. The birds started dying in August, a few at a time. About four days after the last chickens had keeled over, Sakuntala got sick with a cough and sore throat. She felt feverish. Her aunt took her to a local health center for medicine, but the condition worsened. Days later, the girl was admitted to a district hospital with a high temperature, difficulty breathing, and low blood pressure. An X-ray revealed pneumonia in her lower right lung.

When her mother, Pranee Thongchan, learned of Sakuntala's deteriorating condition, she rushed to the bedside. Pranee, almost a girl herself at age twenty-six, lived more than two hundred miles away in a Bangkok suburb, where she worked in a garment factory. Pranee had asked her husband, a cabbie, to drive her back to the province to see her daughter. She reached the hospital at midnight. There she cradled the limp body of her little girl, repeatedly kissing her and wiping her mouth. Though the girl kept coughing, a duty nurse reported that Pranee kept her face "attached" to that of her daughter, spending the night beside her "cheek to cheek."

The next afternoon, with antibiotics failing to make a difference and her body descending into shock, the girl was transferred to the province's main hospital. When she arrived, she was bleeding heavily from her lungs. Blood oozed from her nose and mouth. Three hours later she was dead.

Pranee brought her daughter's body to a Buddhist shrine near her parents' home in Khampaeng Phet for three days of funeral rites. On the third day, Pranee herself began complaining of a headache and fever. Pranee went to the district health center for medicine. When she returned to the Thai capital, she felt even worse. It was getting harder and harder to breathe. Ten days after her daughter died, Pranee checked herself into the Bangkok hospital. By then it was too late. The infection had invaded both her lungs, and nothing could save her.

Alone in the car, Dowell and Kumnuan agreed that her case looked a lot like one of human transmission. "If there is transmission in this way, many people will be interested in this case," Dowell told his colleague.

Kumnuan kept working his cell phone to get more details. He called his subordinates, who were already in the province. Now they were telling him that other family members might also be sick.

"We need to get samples," Dowell urged. These specimens would not only confirm the virus but could also show whether it was mutating.

Had they taken samples from the mother? Not yet, Kumnuan answered. Kumnuan called back to Bangkok. He discovered that Pranee's corpse had already been embalmed. At that very instant, the body was at a Buddhist temple in the capital, about to be cremated. Kumnuan ordered his officers on-site to do whatever it took to hold on to the body and hurriedly dispatched a specialist who could conduct a limited autopsy on the fly. He got there just in time, snipping out a specimen of lung tissue. It later tested positive for the virus.

When Dowell and Kumnuan finally arrived at Kamphaeng Phet Hospital, Sakuntala's thirty-two-year-old aunt had just been brought in. She, too, had chills and trouble breathing, and she was having X-rays taken. They went to the radiology department to have a look. Sure enough, her lungs were clouded over. That raised an alarm. They asked to see her. The woman's condition was serious—her temperature had spiked at over 103 degrees, and samples she gave that day would later test positive—but unlike in the previous cases, would not be fatal. She mumbled to Dowell and Kumnuan that she'd been the one who buried the sick chickens in the yard, wrapping her hands in plastic bags for protection. But that had already been more than two weeks earlier, beyond the incubation period for flu. She also told them she had cared for her dying niece, staying at her hospital bedside until the very moment Pranee had arrived. That was the telling detail.

The following Monday, September 27, the veteran virologist Prasert Thongcharoen chaired a closed-door meeting convened by Thailand's Ministry of Public Health to review the cases of Sakuntala, her mother, and her aunt. In attendance were government health officials and medical experts from WHO and the CDC. Eight months had passed since Prasert blew the whistle on bird flu in Thailand. Now it was his mission to have his government and international health agencies formally acknowledge what scientists increasingly believed: The virus could spread among people.

This wasn't the first probable case of human transmission. But the evidence this time was incontrovertible. Sakuntala's mother, Pranee, hadn't even been in the same province when her daughter got sick. There was no way they could have caught the bug from the same chickens. In fact, there was no poultry at all in the Bangkok apartment where Pranee lived, nor in the factory where she worked. She had certainly contracted the virus at her daughter's bedside, and that was the same way the aunt had likely caught it. "It was a clear indication that H5N1 could be transmitted from person to person," Dowell said later. "Even though a number of us who had studied H5N1 closely over time thought that had probably already occurred, there was a widespread perception that the virus couldn't be transmitted person to person."

Dowell told me he faulted WHO and his own institution, the CDC, for too long leading the public to believe that bird flu could not be passed among people. Even if the transmission was limited, it was of grave concern. This was precisely how the virus could become proficient at spreading, he explained. It was through the process of passing from one human to another that a mutating strain could select the genetic attributes required to become a mass killer.

Once the evidence had been presented, Prasert coaxed his colleagues to accept the inevitable conclusion. He had the rare combination of independence, savvy, and scientific credentials to make it happen.

A day later Thailand's health ministry announced that the Kamphaeng Phet cluster had been "probable human-to-human transmission" of bird flu.

"For the political leadership in Thailand to say there was person-to-person transmission, that hadn't happened before," Dowell recounted. "It's a testament to Prasert and the influence he was able to wield in Thailand."

WHO released a statement about the Thai cluster that same day, copying the government's language and for the first time conceding "a probable case of human-to-human transmission." The virus had crossed a threshold, and so had its antagonists.

But even as they accepted how close H5N1 had now come to an epidemic strain, they had little inkling of what was about to happen. The death of yet another Vietnamese youngster right after New Year's Day 2005 would mark a new, even larger wave of infections. The virus would extend its reach in the new year as never before. It would soon strike beyond Asia, infecting new continents, multiplying its victims, and confronting the flu hunters with the prospect of imminent pandemic.

# PART TWO

PART TWO

# CHAPTER FIVE

# Livestock Revolution

When Prathum Buaklee dropped out of fourth grade to plant rice like his father and grandfather, he could not envision the revolution that would roll across the wetlands of central Thailand, lifting his family out of destitution and ultimately sending his own sons on to university in Bangkok. The royal capital, though only seventy-five miles southeast of Prathum's village of Banglane, seemed like another continent in the 1950s. Those sons of Suphan Buri province fortunate enough to escape its hardships first had to find their way to the Tha Cheen River, which slices through the swampy, low-lying plain. Roads were few, little more than muddy tracks rutted by the wheels of cattle carts. So local journeys were often made in wooden rowboats that glided through weedy marshes and along a labyrinth of canals skirting the glistening emerald paddies. Once the travelers reached the river, they would hitch rides on the lumbering, two-story rice barges that hauled the province's harvest southward. Departing after the worst of the midday heat, they would arrive in Bangkok at dawn the next morning.

Nearly everyone who remained behind grew rice. It was a hard life, long days under the searing, tropical sun, and the rewards were modest. "It wasn't enough. Just barely enough to make a living," Prathum recalled, a deep furrow cutting across his broad forehead like freshly tilled earth. He erected a small, traditional house, a leaky hovel of clapboard and corrugated metal on stilts, and bought himself a bullock cart.

At first, change came slowly to Suphan Buri. In the drier, upland area to the north and west of the province, villagers started cutting down the bamboo forest in the mid-1960s and planting sugar cane. Day by day, the jungle shrank until the cane fields eventually nestled against the base of the mountains. Long-distance bus service was introduced, putting Bangkok only four hours away along a rocky, bone-jarring road.

Then, two decades later, chicken made its debut. A pair of Thai poultry companies, including the Charoen Pokphand enterprise that would ultimately become the country's premier multinational corporation, came to Suphan Buri, urging rice farmers to raise chicken instead. The companies offered them chicks, feed, and guaranteed prices for mature broilers. Some of this activity was driven by Thailand's campaign to boost poultry exports. But far more profound changes were also at work. The kingdom had embarked on an ambitious course of economic development, tapping its wealth of natural resources, cheap labor, and open investment climate to become a low-cost manufacturing dynamo. As its shirts, shoes, and consumer electronics crowded American and European shelves, Thailand staked a claim as one of the new Asian tiger economies, recording annual growth rates of nearly 10 percent. This translated into rising incomes for many Thais, especially in the cities, and the new, burgeoning middle class had new, urban tastes. They demanded a better diet, in particular one rich in animal protein. Nowhere was this truer than in the boomtown of Bangkok.

Suphan Buri was strategically located to meet this demand. In the late 1970s, the government had built a paved road linking the province to the capital. Now the Bangkok market was barely two hours away. Many peasants took advantage of cheap land prices to expand their holdings and establish chicken farms. By 1987, Thais had doubled the average amount of chicken they ate. Yet production across the country was growing so fast that prices actually declined, making chicken an even cheaper source of protein than fish or pork and fueling demand further. The consumption of chicken would soon double again.

But the new middle class yearned for variety, and that also meant soaring demand for eggs.

Even as Prathum continued to toil in the rice fields, fellow villagers in Banglane were starting to experiment with hen farms. In 1991 Prathum followed suit. He began with three hundred laying hens, soon adding several hundred more. When his flock grew into the thousands, he abandoned the paddies altogether.

"I never imagined the changes when I started out," he told me, chuckling softly, creases deepening at the corners of his eyes. He spoke with the exaggerated inflection of a Suphan Buri native. Even today, this distinct accent marks people from the province as something of country bumpkins, at least in the reckoning of their cosmopolitan Bangkok cousins. But they're hardly poor yokels. Prathum eventually bought twenty acres of land, more than tripling the size of his holdings, and erected seven open-sided sheds, each stretching about forty yards under pitched metal roofs. His flock reached fifteen thousand birds. And with average Thai consumption of eggs doubling in just a decade, Prathum's hens were indeed laying gold. "We got a better income so we could do whatever we wanted," he continued, gently shaking his head with wonder and then bowing it slightly to acknowledge the good fortune. "I feel grateful to the chickens. Chickens are like human beings. You take care of them and they'll take care of you."

Three years after he started chicken farming, this broad-shouldered peasant who had once been unable to afford even a motorbike bought a used Ford pickup. A few years later, after the increasingly prosperous village put in paved roads, he added a second, a new one. He knocked down his old shack, replacing it with an airy wood-frame dwelling three times as large. He furnished it with a refrigerator, color television, and air-conditioning. To give his teenage daughter privacy, he later built her a separate room, her territory marked by a Britney Spears poster on the outside of the door. He then went on to construct a second house, a retreat on the edge of some neighboring paddies, and started taking vacations with his family, renting a van twice a year and driving to the mountains of northwestern Thailand. For each of his three children, he bought a new computer. One son went on to study veterinary science at the university in Bangkok; the other, computer engineering.

But then, a dozen years after he answered their calling, chickens changed his life again. On a warm December morning in 2003, truckloads of livestock officers swept into Banglane. It was no surprise. Prathum knew that flu had broken out in a neighbor's poultry shed just beyond a nearby canal.

In less than a generation, a livestock revolution had brought unthinkable wealth to dirt-poor peasants, not just in Thailand but across much of East Asia, and dramatically enhanced the diets of tens of millions of people. But now this fundamental transformation of Asian farming was posing a threat unprecedented in the history of human economy. By packing together so many birds, often in close quarters with people, pigs, and other livestock, farmers like Prathum have created ideal conditions for a flu pandemic. The sheer number of birds has opened the door for the disease to take hold. The proximity of other creatures heightens the chance it will jump species, swap genes, and mutate. Cramped together, these flocks are dry tinder awaiting a conflagration that could race across farms, provinces, and national borders, burning through the unrivaled concentrations of humanity in their midst.

This far-reaching economic change is but one of the factors making East Asia so treacherous for those struggling to avert a pandemic. The same forces of globalization that birthed the Asian tiger economies can now speed the flu virus around the globe within a day. Traditional Asian practices, from cockfighting to live poultry markets, have acquired a sinister cast, defying efforts by WHO and its allies to reform them before they seed a pandemic. Confronted with these hostile realities on the Asian terrain, the world would hope for a demonstration of political will equal to the threat. Instead Asian governments have repeatedly hushed up their outbreaks until death's reach caught them in the lie.

When the livestock officers descended on Prathum's farm, he tried to turn them back. He vowed his hens were healthy. But he realized the battle was lost.

"Everyone has to abide by the government's decision," a senior officer urged him. "Go with the flow."

Prathum shuffled through the sheds, counting his birds so he could apply for compensation.

"Where do you want us to dig the hole?" the officer asked.

Prathum motioned to the edge of his property and left. He couldn't watch.

Since animals were first domesticated ten thousand years ago, they have promised humans a richer, fuller life but all too often delivered death. As scientist Jared Diamond notes, the peoples who first drew animals into their daily lives were the first to fall sick, infected by germs descended from those afflicting their livestock. Though these pioneers later developed a measure of immunity, mankind has continued to be ravaged by such offspring diseases. Many of the most prodigious killers of the modern era, including measles, tuberculosis, smallpox, and, of course, flu, have evolved from animal pathogens.

The majority of illnesses that now strike humans are cross-species zoonotic diseases. Of the 1,415 human pathogens that have been catalogued, about 60 percent also cause disease in animals. These microbes can hopscotch among species and mutate along the way, acquiring new, more lethal characteristics. An even higher proportion of previously unknown human diseases, about three-quarters, originate in animals. These maladies include recent arrivals like SARS, which passed from infected civets in China to humans before spreading to thirty countries in 2003, and West Nile Virus, which first appeared in the Western Hemisphere in 1999, before going on within a decade to sicken people across much of the United States and become endemic in the country's wild birds. Indeed, the emergence of new, zoonotic diseases has ominously accelerated since the 1970s. "Similar to the time of animal domestication, which triggered the first zoonoses era a number of millennia ago, a group of factors and driving forces have created a special environment responsible for the dramatic upsurge of zoonoses today," writes the National Academy of Science.

Chief among these causes is development, which is extending human settlement into new habitats and bringing people into contact with animals as never before. In late 1998 a mystery illness erupted in the Malaysian district of Nipah, infecting 265 people and killing more than 100. While local health officials initially identified the disease as encephalitis because it often caused inflammation of the brain, investigators later concluded it was an entirely new pathogen. They discovered that the virus was carried by fruit bats, which gathered in trees on newly developed pig farms. The bats infected the pigs and the pigs infected the farmers. To break this chain, 1.2 million pigs were ultimately slaughtered.

But the classic example of the unintended consequences of progress is not a new one: bubonic plague. Some scholars have posited that the opening of trade routes between China and Europe in the Middle Ages was responsible for conveying the Black Death from its source in Asia's Gobi Desert to its killing fields in the West. The plague bacterium had found itself a permanent home in the burrowing rodents of the Asian steppe. Marco Polo himself had remarked on the great number of what he called "Pharoah's rats" that he encountered in the Gobi Desert. Caravans of the mid-fourteenth century snaked through their habitat, steadily carrying infected rats and fleas onward toward the Crimea and Europe's doorstep.

Some researchers have contested this account, saying evidence of plague in China centuries ago is thin. But another recent pandemic makes an even more convincing case for the fateful relationship between progress and plague and for East Asia's starring role in this drama. In the late eighteenth century, plague erupted in southern China, not far from the Burmese border. During the preceding decades, hundreds of thousands of migrants had streamed into a largely undeveloped corner of Yunnan province, lured by a boom in copper mining. This explosive growth transformed the area from a rural hinterland into an increasingly urban outpost and exposed the miners, merchants, transporters, and various other fortune hunters, laborers, and camp followers to plague bacteria long harbored by local mice and voles. Caravan trade in copper, as well as other minerals, salt, cotton, tea, and grain, dispersed the disease around the province. With the acceleration of long-distance trade in opium grown in Yunnan, plague spilled beyond the provincial borders. This lucrative commerce carried the epidemic inexorably eastward, by land, river, and sea, past the border regions north of Vietnam until the Pearl River delta and Hong Kong fell prey in the last decade of the nineteenth century.

When the epidemic came ashore in the spring of 1894, it ravaged Hong Kong. Corpses were left abandoned in the streets of the British colony. Stores and houses were shuttered, draining the once teeming quarters of life but for the English infantry. The soldiers went from home to Chinese home in search of the sick and dead, forcibly disinfecting furniture, sheets, and kitchenware and carrying off the ailing to a great ghostly ship, the *Hygeia*, moored three hundred yards off the waterfront. "Little wonder, then, that this malevolent-looking hulk, pressed into service at the start of May 1894 as a floating plague hospital, should have become an object of terror," recounts author Edward Marriott. Local Chinese resisted the raids, and some doctors took to carrying revolvers for protection. It was whispered that the *Hygeia* was no hospital but a sinister laboratory where the English were concocting a cure from the livers and other organs of patients. Fearing the abduction of their children, mothers pulled them from classes, and by the middle of May, half of Hong Kong's schools had closed. About eighty thousand Chinese fled the colony altogether. And though international shipping companies urgently rerouted their vessels to bypass what had been the world's fourth-busiest port, the plague would not be denied, eventually spreading as far as San Francisco.

Six years earlier, in 1888, another epidemic had struck Hong Kong. Its stay was less tumultuous but its symptoms nonetheless severe. James Cantlie, a fellow of the Royal College of Surgeons posted in the colony at the time, reported that patients suffered from headaches, backaches, and sore eye sockets and limbs. At times the pain was agonizing. Nearly all had runny noses, and many suffered from coughs, diarrhea, and vomiting. Some complained of jaundice, profuse rash, and mottled skin. Their fever would usually spike by the third day, approaching 104 degrees. When the disease first broke out, Cantlie misdiagnosed it as a form of "tropical measles." Others called it dengue fever. But by the time Cantlie reported his findings in 1891, he knew what it was. In the intervening years, an influenza epidemic had sprinted around the world. The Europeans dubbed it the Russian flu, because it came from the east. The Russians in turn called it the Chinese flu. Cantlie told readers of the *British Medical Journal* that the Hong Kong outbreak had in fact been the first recorded appearance of this flu pandemic and its original source. Moreover, after conducting research among the Chinese, he concluded that flu was endemic in China.

Writing a century later, another Westerner who devoted his career to medical inquiry in Hong Kong determined that China was the "epicenter" of all influenza viruses. Kennedy Shortridge has been one of the world's premier scholars of flu, a lanky Australian with thick, graying eyebrows and a deep melodious voice. He spent three decades as a microbiologist in Hong Kong before retiring to New Zealand. His research pinpoints southern China, and in particular the Pearl River delta of Guangdong province, adjacent to Hong Kong, as the cradle of the world's flu. He noted that the natural hosts for flu viruses are aquatic birds, harboring the microbes in their guts, with ducks playing a unique and leading role. Before the rise of China's Qing dynasty in the mid-seventeenth century, duck herders grazed their flocks along riverbanks and in canals and other waterways. But that changed in response to a mounting problem with pests in the rice paddies of southern China. Ducks were introduced into the paddies and fed on the insects and snails. When rice plants started to sprout, the flocks were temporarily rotated elsewhere. This elegantly balanced agrarian system made ducks far more profitable, and their population soared. With as many as five rice harvests annually in Guangdong, ducks became a year-round presence in the densely populated villages of the delta, and duck droppings, often larded with virus, became ubiquitous. Pigs snuffed up the fecal matter, offering themselves as a natural laboratory for gene swapping, because they can contract both bird and human flu viruses at the same time. Contagion was everywhere, Shortridge observed. It was a recipe for repeated epidemic.

The flu pandemic of 1957 is known as the Asian flu. Western countries learned of it after it was identified in Singapore. But Shortridge told me the strain had actually been isolated earlier in China's southern Guizhou province. The 1968 pandemic was dubbed the

Hong Kong flu. Yet Shortridge asserts that this, too, arose first on the mainland, specifically Guangdong, but the mainland Chinese were too preoccupied with the Cultural Revolution to take note. Hong Kong was just about the only conduit out at the time. It caught the bug and lent its name.

The origin of the 1918 Spanish flu is a matter of greater dispute. All agree the epidemic did not start in Spain. It only took this name because the Spanish were willing to report it. Unlike the United States and the European powers embroiled at the time in World War I, Spain was neutral and did not censor news deemed to undercut morale. Spain broke with their policy of censoring news about the outbreak. The flu made headlines in Spain, and eventually the press in other countries picked up reports of this "Spanish" flu.

American author John Barry has made a strong case for Haskell County, Kansas, as the origin of the pandemic. He cites medical reports as evidence of an extraordinary flu outbreak there in early 1918 and describes how the county's young men would have reported to an army camp three hundred miles to the east before being deployed with their germs to the European front. British virologist John S. Oxford, by contrast, has postulated that the scourge first arose at a mammoth British army camp in northern France, where many soldiers had been treated in 1916 for what was then diagnosed as acute purulent bronchitis. Oxford's review of those clinical findings concluded that the outbreak was actually pandemic flu.

But Shortridge maintains that even this great influenza of 1918 has a Chinese pedigree. Part of his proof is in the antibodies. Citing the medical accounts of an American missionary working in Guangdong at the time, Shortridge notes that Chinese children born after 1907 appeared to have a heightened immunity to the virus when the full-blown epidemic hit, suggesting they had already been exposed to a less virulent version of the same strain. "The virus had been smoldering in southern China for at least eleven years before it appeared," he told me. Skeptics of his theory point out that the first wave of pandemic was recorded in the United States and Europe early in 1918, several months before it was documented in China. Yet Shortridge says this disregards the little-appreciated nature of flu in tropical climes like southern China. There, flu is primarily a summer malady, not a winter one, and would not have fully manifested itself until the middle of 1918 even if the pandemic strain was already circulating locally. So how, then, would the virus have found its way to Europe? Shortridge says he discovered a possible explanation by accident. While listening to a program about World War I, he unexpectedly heard the sound of Chinese. It was the taped voices of economic migrants who had set off for the European front to dig trenches for the Allied forces. He recognized their dialect. It was Cantonese, the dialect from around Guangdong.

Four centuries ago, it was a change in farming techniques that consolidated southern China as the world's influenza epicenter. The introduction of ducks into paddies boosted agricultural productivity and set the conditions in which novel strains could smolder. But in the last generation, it's all been about demand. Much of East Asia has witnessed a population explosion of chickens, ducks, and pigs in response to the region's rapidly rising incomes. More money has meant a greater appetite for meat, milk, and eggs to complement and even replace the traditional staple crops. And nowhere has more money come more quickly than in East Asia. These countries, often benefiting from open market policies and tremendous Japanese investment, have achieved unrivaled growth as manufacturing exports have eclipsed rubber and rice at the heart of the economy. Steel and glass have thrust into urban skies from Shanghai and Guangzhou to Bangkok and Jakarta, attesting to the region's ambitions.

Since China began adopting market reforms in 1978, it has consistently recorded annual growth rates of more than 10 percent, raising living standards and reducing poverty as never before in history. The Beijing government turned much of this raw energy southward toward the marshes and paddies of the Pearl River delta. By establishing a special economic zone in Guangdong, China unleashed what author Karl Taro Greenfeld labeled the "greatest mass urbanization in the history of the world." The province became the world's workshop, "where more of everything is being made than has ever been made anywhere at any time." China now manufactures enough televisions to replace the world's supply every two years and a quarter of everything sold at Walmart. Guangdong became China's richest province, the boomtown of Shenzhen its richest city.

Yet for a decade after 1985, Thailand actually outpaced China and registered the fastest growth on Earth. No longer was the sex trade Bangkok's main calling card. The capital built cavernous shopping malls, a hot fashion industry, and a sleek commuter Sky Train to whisk its young professionals among their high-rise office towers. Then it was Vietnam's turn, emerging for a time as the fastest growing country in Southeast Asia by capitalizing on the Communist Party's *Doi Moi* economic reforms. As growth rates topped more than 8 percent a year, storefronts along the romantic, tree-lined streets of Hanoi overflowed with iPods, computers, and digital cameras, and young Vietnamese plotted an even brighter future, expressing far greater admiration for Bill Gates than for anyone in their Politburo. Indonesia, in the meantime, diversified an economy long dependent on exports of oil, teak, and minerals, gaining recognition as a major newly industrialized country. The government in Jakarta eradicated much of the country's poverty while motorbikes and cell phones became de rigueur even for many in the working class.

The Asian financial crisis of 1997 temporarily knocked the wind out of this fabulous progress. But the region's economies rebounded, albeit some faster than others. Incomes and ambitions resumed their ascent. Malaysia, always one of the region's best performers, pressed ahead with the completion of its Petronas Twin Towers in downtown Kuala Lumpur and boasted they were the tallest buildings on the planet. (As measured to the tips of their spires, a controversial standard, they were.) The towers were a pair of exclamation marks rising above Asia's transformed landscape.

To a casual visitor, the agricultural changes that have accompanied this era of remarkable growth may not be as visible as the city lights. But as economist Christopher Delgado from the International Food Policy Research Institute and his fellow authors wrote, "The demand-driven Livestock Revolution is one of the largest structural shifts ever to affect food markets in developing countries. . . ." The revolution is not limited to East Asia. It has been manifest across much of the developing world as rising incomes, rapid population growth, and the broader diet that comes with urbanization combine to stoke demand for animal protein. During the two decades that followed 1980, people in developing countries doubled the average amount of meat they ate. By 1995 the volume of meat produced in developing countries for the first time surpassed that in developed ones.

But this is mostly because of China and Southeast Asia. China alone has accounted for more than half the developing world's total increase in meat output. A large majority of that has been poultry products and pork, with the production of chicken meat growing fastest. The radical expansion of flocks that began in the 1970s and 1980s continued into the following decades, barely pausing for the East Asian financial crisis. From 1990 through 2005, China's production of chicken nearly quadrupled, as did that of duck and goose. The amount of pork more than doubled.

Southeast Asia's record ranks second only to that of China. During the same fifteen-year period, Indonesia more than tripled its production of chicken meat, Vietnam and Malaysia more than doubled theirs, and Thailand, which had registered a breathtaking growth of 10 percent annually in earlier decades, saw its output of chicken slow, increasing a mere 60 percent over this period. Malaysia more than doubled its output of duck and goose while Vietnam more than tripled its pork. To get a sense of the sweep of this revolution, consider the case of the Indonesian egg. In 1970, just as Indonesia's long-slumbering economy was preparing to embark on a generation of sustained growth, the government statistics agency reported that the annual production of eggs was 59,000 tons. Three decades later, the total was 783,000 tons—a thirteenfold increase.

This transformation has literally put a chicken in every middle-class pot. Many among the urban poor have also secured better diets as meat prices dropped. (Consumers, fortunately, did not face the kind of inflated prices for grain and vegetables due to rising demand for animal feed that some economists had predicted.) But across vast swaths of rural Asia, the record is more mixed. Some small-time farmers have proven unable to compete with new industrial producers and lost their livelihoods. Others, by contrast, have found that livestock, one of the few sectors they could afford to enter, was their ticket out of poverty. For Prathum, the revolution was a bonanza—until the virus discovered the same thing.

Prathum's wife said the livestock officers stormed into Banglane like marauding communists. "They came by the hundreds in trucks, bringing soldiers and prisoners to kill the chickens. We argued for some time. But they weren't listening to us," Samrouy Buaklee recounted. She raised her leathery hands in exasperation and then wiped her deep brown eyes with the checkered scarf around her neck. "It broke my heart. I felt that the chickens were like my children."

Samrouy had retreated in tears to the house deep in the rice paddies and remained sequestered there, alone, for two days. When she returned, she noticed the silence. It's always the silence. Over and over, farmers who lost their flocks told me it was the absence of the cackling and cooing they found hardest to bear. The village had gone dark. Farmhouse lights that once flickered on in predawn hours as villagers awoke to tend their flocks remained extinguished. The roads were abandoned. "No one walked around," her husband recalled. "Everybody sat at home and nobody talked. With the chickens gone, we didn't know what to do with ourselves."

After more than half a year, Prathum decided to restock, rebuilding his flock though not his confidence. When I met him, his brown eyes had grown heavy, and bags hung low on broad, sunbaked cheeks. "Even if we're afraid of the disease returning, what can we do? Nothing. We can't run away," he said softly. "It's my job. If I don't do chicken farming, what else can I do?"

Prathum left the question hanging. He rose from the wood crate where he'd been resting in the barn and emptied a sack of chicken feed into a wheelbarrow. Emerging into the morning light, he pushed it down a short concrete causeway jutting into the fishpond and trudged past a pair of spirit houses, those colorful, birdhouse-size shrines on pedestals he had once hoped would keep the local spirits content. At the end of the causeway, three open-sided chicken sheds on stilts extended across the green water. Prathum started with one on the left, the hum from hundreds of excited hens rising to greet him. He stepped nimbly along the aging wood planks that ran between the cages, his meaty hands shoveling grain from a bucket into the long feeding trays. The plaid shirt hanging from his stocky frame was soiled and his bare feet were caked with dirt. The planks were stained with droppings, the air rank with a cocktail of feathers, feed, and feces.

As a concession to new government rules, Prathum had draped fishnet along the sides of the two sheds. This was meant to keep out wildfowl, which could be carrying bird flu. But mice had already gnawed holes in the netting, and a few crows and swallows were darting about under the corrugated metal roofs. That was the extent of Prathum's effort to prevent contamination and stem another outbreak.

The most important line of defense against a human pandemic is not at the hospital or vaccine lab but at the farmyard gate. A single gram of bird feces can contain up to 10 billion virus particles. A speck on a heel or a pant leg or a bucket or a tire can introduce an infection capable of decimating a whole flock.

Health officials have long made clear how to prevent epidemic contagion from spreading among flocks or from farm to market and on to other farms. The first principle is to severely restrict access to poultry flocks. This means keeping chicken sheds off-limits to most visitors. Those raising birds of their own must be categorically banned. The second is strict hygiene. Anyone entering a shed should wash his hands and don sanitized shoes. Poultry workers should change into clean, disinfected clothes and take them off when they leave so they can be washed. Feeding pans and cages should be cleansed daily. Equipment, such as pallets and egg crates, are easily contaminated and should never be shared among farms. Vehicles that have visited other farms could inadvertently be carrying the seeds of disaster and should be kept at a distance. Other animals must be barred from the chicken sheds.

When Prathum's black dachshund trotted after him into the

henhouse and then curled up for a nap beneath the cages, I knew there was trouble.

Nirundorn Aungtragoolsuk, a director of disease control in Thailand's livestock department, later confirmed as much. He told me the government had adopted strict regulations, including a requirement that poultry workers shower with disinfectant before entering a farm and vehicles be sprayed with disinfectant before arriving on premises, but these applied solely to the large, export-oriented operations. The regulations were not meant for most farms, like Prathum's. "They have done it their way for a long time and we cannot change it overnight," Nirundorn said.

It took Prathum half an hour to finish feeding the hens in the three sheds. He returned to the barn, sweat glistening under his thinning hair, and hopped on his Honda motorbike. With a sack of feed in the sidecar, he buzzed up his gravel driveway, across the road, and down a dirt track that paralleled a canal on the far side. His other dog, a white crossbreed, had joined the dachshund, and now the pair gave chase, scampering behind Prathum until he reached two more chicken sheds suspended above another pond. As he resumed his feeding rounds, the dogs followed him inside.

A few moments later, as he emerged to fetch more feed, a silver Isuzu pickup coated with dirt pulled up right at the entrance to the sheds. It was the neighbors. The husband, Monchai, had bad teeth, and his wife, Boonsveb, had big hair. But they also had three times as many chickens as Prathum. They had culled the whole lot when the flu erupted and replaced them all. They told Prathum they had an uneasy feeling about another outbreak and wanted to compare notes. The talk turned to the question of whether they should erect modern, all enclosed, climate-controlled sheds.

"Of course that would be better," the wife said. "It would keep out disease. But it's expensive."

"Yeah, that's the problem," Prathum agreed. "Who can afford it?"

"You certainly can't afford it," the husband quipped, needling Prathum. "You can't even afford enough staff. You have to do the farming yourself."

As if on cue, Prathum refilled his bucket and vanished deep into the chicken shed again. The husband, wearing cracked, dirty sandals, accompanied him inside. The dogs took up the rear as hundreds of red-crested heads poked out of the cages, viewing the procession.

Shortly after the neighbors left, another Isuzu pickup, this one red, rumbled down Prathum's gravel driveway, pulling up in a cloud of dust just outside the barn. The cab door opened. Out got Nikon Inmaee, an egg vendor with a narrow face and short, wavy hair. Prathum had collected the eggs in the hours just after dawn, and now they were waiting, packed into plastic trays stacked ten high amid dirt and dead grass on the barn's concrete floor.

Prathum helped Nikon gingerly hoist the trays into the truck bed. Three days a week, Prathum's eggs were ferried to Bangkok, but this batch was headed for a closer market, about fifteen miles away. While Prathum calculated the tab on a small pad, Nikon returned to the back of his truck and began pulling out a separate set of empty trays. He deposited dozens of them in the barn for use later in the week. They were still soiled from the market. It was like addicts swapping dirty needles.

There was a time in the United States when chicken was a luxury, an indulgence for those weary of more affordable dishes like lobster and steak. Chicken was precious because it was relatively rare. In the nineteenth century, raising poultry was little more than a hobby for farmers' wives, and in 1880, when the U.S. Census started counting chickens, they numbered only 102 million nationwide. By 2006, that number were butchered nearly every four days.

This American revolution would have been impossible but for a series of advances in animal husbandry, starting with the debut of commercially sold chicks in the late nineteenth century. Next came the development of artificial hatcheries, which lowered prices and brought chickens to selling weight faster. Companies specializing in feed emerged. Vitamin D was introduced to fight rickets. Broilers and later layers were shifted indoors, where temperature, lighting, and diet could be precisely calibrated, and then the birds were raised off the ground and confined to tiers of wire cages, where care and feeding was even easier. But the watershed was the introduction in 1971 of a vaccine for a poultry plague called Marek's disease, which was killing 60 percent of the birds. Chicken prices plummeted, further fueling American demand already on the rise for familiar reasons: population growth, increasing income, and urbanization.

In the United States, where chicken was fast replacing beef as the animal protein of choice, safety measures to prevent disease followed quickly. The U.S. Department of Agriculture launched an aggressive campaign to educate farmers about biosecurity. Many family farms were swallowed by integrated agriculture companies, which insisted on stricter practices to protect their operations. In Europe, which was experiencing a similar chicken boom, many farmers had turned to banks for financing and inked contracts with feed mills. Both demanded measures to safeguard their investments.

Asia, for the most part, has yet to follow suit. So today, birds in China and Southeast Asia are amassed in once unimaginable densities, often weakened by their stressful confinement and exposed to the whims and wrath of viruses like influenza. The tremendous amounts of feed, water, and human traffic required to maintain these flocks offer a generous avenue to infection. In the unnatural setting of intensive agriculture, chickens are more vulnerable to contagion because they are pressed together and housed atop one another's droppings, which are a main way, if not *the* main way, birds transmit the virus. A year after bird flu's arrival in Thailand, a study found that Thai commercial farms were at a significantly higher risk of infection than the small, informal flocks of several dozen poultry that villagers have raised in their yards for centuries.

It's not only the *risks* of infection on large farms that are greater. So are the consequences. The lack of genetic diversity in many commercial flocks means that a virus that infects one bird can likely infect them all, offering abundant opportunities for microbes to reproduce. The chances for the virus to mutate as it skips from bird to bird are multiplied many times over. "Once an influenza virus invades a commercial poultry farm," scientists warned, "it has an optimum number of susceptible poultry for rapid viral evolution."

But researchers have also found there's something even more peril-

ous than a country of dense commercial chicken farms. That's one like
Thailand, a country in transition, where commercial farms operate in
the midst of extensive traditional flocks. These small holdings represent
a vital link in the chain of infection. For the flu virus to migrate from its
natural reservoir in waterfowl, it needs an initial toehold in domestic
poultry. Asia's traditional backyard farms, with their freely grazing birds
and even fewer safeguards, offer just this opening. They are like kindling
wood around the larger commercial farms. And the larger commercial
farms, which have mostly sprung up near their markets, are concentrated around another vulnerable population: the unprecedented accu-
mulation of humanity in the metropolises of East Asia.

As Jan Slingenbergh of the UN Food and Agriculture Organization
and his fellow researchers write, "Agricultural practices have become
the dominant factor determining the conditions in which zoonotic
pathogens evolve, spread and eventually enter the human population."
More pointedly, the FAO said in 2008 that the rapid development of
Asia's poultry industry without due regard for animal health created a
"virtual time bomb" that "exploded" with the outbreak of H5N1.


Sangwan Klinhom was a Thai country singer in the parts around
Suphan Buri. His resonant tenor earned him a following, but little
money. The tips couldn't even pay the rent. "If you're a singer, you're
very poor," he explained. "Some die without a coffin." So he eventually
abandoned the circuit of farmyard weddings and cheap beer joints for
the roving life of a duck herder.

When I encountered him in the shade of a coconut tree, Sangwan
was rolling a homemade cigarette fashioned from a palm frond. He
would occasionally glance up to check on his flock, nearly a thousand
khaki Campbell ducks pecking and scavenging in the mucky waters
of a rice paddy several miles north of Banglane village. Despite the
intense midday heat, he wore a heavy brown knit cap with a blue
pompom to keep the sun off his head. His brow was deeply furrowed,
his jowls weathered. Beneath thick, graying eyebrows, his deep-set
eyes were bloodshot from sun and stress. Once again, he was singing
a plaintive tune.

The practice of grazing ducks in rice fields, which initially developed in southern China, had long since spread to the wetlands of Southeast Asia. Herders like Sangwan followed the rice harvest, trucking their flocks from province to province in pickups and feeding the ducks for free on residual grains, insects, and snails in the muddy water. "The ducks give you anything you want," he told me. "If you want something, you wait a bit and you get it. I didn't even have a house before." But now this barefoot nomad and countless others like him were being pressed by Thai officials to renounce their wandering ways and shut their flocks up in closed shelters. Ducks had been fingered as silent killers.

Researchers had discovered that ducks were spreading the novel flu strain like never before while no longer displaying any symptoms of their own. Wild waterfowl had long been recognized as a natural host for flu viruses, carrying the infection without getting sick. As the pathogen grew more virulent, it initially turned on the cousins of these wild birds, domesticated ducks, and caused widespread die-offs. For a while, this helped tip public health officials to proliferating poultry outbreaks that could endanger people. But in 2004, the virus abruptly changed its modus operandi a second time. Infected ducks once again showed no symptoms, according to an international team of scientists based at St. Jude Children's Research Hospital in Memphis. But now these infected birds spread the virus in larger amounts and for longer periods, in some cases a week longer than before. The virus also survived for more time in the surrounding water than it ever had. The duck had become "the Trojan horse" for Asian flu viruses, the researchers warned darkly.

When investigators in Thailand tested flocks of free-range ducks, nearly half proved to be infected with flu despite few signs of illness. Scientists warned that traditional duck farming posed a tremendous risk not only in Thailand's central plains but also in the Mekong River delta of southern Vietnam and the Red River delta of northern Vietnam.

Separate studies of the bird flu epidemic in Thai poultry had also deeply implicated free-ranging ducks. The research showed that outbreaks in the chicken population were concentrated in areas where ducks commonly graze, primarily wetland areas of intensive rice cultivation. Suphan Buri was singled out as a hot spot for disease. By contrast, provinces with high concentrations of chickens but few ducks largely escaped the brunt of the epidemic. The authors suggested that paddies were a likely meeting point where migratory water birds relayed contagion to ducks, which in turn infected chickens before shuttling it to other fields and provinces.

Sangwan said his rambling took him through the rice paddies of more than ten provinces over the course of a season. Every two or three days he moved on, generally drifting southward with the harvest. Only hours earlier, after exhausting the pickings in a nearby field, he had herded his flock to a new paddy, where young rice plants were just starting to poke through the still surface. "I marched them here like little soldiers. 'Keep walking,' I told them. 'Keep walking.'" He gestured with his open palms to show how he nudged them along, a smile settling on his stubbly face and crow's-feet deepening at the corners of his eyes. The ducks had filed down the grassy banks into the water, waddling and ruffling their tail feathers. A flotilla set sail with a whoosh toward a low line of palms on the distant shore. Sangwan had claimed a rare sliver of shade on the dike. He lay down his long bamboo rod and stretched out his scrawny legs.

At the end of the day, Sangwan and his wife would line the birds up again and march them back to the campsite. The ducks spent their nights in a temporary enclosure of plastic sheeting. Sangwan looked for a dry patch of earth to pitch his tent. "I've gotten used to living in the open fields," he said. "I love spending the time with the ducks rather than in a house, where you have to hear a television and people talking and traffic on the street." In the hours before dawn, he would listen to his charges rustle as they scouted for comfortable nooks to lay their eggs. "That's a nice sound," he mused. "That's the sound of making money."

Sangwan had turned to herding two decades earlier as the livestock revolution was accelerating, doubling and redoubling Thailand's duck production. As a younger man, he had dabbled in construction, growing rice, and raising vegetables. He took up singing after winning a local contest. Later, dead broke, he persuaded his uncle to teach him about ducks. He learned how to call to them in an authoritative voice so they'd respect and obey him. But alone on the dikes, Sangwan still serenaded his flocks with ballads of rural heartbreak.

Lowering his cigarette to his side, his melancholy voice began to carry across the glistening paddies, rising above the soft swooshing sound of birds foraging in the water.

> *I am looking at the rice fields at harvest time.*
> *I feel so lonesome thinking of you, my darling.*
> *I used to hold my sickle harvesting with you each year.*
> *But now things will never be the same . . .*

He returned the cigarette to his lips and took a long drag. His sunken cheeks slipped even deeper into shadow. Then he continued.

> *Oh, my dear, did you forget your promise?*
> *You asked me to wait for you these three years.*
> *But you seem to have forgotten our homeland.*
> *Oh, where are you right now?*

He paused again, briefly, eyes lowered.

> *Have you been seduced by life in the big city?*
> *Have you forgotten our land of farms?*
> *Or are you ashamed because someone cheated you in love?*
> *Is that why you're not coming back home to me?*

When he finished, Sangwan drew a bag of tobacco from the pocket of his baggy shirt and began rolling another cigarette. He fretted that the best days seemed to be over. Thai officials were already threatening to restrict the movement of ducks from one village to another. He could never afford to raise his flock in a closed shelter, he said. The feed bill would bankrupt him and the ducks would rebel.

After the government first floated the idea in late 2004, Sangwan had experimented with confining the ducks to a shed beside his house. It lasted a week. "I felt restless because the ducks couldn't walk around and they didn't have enough food," he recounted. "The ducks were not happy." That was bad news for business because, he confided to me, ducks are like pregnant women. They need to be pampered or they get nervous and lay their eggs prematurely. "I feel like I have a thousand little wives," he said, a grin briefly breaking through. "When the ducks get tense, I get tense."

To protest the proposed farming regime, his wife had led hundreds of peasants to the provincial capital. They besieged a government building for three hours, accusing officials of acting arbitrarily and sowing needless anxiety. "When the government says ducks carry bird flu, it just makes people panic," Sangwan complained, growing agitated. "It's not true that ducks get the flu. For twenty years I've been raising ducks and I've never seen one get bird flu."

In the months after I met Sangwan, the Thai government would bar farmers from transporting their flocks from one region to another and eventually, in 2006, place a total ban on duck grazing. Thailand's initiative sputtered, but the country ultimately achieved more than neighbors like Vietnam, China, and Indonesia, where duck herding remains common. When flu outbreaks unexpectedly erupted across more than a dozen provinces of northern Vietnam in 2007 after a long period of quiet, sickening people in the country for the first time in eighteen months, ducks were implicated. A special investigation blamed the epidemic on a dramatic influx of young ducks into the paddies of the Red River delta. By contrast, many of the estimated 10 million free-range ducks in Thailand were ultimately slaughtered or moved indoors.

But even there, compliance was spotty. Some Thai duck herders continued to follow the cycle of the crops as they had for generations, thwarting efforts to snuff out the disease. It had been several harvests since I met Sangwan when I heard about a group of herders who'd illegally moved three flocks with as many as fifteen thousand birds into the fields of Kanchanaburi province, just west of Suphan Buri. The chickens in several local villages began to die within two weeks. When those near the home of a peasant named Bang-on Benphat started to fall sick and collapse, the forty-eight-year-old man butchered them for dinner. His young son helped pluck the feathers. Both soon developed a fever and lung infections. Bang-on was hospitalized with severe pneumonia. Two days later he died, a casualty of flu.

Prathum sat cross-legged on his back porch, surveying all that he and his chickens had built. His eyes panned across the barn and the sheds, where his amply nourished hens were settling in for the afternoon, past the fish ponds, where a fleeting fin glinted amid the vines of morning glory, toward a line of trees casting long shadows at the edge of his property.

His thoughts returned to those new, modern chicken shelters that farmers were chattering more about. They were called evap houses, short for evaporative cooling houses. They had automatic ventilation and used large fans and water to maintain mild temperatures even during intense tropical heat. Because they were enclosed, they could keep out most contagion. "Even insects can't get in," he noted, impressed. But the cost was tremendous. He would need a loan and have to quadruple the size of his flock to make the numbers work. He would need at least five years to break even. No need to be hasty, he reasoned.

"I'm not worried right now," he put it to me. "We haven't heard anything lately about the epidemic. Maybe the disease left with our last lot of chickens. The new ones all look healthy."

His wife appeared in the doorway with a watermelon. She wasn't buying his cool assurance. "I'm definitely afraid the disease will come back to this area," she offered. "Some people say the disease came with the wind. Some say it came with birds. We have no clear idea. And deep down, he's still worried about it, too." She glared at Prathum, then laughed.

"Yes, I'm still scared," he confessed. "But I try not to show it. What can I do? We'd never had bird flu before. It just came. I'm hoping it won't come again."

Prathum took the watermelon from his wife. He grabbed a knife from the bench and started carving the fruit.

His sons were urging him to invest in an evap house, he told me without looking up. It was all that fancy university education. His older son, the one studying veterinary science, he'd even visited several evap houses to check them out. But Prathum had seen enough change in his life.

"I may not be able to learn as fast as young people," Prathum said. "I'll retire after a while and pass the farm on to my son. Then, he can do what he wants."

# CHAPTER SIX

# From a Single Spark

Professor Yi Guan gingerly placed the small cooler box with its mysterious contents into his black canvas satchel. He covered the box with a towel, then a newspaper, to conceal it from prying eyes. He wasn't quite sure what he had. Whatever it was, it had already proven to be a ruthless killer. The cooler box contained about two dozen vials, and lurking inside each one, Guan feared, was enough biohazardous material to start a global epidemic. But the specimens could also be the world's salvation—if only he could get them back to his laboratory in Hong Kong.

Guan slung the strap of the satchel over the shoulder of his gray suit jacket and headed for the door of the hospital. The medical staff at the Guangzhou Institute of Respiratory Diseases had nervously collected the mucus specimens from the noses and throats of patients stricken by the strange plague now burning through China's Guangdong province. The institute director, an esteemed scientist named Dr. Nanshan Zhong, had agreed that Guan could take the samples back to Hong Kong University for identification. But Guan had no such permission from the Chinese government. If he was stopped, he had no papers to show. If the vials were discovered, they could be confiscated and Guan detained. He could be held until his Beijing contacts vouched for him, if he was fortunate. If not, he could be accused of stealing state secrets or espionage and sentenced to life in a labor camp.

Chinese officials were determined to keep the severity of the epidemic under wraps. That very morning, February 11, 2003, Guangzhou's vice mayor had announced that the city was facing an outbreak of unusual pneumonia but it was under control and no extraordinary measures were required. But WHO was already picking up rumors of a far more serious outbreak involving a "strange contagious disease" that "left more than 100 people dead in Guangdong Province in the space of one week."

Guan suspected avian flu. Three months earlier, in November 2002, the wild birds of Hong Kong had started to die, first in the New Territories bordering Guangdong Province, then at a park in the teeming downtown of Kowloon. Samples from the outbreaks tested positive for the virus. Guan's suspicions hardened in February when bird flu was detected in a Hong Kong family. They had been traveling in China's Fujian province for the Chinese New Year when a young daughter came down with a severe respiratory illness. She had perished before the family returned home and was never tested for the virus. Soon her father and brother also fell sick and were hospitalized in Hong Kong. The father died. Both tested positive. It was the same H5N1 subtype that had first struck Hong Kong in 1997. They were the first confirmed cases anywhere since then.

As a fledgling researcher, Guan had helped investigate the 1997 outbreak. He had been part of the team that uncovered the widespread infection among Hong Kong's poultry, crucial information that helped energize the city's decisive response. He believed a pandemic had been averted. Now he was trying to repeat the feat.

As he left the Guangzhou institute, an aging seven-story gray cement edifice along the Pearl River, and set out to catch his Hong Kong–bound train, Guan felt time was running out. He feared that the next time the virus departed the province, it wouldn't be in securely sealed vials nestled inside a carefully prepared cooler box but unknowingly in the lungs of a victim. Once it escaped southern China, he was afraid, moreover, that the pathogen would spread to dozens of countries. Finally, he was sure it would then take only days to reach the far side of the planet.

Guan was tragically prescient on all three counts. The transformation of Asia over the previous generation had not only been internal, amplifying the hazards of an animal-born epidemic; but it had also redefined the region's ties with the rest of a globalized world. And in this age, the magnitude of a pandemic threat was growing as the distance between its origin and the rest of the world was shrinking.

Guan, however, was wrong about one thing.

When Yi Guan was six years old, growing up in the impoverished Chinese province of Jiangxi, his sister changed his name. He had been born Qiu Ping Guan. Qiu meant "autumn," the season of his birth. Ping meant "peaceful." Guan was the family name.

He was the youngest of three boys and two girls raised in the remote countryside about 180 miles from the provincial capital. In 1966, when Guan was four, Chairman Mao Zedong launched China's Cultural Revolution, a decade of violent upheaval targeting those considered as capitalists, intellectuals, or vestiges of the former ruling class. Guan's mother was descended from property. Though spared the worst excesses, his family was forced to subdivide its six-room house to make space for others. His father, an engineer, was sentenced to reeducation and put to work threshing flax plants to extract an ingredient for wine.

One day, as Guan was preparing to enroll in first grade, his adult sister called him aside.

"Come on, brother. I need to talk to you about something," she said. She seemed unusually earnest.

"What do you want to talk to me about?" the young Guan asked.

"I want to give you a new name," she responded. "You are the only boy in our whole family who has the hope to become successful. So I'm changing your name. It's becoming Yi."

She wrote the name on a piece of paper. Guan couldn't understand the significance. His sister said one of its meanings was "extraordinary."

"I picked this meaning to make you remember you must become outstanding, extraordinary," she told him. "That is your duty." She took him to school and registered him under his new name. It was a heavy burden, Guan later recalled. But he took his charge seriously.

As part of his radical remaking of Chinese society, Mao had shuttered the colleges. But just as Guan was preparing to graduate from high school, China announced they would reopen. For nine months he crammed for the entrance exam. Less than 1 percent of high school students would make the cut, Guan recounted. He would be among them.

Guan went on to study medicine and specialize in pediatrics, winning a place at an elite Beijing institute where he hooked up with a senior scientist specializing in infectious diseases of the respiratory system. He was later offered a slot in the PhD program at Hong Kong University and, after that, a chance to go overseas. He continued his research with one of the world's top flu scholars, Dr. Robert Webster, at St. Jude Children's Research Hospital in Memphis, Tennessee.

Several years had passed when, on a Saturday morning in late November 1997, Webster called him. Guan had returned to Memphis hours earlier after defending his doctoral thesis in Hong Kong and visiting his aging mother for the first time in two years. "Don't open your baggage," Webster ordered him. Guan was to turn around and go back.

"What happened?" Guan asked.

"While you were in the sky crossing the Pacific," Webster said, "they had three cases of H5N1."

"Really?" Guan was shouting excitedly over the telephone. "Really?"

Webster instructed him to get his travel documents ready and prepare the biological materials he would need to transport to Hong Kong. They were going to join the virus hunt.

Guan never gave up the chase. He soon moved back to Hong Kong to become a researcher in microbiology at the university and quickly went to work sampling the city's birds. Before long, he would emerge as one of the world's great collectors of flu viruses. Even as memories of the 1997 outbreak were fading, he was compiling data on myriad strains and amassing thousands of samples from birds in Hong Kong and southern China. In the summer of 2000, he had extended the net to Guangdong, establishing a virology lab at Shantou University Medical College. The facilities there had been idle for a decade. Guan spent a week cleaning the lab. He scrubbed the floor and washed the research bench and its protective hood. Only then did he set off to collect specimens from nearby poultry markets. Within two years, he had set up a network of field researchers that was gathering samples from birds in four provinces of southern China.

Webster, his mentor, had helped recruit Guan to the post at Hong Kong University. Webster was convinced that the novel flu strain simmering in southern China posed a grave danger to the world and wanted someone, preferably a Chinese virologist with Western training, who'd be nothing less than bullheaded in tracking the evolving threat. "Yi doesn't know the word *no*. He doesn't take *no* from anyone," Webster put it to me. "He believes in what he's doing and he's intellectually driven to do these things. He talks a million miles an hour, and a lot of it is not totally focused, but his overall mission is focused. You've got to have someone who is hard-driving to get out there and be able to interact with the people and understand the region, and he was the perfect person to do the surveillance."


Rumors of a bird flu epidemic among the Chinese of Guangdong first surfaced in November 2002. WHO's influenza chief, Klaus Stohr, who would later mobilize the agency's flu hunters after bird flu exploded in Vietnam in early 2004, was at a medical conference in Beijing when a health official from Guangdong stood up and described an especially nasty outbreak of respiratory disease among people of his province. "He talked about deaths, very severe disease and deaths," Stohr recounted. Chinese doctors had been unable to identify the precise cause, but they said it looked a lot like flu. Stohr was inclined to agree. "I just put two and two together, and it added up," he recalled. "I thought this must be H5N1 coming back in precisely the way we had feared. It was our worst nightmare, and the world's."

But when WHO subsequently pressed Chinese officials for more details, they offered a terse, dismissive reply. It was indeed flu, they reported, but just routine flu and everything was under control. In essence, "Now buzz off."

By the waning days of 2002, with wildfowl in Hong Kong starting to drop, Guan and his fellow researchers suspected that whatever was killing the birds was also afflicting the patients in Guangdong's hospitals. So on Christmas morning he came to Kowloon Park, an exquisitely maintained expanse of manicured greenery, flower beds, and faux waterfalls at the heart of central Kowloon, just off a stretch of Nathan Road known as the Golden Mile for the bountiful commerce of its shops and boutiques. Toward the center of the park, fringed by palms and shade trees, was the man-made lake where several dozen species, including flamingos, ducks, geese, and teal, frolicked in the water and sunbathed on the banks. Guan laid out his gear. Meticulously, he clasped a small vial in his curled pinky, leaving the rest of his fingers free. As a colleague restrained the first bird, Guan slowly inserted a Q-tip-like swab into its cloaca and withdrew a specimen. Guan sampled at least a dozen birds this way. Most later tested positive in the lab for avian flu.

He and fellow researcher Malik Peiris also continued to stalk the strain into the Mai Po Marshes of the New Territories. There the mudflats and mangroves offered a refuge unique in Hong Kong for hundreds of species of wild birds. The scientists drove up before dawn. It was a cold, damp morning. Though they were wrapped in heavy coats, the chill penetrated Guan's bones. He pushed a small boat into the dark water, mud soaking his sneakers, and then rowed across the narrow inlet. He came ashore on Duck Island, a sliver of land that fittingly boasted more than twenty species of ducks. It was too hard to catch these birds. So for an hour and a half, Guan scoured the ground for their droppings. Live virus would be lurking inside.

The specimens collected from scores of birds in Hong Kong suggested that the mystery outbreak in Guangong's hospitals was H5N1. But by February 2004, Guan and his fellow microbiologists realized there was no substitute for actual human samples. Someone would have to go to Guangzhou to get them.

"Why not Yi?" Webster asked. "It's not everyone who's going to want to go into that room and risk his life." Guan was impetuous and courageous, and it was obvious to him that this was his moment.

When Guan arrived at the Guangzhou Institute of Respiratory Diseases, he told the director, Nanshan Zhong, that the pathogen was most probably influenza. "It is possible this is the early stage of a pandemic. If we don't deal with it carefully, this will be a disaster," Guan warned.

To contain it, medical experts had to determine precisely what it was. Guangzhou didn't have the necessary lab facilities, Guan concluded, but Hong Kong did. Zhong concurred.

With the vials stashed in his satchel, Guan hailed a taxi outside the institute gate on the afternoon of February 11 and set off into rush-hour traffic. He wanted to make the 6:30 P.M. express train to Hong Kong. If he did, he could turn over the cache of vials to his lab staff in time for them to begin the process of culturing the virus samples that very night. Guangzhou East Railway Station was teeming with travelers. The cavernous hall echoed with the announcement of trains departing for destinations in the Chinese hinterland. Guan headed toward the terminal for the Kowloon-Canton Railway, which would whisk him to Hong Kong. Police officers, some alone, some in pairs, meandered through the crowd. Guan avoided eye contact.

He ascended the escalator. At the top were the immigration counters. Guan liked to tell himself it was all one country, Hong Kong and the mainland. That was, after all, the official Chinese government line. And by that logic, he wasn't smuggling the samples abroad. But in many practical ways, China still treated its border with Hong Kong as an international frontier. Guan showed his passport to the blue-uniformed immigration officer, who waved him through. When he reached customs and saw the X-ray machine, he momentarily considered looking for another way around. Then he thought better. "If you try to avoid that," he told himself, "there will be more trouble." He placed his satchel on the belt. Seconds passed before it reappeared on the far side. The white-uniformed customs officer didn't say a word. Guan retrieved his bag and continued toward the waiting room.

There were families with large suitcases, and businessmen returning home after a day trip to their factories and suppliers. They were already queuing up when the train was announced. Guan joined the line. An immigration officer was conducting a final passport inspection. "The quieter you are, the safer you are," Guan reminded himself. Usually he was a dervish of activity, a fast-walking, fast-talking impresario of scientific notions who pressed his theories, passions, and grievances on listeners in a shotgun spray of sentences, a chronically restless soul who found it nearly impossible to remain seated or stand still unless, of course, he was smoking a cigarette out an open window. Yet in his plain gray suit, inexpensive haircut, and large, silver-framed aviator glasses, he could melt into the undifferentiated mass of commuters if he could just feign the right air of indifference.

"The more you keep quiet," he repeated to himself, "the safer you are." The immigration officer asked for Guan's passport. He flicked his cigarette to the floor and produced the document from his jacket with an affected look of weary annoyance. The officer returned the passport and moved on.

Finally the line moved. The passengers filed downstairs to the red tile railway platform and onto the train. Guan claimed his seat. He placed the satchel carefully on the overhead rack. Then he took out his cell phone and called his lab. "Are you ready?" he asked. "The samples are on their way."

But as they were analyzed over the coming days, the samples stumped Guan and his fellow researchers. The virus wasn't H5N1. It wasn't flu at all. It would later be identified as SARS, and that was fortunate. Because in the age of globalization, flu would have been much worse.


Half a year after the SARS epidemic had subsided and life had returned to Hong Kong's deserted streets, the city's legislative council would conduct an inquiry in January 2004 into the government's handling of the crisis. The outbreak had killed 299 people in Hong Kong alone. Nearly six times that number had been infected. Amid stinging criticism, Hong Kong's secretary of health, welfare, and food and the chairman of the hospital authority would lose their jobs.

Margaret Chan, the city's health director who had so ably steered

Hong Kong through the bird flu outbreaks of 1997, had also skippered its emergency response to SARS. As she testified before the legislative council, she broke into tears. "We tried to do our best," she assured the members before being overcome by emotion, forcing them to briefly suspend the hearing. During her testimony, Chan told the council she had tried several times in early 2003 to confirm the press reports of an epidemic brewing in neighboring Guangdong province. On February 11, 2003, the very day that Guan was making his clandestine run to the Guangzhou institute, Chan and one of her departmental consultants had repeatedly phoned health officials in Guangdong about the rumors. No one answered their calls. "Usually, with other infectious diseases, there was no problem with communication," she testified. She added that a Guangdong official later told her "there was a legal requirement for infectious diseases at that time, that infectious diseases were classified as state secrets. That is why they cannot share the information."

The council went on to censure Chan, who by that time had resigned from the health department for a post at WHO in Geneva. She was faulted in part for leaving Hong Kong vulnerable "in that she did not attach sufficient importance to 'soft intelligence' on the [acute pneumonia] epidemic in Guangdong." The report suggested she could have dispatched a team to the mainland to investigate.

Ultimately, it was a doctor named Liu Jianlun who had brought the disease to Hong Kong's attention. Liu was a retired kidney specialist from southern China. At age sixty-four, he still worked part-time in an outpatient clinic at a hospital in Guangzhou. It was this hospital that first treated one of the earliest victims, a forty-four-year-old seafood seller from the suburbs who came in January 30, 2003, with a severe cough and fever. This patient stayed only two days before being transferred to another hospital. But in that remarkably brief time, he infected at least ninety-six other people, including ninety health-care workers.

Liu himself started feeling sick two weeks later. He worried that he had contracted whatever horrible illness was besieging his hospital. But his chest X-rays looked clear. So he dosed himself with antibiotics and set out with his wife on a three-hour bus ride for a nephew's wedding in Hong Kong. When they arrived, he felt well enough to go shopping and enjoy a long lunch with relatives. Late that afternoon, on February 21, Liu and his wife checked in to their room on the ninth floor of the Metropole, a three-star hotel in Kowloon with a large swimming pool on the roof and a karaoke bar in the basement. Their room was what marketers call cozy, with two single beds and a pale olive carpet. From the window of room 911, they could see a Shell service station, an Esso service station, a YMCA guesthouse, and beyond, one of the most congested quarters in all Hong Kong.

Long before the term *globalization* was coined, Hong Kong was the definition. As a colonial entrepôt, it evolved into a bridge between Occident and Orient, a global financial center, and one of the world's busiest ports. The ninth floor of the Metropole was true to form. Staying on the floor that same evening were also three women from Singapore, including a former flight attendant on a shopping excursion, four Canadians, among them an elderly woman from Toronto visiting her son, a British couple on the way to their native Philippines, a young German tourist headed for a two-week vacation in Australia, and three Americans, including, right across from 911, a Chinese-American garment merchandiser bound for Hanoi to meet his denim suppliers. Sometime in the course of that evening or early the next morning, before Liu checked out of the hotel with a searing headache and dragged himself five blocks to a hospital, he managed to infect all thirteen of those neighbors. Three others at the Metropole also caught the virus. Based on intensive sampling, health investigators later theorized that Liu, upon returning to the ninth floor from dinner, had thrown up on the teal-colored carpet outside the polished, wood-trimmed doors of the elevator. Someone, perhaps his wife, cleaned up the mess. An invisible mist of infectious particles wafted along the corridor.

Liu would become Hong Kong's first case. Before he died, he told the doctors and nurses caring for him about the disease raging in his own hospital.

The Singaporean hotel guests returned home, where they were all hospitalized, and one, the woman on the shopping spree, in turn sparked an outbreak that sickened at least 195 people in her own country. The doctor who treated the initial Singaporean case later flew to New York for a medical conference and on the way infected a Singapore Airlines flight attendant.

The Chinese American merchandiser continued to Hanoi, where, before succumbing to his sickness, he seeded a Vietnamese outbreak that infected sixty-three others. A French physician in Hanoi who cared for a stricken colleague later carried the virus home to Paris, along the way infecting three others on the Air France flight.

Just four days after Dr. Liu had boarded the bus in Guangzhou, the elderly Canadian woman from the Metropole was back in Toronto, halfway around the globe, and feeling ill. Before she died, this grandmother passed the virus to four family members, ultimately igniting a cluster of 136 cases in Canada. One of those stricken in Toronto was a nurse from the Philippines, who later flew home to help find a faith healer for her cancer-stricken father and instead infected her family with the killer virus, volleying the illness right back to Southeast Asia.

This is how an epidemic becomes a pandemic. This was the first great wave of a still-unnamed virus washing over the world. More would follow.

Of all those sickened at the Metropole, the one who went on to infect the most people directly wasn't a guest at all. He was a twenty-six-year-old airport freight handler from Hong Kong itself who visited a friend staying on the ninth floor. About two days after this young worker started feeling lousy, he went to the emergency room, where he was diagnosed with a respiratory infection and sent home. Almost a week later, as his condition worsened, he was admitted to Prince of Wales Hospital. As the main referral hospital in the New Territories, this modern medical complex has the feeling of a bus terminal at rush hour, with crowded corridors and long lines at the reception windows. The young man was placed in Ward 8A. He went on to infect at least 143 others, all at the hospital.

From here, the virus again exploded into the world. Nearby in Ward 8A was an elderly Chinese man being treated for an unrelated salmonella infection. His seventy-two-year-old kid brother visited him often. On March 11, the younger brother developed a fever and three days later came down with a cough and chills. Though a physician urged him to get hospital care himself, the man insisted on flying back home to Beijing as planned. On March 15 he did so.

"It was like seeds thrown into the wind," a doctor in Beijing later remarked. "Who knows where they will land?"

Air China flight 112 was nearly full that Saturday. The Boeing 737 had 112 passengers and 8 crew members. The ailing seventy-two-year-old was slumped in seat 14E. He looked pale, his brow was drenched. He couldn't quiet his coughing and kept hacking until his handkerchief was soaked. He went to the galley to ask a flight attendant for water to take some pills. In the three hours it took to reach Beijing, he infected 22 passengers between rows 7 and 19 and 2 flight attendants. The first victim would spike a fever within days. By the middle of the next week, three-quarters would develop what initially felt like a bad head cold. Five would later die, as would the elderly traveler himself.

The two flight attendants were from Inner Mongolia, a region of northern China. When they returned home, they sparked an outbreak that accounted for most of its 290 subsequent cases. From there, the virus leapfrogged across the border to the independent country of Mongolia.

Ten of those infected on CA-112 were members of a Hong Kong tour group.

Four were employees of a Taiwanese engineering firm, who eventually carried the virus home with them.

Another was a young woman from Singapore, who later flew home and was hospitalized there.

Yet another was a Chinese official who journeyed to Bangkok. As he headed back to Beijing from that subsequent trip, on a Thai Airways flight, he in turn infected a Finnish official of the International Labor Organization, who had been seated next to him.

It had only been five weeks since Dr. Liu got on the bus. Of the 8,098 cases of SARS ultimately detected, more than 4,000 could be traced back to his overnight stay in the Metropole.

———

When the Spanish Lady came calling in 1918, no place was too remote to elude her entreaties. She even found the islanders of the Pacific. In prior centuries, when seafaring ships were driven solely by the wind, an epidemic disease brought on board would have time to burn itself out before coming ashore on these distant islands. They were beyond reach. Maritime technology changed that. In October 1918, a U.S. Navy transport called the *Logan* sailed from Manila with an infected crew and put in at Guam. Nearly everyone on the island fell sick with flu. About 800 died. Another vessel, the *Navua*, set out from the stricken port of San Francisco and in mid-November docked in Tahiti. Three thousand Tahitians caught the disease, and more than a tenth of the population perished. From the ailing port of Auckland in New Zealand, a steamer named the *Talune* set sail, scattering death at each stop along its tropical itinerary. In Fiji, 5,000 died. In Tonga, as many as 1,600, about a tenth of its inhabitants, died. The *Talune* fatefully docked at Apia, the capital of Western Samoa, in early November. By the time the suffering subsided early the next year, an estimated 8,500 natives had succumbed, more than a fifth of the population.

For many Eskimo villages of Alaska, the plague was even less forgiving. As winter was closing in, the final ship of the season, a vessel from Seattle called the *Victoria*, moored in the port of Nome on Alaska's Seward Peninsula and deposited its lethal cargo. From there, sped by the wanderings of white missionaries, influenza advanced along the frozen tundra, penetrating the coast to the north. It killed every last Eskimo in the village of York, about 150 miles from Nome by dogsled. The inhabitants of nearby Wales, the westernmost point on the North American continent, joined in a funeral for a boy from York. Soon more than half those from Wales were also dead. At another outpost, Teller Lutheran Mission, disease erupted after a pair of visitors from Nome had joined a local church service. The first native fell sick two days later. Soon corpses stacked up inside the igloos. All but eight of the village's eighty residents perished and were buried beneath the permafrost. One was a woman who ultimately helped crack the genetic code of the Spanish flu after researchers excavated her grave seventy-nine years later and retrieved a sample of infected lung tissue from her well-preserved body.

The global reach of pandemic flu is thus nothing new. But globalization is. And over the last generation, it has fundamentally recast the threat of infectious disease. As with SARS, the next flu pandemic will spread at the speed of jet aircraft, coursing along an ever-thickening web of international travel, each new thread reducing the time the virus must wait before breaching another frontier.

"As the first severe contagious disease of the twenty-first century, SARS exemplifies the ever-present threat of new emerging infectious diseases and the real potential for rapid dissemination made possible by the current volume and speed of air travel," said Mark A. Gendreau, a senior attending physician at the Tufts School of Medicine, in testimony before the U.S. Congress. Margaret Chan was even blunter: "SARS was a wake-up call for all of us. It spread faster than we had predicted." Within six months, it reached more than thirty countries on six continents.

More people are traveling more places than ever before. Though Hong Kong remains an exceptional crossroads, Yi Guan rightly suggests that the world increasingly resembles the ninth floor of the Metropole.

"Today you are in England, tomorrow in New York, and the third day you might be in Hong Kong," Guan noted. Imagine how many people an infected traveler encounters along the way. "The case lands in London or New York or Hong Kong. Maybe ten thousand people have connecting flights in that airport within two hours. It spreads to the whole world. Globalization accelerates the transmission speed, maybe by a hundred times."

A century ago, he continued, a novel flu strain could take more than a year to circle the world. "Now, currently, does it take one year? I don't think so. Maybe one month," he said.

Over history, each advance in transport and trade has sped disease on its way. The Black Death of the Middle Ages spread faster by merchant ship on the Mediterranean than by horseback on the Asian steppe. The last of three cholera epidemics in nineteenth-century America was the swiftest, exploiting the country's new railroads. Even since 1968, the date of the last flu pandemic, change has been dramatic. Air traffic has increased about tenfold since then. Using data on the volume of travelers at fifty-two major cities around the world, a team of American researchers projected how long it would take a flu pandemic to spread and compared it to the Hong Kong flu of 1968. They found that the same virus, if it had erupted in 2000, would have struck cities in the Northern Hemisphere nearly four months earlier. And while the Hong Kong flu required almost a year to sweep the globe, in 2000 the virus would have peaked in every one of the cities in half that time. A separate team of researchers in Britain, using a different statistical approach and more recent data, from 2002, concluded that in some cities in the Southern Hemisphere, the epidemic actually would have peaked a full year faster than it had in 1968.

An accelerating epidemic leaves public health officials little chance to top off their stockpile of antiviral drugs or distribute them. There's less lead time to prepare measures meant to slow the inexorable advance of epidemic—for instance, isolation policies and school closings—or to make sure that strategic infrastructure and crisis manpower plans are in place. Most crucially, scientists expect it will take at least six months to develop a pandemic vaccine and far longer to make sure everyone gets it.

"All of technology cannot keep up," Guan warned. "To manufacture a vaccine takes months. The transmission of disease is by the hour now."

No matter how many ways Guan and his colleague Malik Peiris tried to find a flu virus in the specimens smuggled back from Guangzhou, they couldn't. For that matter, they couldn't isolate a virus of any sort at all. In the lab, they tried to grow the puzzling pathogen using chicken embryos, dog cells, monkey cells, and even human larynx and lung cells. Nothing. But each disappointment refined the search. Each time they failed to corner their quarry in the Guangzhou samples—for weeks, the only ones outside the hands of the Chinese government— the Hong Kong University team weeded out false pretenders, bringing the researchers that much closer to the golden moment of discovery.

When it came, it was Peiris who made it. His lab isolated a pathogen called *Coronavirus* in a new specimen taken at a Hong Kong hospital from the dying brother-in-law of Liu Jianlun. Precisely one month after Dr. Liu had checked into the Metropole, Peiris sent an e-mail to a global network of laboratory scientists announcing that he had found the cause of the disease now named SARS.

The discovery was an unprecedented coup for WHO. Peiris was part of a virtual laboratory network that Klaus Stohr had assembled in mid-March 2003 for the SARS hunt. He had recruited eleven premier labs from nine countries for a rare collaborative effort, appealing to many of virology's brightest and most competitive researchers to set aside their egos and their lust for scholarly publication. Instead they compared notes, speaking daily by teleconference to review their progress. Crucial findings were shared through a secure Web site. WHO also established parallel networks, so epidemiologists could analyze how SARS was spreading and clinicians could consult about how to treat it.

The overriding fear was that this killer could become endemic, like HIV-AIDS, before the world had time to diagnose the threat, contain its spread, and eradicate it. WHO rallied scores of disease specialists from inside the agency and out, dispatching them to East Asia. Keiji Fukuda, for one, spent eight weeks in mainland China and Hong Kong. It was what Fukuda saw in the wards of Prince of Wales Hospital that prompted WHO to sound its first global alert about this severe, unidentified pneumonia in mid-March 2003 and urge that patients be isolated. A second, stronger alert followed three days later after Mike Ryan, WHO's global alert coordinator, was awakened with news that an infected physician had boarded an airplane in New York bound for Singapore. The man was bundled off the airplane during a stop in Frankfurt by German emergency medical staff in orange hazmat suits. Within hours, WHO had begun taking measures to curtail the international spread of SARS.

This was the agency at its best. "The quality, speed and effectiveness of the public health response to SARS brilliantly outshone past responses to international outbreaks of infectious disease, validating a decade's worth of progress in global public health networking," according to an assessment by the U.S. Institute of Medicine. "The

World Health Organization (WHO) deserves credit for initiating and coordinating much of this response."

Yet even after the *Coronavirus* had been isolated and containment efforts put in place, the source of the disease remained a mystery. WHO investigators suspected a link to wild animals. Some of the earliest cases in Guangdong had been in restaurant employees who prepared exotic fare, often from small imported mammals, to sate southern China's appetite for what locals called "wild flavor." To choke off the epidemic, researchers would have to determine which creature was the culprit. Someone would have to literally stick a needle into the heart of an animal and a swab up its anus. Once again, the mission would fall to Guan.

In early May 2003, he crossed to the Chinese city of Shenzhen, just beyond the narrow river that serves as Hong Kong's border with the mainland. Once a fishing village, Shenzhen had been designated a special economic zone in 1979 to attract foreign investment. The gold rush had transformed it into an audacious boomtown with a population rivaling New York's and skyscrapers to rival Hong Kong's. It had become China's wealthiest and fastest-growing city and the quintessence of excess. In the city's storied restaurants, the new rich spent hundreds, even thousands of dollars to dine on nearly any form of life they hankered after. At Dongmen Market, the hungry and the adventurous perused wire cages stacked high with writhing snakes, barking raccoon dogs, growling ferret badgers, turtles, hares, palm civets, hog badgers, house cats, scaled pangolins, rabbits, beavers, and the miniature Asian deer called muntjac.

By the time Guan set foot on the slick, bloody floors of the covered market, he had lost count of how many thousands of birds he'd sampled over the years, looking for flu. But he'd never collected specimens from the kind of grim menagerie that now confronted him. Many of these animals were carnivores with claws and fangs. The merchants, engaged in a shadowy yet highly lucrative trade, could be equally vicious. Guan had won prior permission from Shenzhen health officials for this expedition. At least that would keep the police off his back.

Dongmen Market was huge. It sprawled across an entire city block, consuming the ground floor of a mammoth clothing-and-textile center. Stalls disappeared into the twilight of scattered fluorescent bulbs dangling from the metal ceiling. The odor was oppressive. "Where do I start?" Guan asked himself. He had applied his full deductive powers to the question even before he arrived. Whatever creature was the source of the virus, it had to be a mammal, he reasoned. That would explain why the microbe was so quick to become transmissible among humans, which of course are also mammals. So no turtles, snakes, or, for once, birds. The creature would also have to be fairly common. If it was too rare, the virus might have burned itself out before it had a chance to cross to people. Guan narrowed the list to eight species. He was especially interested in Himalayan palm civets, also known as masked civets because of the black and white stripes that run from forehead to nose and white circles around their eyes. About two feet long and ten pounds in weight, these catlike creatures have long been a popular Chinese delicacy.

He approached the traders. He explained that he and his team were looking to take a few samples from the animals: a throat swab, a rectal swab, and some blood. To get the blood, he would have to jab a needle into the heart of each beast. It would be too hard to find a vein through all the fur.

The dealers wanted no part of it. They were afraid Guan might kill or otherwise harm their lucrative creatures, perhaps somehow rob them of that raw bestial energy that made them so coveted by customers. But if Guan was willing to buy the animals, well, then they could do a deal.

Guan pulled a thick wad of Chinese banknotes from his pocket. He had thousands of dollars worth. Yes, he'd pay, he told them. But not full price. Here's how it was going to be: He would give them one hundred yuan to sample an animal, about twelve dollars each. If the animal died within a day, the trader could notify Shenzhen's disease-control officers and be compensated in full. The traders agreed and crowded around, eager for easy money.

There wasn't enough space at each stall to take specimens, and it was too dark to see. In any case, Guan didn't want to scare off anyone's business. So once he made his selection, he had the merchants lug the cages to the muddy alley outside. There, amid all the hustle, among the army of porters hauling crates of poultry and produce, exotic roots, mushrooms, and broad bushy vegetables, in between the handcarts, trolleys, and bicycles stacked with boxes, Guan spread a plastic tarp, put down his gear, and prepared to operate. He slipped on a white lab coat and mask. He donned thick protective gloves. The curious quickly crowded around. Guan and his colleagues asked the security guards to push them back. "There's virus," he warned.

Before they could begin, Guan's team had to anesthetize each animal, pump it full of ketamine. That meant coaxing an often hostile creature out of its cage and plunging a needle into its flesh. "It's very, very dangerous," Guan advised. "They can bite you. The civet, his head can spin around 360 degrees and you never expect it. To catch it by its back, it's too hard to do that." Some beasts cowered, some lunged. So to restrain them, the researchers used a special tool fashioned from a long tube with a Y-shaped attachment that fit around the animal's neck. With its head thus pinned down, Guan and his colleagues wrestled the critter to the ground and injected the anesthesia.

The subject soon went limp. Guan stuck a needle into its heart, filling a vial with blood. "My medical training helped me a lot," he recalled. Next he inserted a swab into the animal's throat and finally into its anus. Over the course of two days, he jabbed and swabbed twenty-five animals, including a half-dozen civets and assorted beavers, hog and ferret badgers, muntjacs, raccoon dogs, and domestic cats.

Back at the university lab, he quickly found the evidence he was looking for. He isolated the SARS *Coronavirus* in samples from the civets and one raccoon dog. These animals plus a ferret badger had antibodies that indicated they had been infected. There was no way to know whether these species were the ultimate origin of the pathogen or had caught it somewhere else. But it was now clear how the virus was spreading to people. The wild-game markets had to be shuttered.

Based on Guan's research, WHO urged China to close them down. This time, at least for a while, China listened. New infections ceased.

On July 5, 2003, the world officially defeated what Shigeru Omi, WHO's regional director for East Asia, later called "the first emerging disease of the age of globalization." The agency reported that the SARS epidemic was over. The last chain of transmission had been severed in Taiwan, with no new cases detected there since mid-June. The global death toll had been kept below eight hundred.

This epidemic had been contained, first and foremost, through the successful isolation of those infected. Some countries, like Vietnam and Singapore, had tackled it more aggressively than others, adopting measures like temperature screening for airplane passengers and hospital visitors, isolation rooms at airports, designated wards for suspected cases, and rapid tracing of those who'd been exposed. But in the end, the key everywhere was preempting the virus before it could infect again.

Had the pathogen been a novel strain of flu, the strategy would have failed.

Pandemic would have followed, potentially with millions of deaths, and not because of any human miscue. The difference is wired into the biology of these two viruses.

On many scores, there is an uncanny resemblance between influenza and SARS. They are both cruel respiratory afflictions that originate in animals and cross to the humans who prey on them. Like SARS, flu has often if not always emerged out of southern China, where live markets have proven central in amplifying and spreading the disease. Yet flu is far more sinister.

Flu, for starters, is a more nimble virus that spreads with an ease unmatched by other respiratory diseases. "Flu replicates far more efficiently in humans than SARS," Guan reminded me. "After adapting to humans, SARS can spread quite quickly. But if compared to flu, it is still quite slow."

In analyzing outbreaks, researchers focus on what they call the basic reproductive number. That figure represents how many other people are typically infected by each sick person. Obviously, the higher the number, the more infectious the disease. An early study of SARS concluded that its reproductive number was lower than that for other respiratory viruses. Researchers said this accounted for why SARS could be contained. Later research compared SARS and flu and found that the transmissibility of flu, as represented by its reproductive number, might be three times greater or more.

SARS was also a soft target because its victims became contagious only after developing a fever and other symptoms. An analysis of SARS patients in Hong Kong found the amount of virus in their nose and throat remained low for the first five days after they started feeling sick, only peaking on the tenth day. When researchers looked at the actual pattern of cases, they reached a similar conclusion that people were rarely contagious in the first few days after they came down with symptoms. Typically it took almost a week or more. For public health authorities, this pattern was a blessing. It gave them ample chance to identify and isolate victims before they disseminated the virus further.

"It is difficult to escape the conclusion that the world was very lucky this time," wrote a team of researchers from Britain and Hong Kong, adding that an anticipated flu pandemic by contrast could have a "devastating impact."

With flu, people may be contagious even before they develop symptoms. One analysis estimated that between 30 and 50 percent of those infected with a novel flu strain would catch it from someone who wasn't yet ostensibly sick. The true figures will depend on the specifics of the strain. But still, in a brewing flu epidemic, there may be no sure-fire way to determine who is contagious and isolate them. Even those infected might be ignorant of their fateful role in spreading the plague until it's too late.

"Once adapted to human-to-human transmission, influenza is highly transmissible, both in the late incubation period as well as early in the disease. Therefore, its spread may not be amenable to interruption with the same public health measures used to contain SARS," wrote Peiris and Guan in the cautious language of scholarly publication.

In person, Guan was more succinct: "From a single spark, you can burn up the world."

I visited Guan several times in the years after SARS. When I last met him at his university office, he was agitated as usual. It was flu, not SARS, that was keeping him up late at night.

"I can see what many people cannot see," he told me. He had always been a bit of a Jeremiah. "For me as a scientist, my record and reputation are fully acceptable to the scientific community. So why do I keep working so hard?"

He paused and reached across the table to pour me a cup of traditional Oolong tea from a white ceramic pot.

He resumed. "The flu virus is not easy to track down. I am building up information so I can know where and how a pandemic might happen. I want to tell the world we can create a different future." He was on his pulpit, urging humanity to prepare for the inevitable mutation of the virus and adopt economic and political reforms that would stem its spread. These should include restructuring farms, markets, and trade and improving how disease is monitored. Most important, he preached, was candor in disclosing outbreaks when they occur. "If not, who will be the losers? The whole globe."

Guan shifted on the edge of his chair and told me he envied Al Gore. As a former U.S. vice president, Gore could find the financing to make his film *An Inconvenient Truth* about the threat of global warming. Guan had been finding it hard to make his own exhortation heard.

"We are all sitting at the same table," he continued. "We share the same benefits, share the tea. Globalization is good for promoting civilization. But if you're part of globalization, you need to take responsibility. If not, it will damage not only one country but many, many countries."

Guan himself had come to epitomize the age of globalization. From Jiangxi, where he grew up reading in the courtyard with chickens for companions, this country boy had worked his way to Tennessee to study with the dean of all influenza researchers and then found a perch at Hong Kong's most exclusive university. There, he was quickly

Jones Ginting with his son. Ginting was the sole survivor of the Sumatra cluster of bird flu cases. Hospitalized for ten weeks, he is shown here days after being discharged and is still unaware that all his siblings are dead.

Agenda Purba, one of the most prominent witch doctors in his remote corner of Indonesia, was asked to treat Jones Ginting after Ginting and his family resisted modern medical care. Purba boasted it was his magical power that brought the outbreak to an end.

Prasert Thongchareon, the virologist who literally wrote the book on flu in Thailand, suspected something was amiss in his country after his friends stopped bringing him eggs. Prasert's lonely campaign to reveal the truth about the virus helped force his government to end its cover-up.

Tim Uyeki (left) and Keiji Fukuda are among the world's premier flu hunters. After the virus erupted in Vietnam, some disease specialists were nervous about investigating but Uyeki and Fukuda were "ready to go right away." *(Paula Bronstein/Getty Images)*

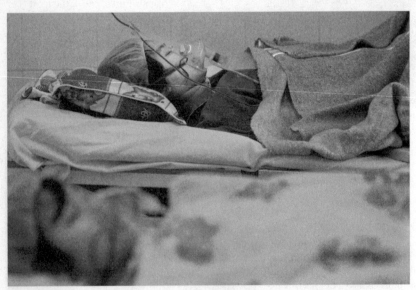

Two Vietnamese sisters, part of a stricken wedding party from Thai Binh province, deteriorated rapidly after contracting bird flu. When WHO reported that this early cluster of fatal cases could be the first instance of human-to-human transmission, the political backlash was tremendous. *(Paula Bronstein/Getty Images)*

Born the son of dirt poor Thai peasants, Prathum Buaklee's pursuit of chicken farming once brought him unimaginable wealth. But the agricultural revolution of which he was a part has also created prime conditions for hatching a worldwide epidemic.

Sangwan Klinhom gave up the life of a Thai country singer to herd ducks. But his new livelihood was put in jeopardy after scientists identified ducks as silent killers that could be responsible for spreading bird flu.

Yi Guan (right) and his "band of heroes" braved political pressure and the risk of infection by sampling tens of thousands of birds across Hong Kong and southern China. The collection of virus samples became in essence an early warning system for pandemic. (Courtesy of Yi Guan)

At a cockfighting ring deep in the rice paddies of northern Thailand, spectators wager hundreds, perhaps thousands, of dollars per bout and then watch anxiously for one of the badly bloodied roosters to succumb. The sport has been part of Southeast Asian culture for centuries.

A Thai trainer teaches an up-and-coming rooster to fight. The trainer's own arms are scarred and swollen from the errant attacks of his pupils. Intimate contact with fighting cocks has been repeatedly blamed for causing bird flu infections.

A Cambodian peddler sells merit birds outside the Wat Phnom temple in Phnom Penh. Buddhist adherents raise the birds to their lips, offer a prayer and sometimes a kiss, and release them as an act of religious devotion.

Indonesian butchers at a live poultry market in Jakarta meet the widespread demand for fresh meat. But these markets, prevalent across much of East Asia, pose a threat to customers and workers and are a nexus where viruses can amplify, swap genetic material, and spread.

Suparno (left), an Indonesian animal-health officer, and his staff vaccinate a hen against bird flu. After giving injections to only a few of the chickens at this Javanese home, the team left, saying it was too hard to catch the rest of the birds.

Hoai Nam (left) and Duc Trung, investigative reporters for the *Thanh Nien* newspaper, posed as poultry traders to expose corruption in Vietnam's disease control efforts. The men each wore a tiny video camera to film illegal commerce at a slaughter-house in Ho Chi Minh City.

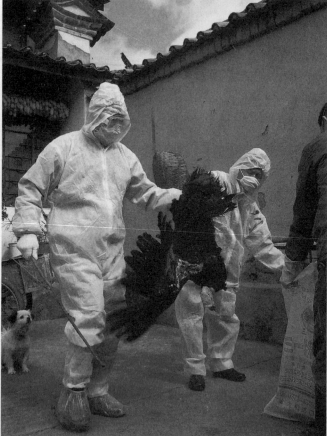

Chinese health workers conduct a mass poultry cull one week after the Beijing government initially confirmed human cases of bird flu. In this one area of Yunnan Province, 100,000 birds were reportedly slaughtered as part of the operation. (*China Photos/Getty Images*)

With human cases accelerating in Indonesia, government officials and volunteers burn nearly five hundred chickens and pet birds in a pit dug behind a public tennis court in Jakarta. Outside the capital, control efforts were more modest. *(Dimas Ardian/Getty Images)*

Margaret Chan, newly elected as WHO's director general, met Chinese Premier Wen Jiabao at the Great Hall of the People in Beijing and told him it was vital that China disclose outbreaks and share virus samples. Chan was the first Chinese national to hold such a high United Nations post. *(Getty Images)*

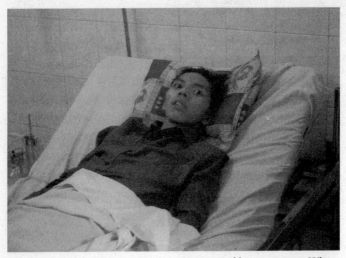

Nguyen Sy Tuan's doctors in Hanoi never expected him to recover. When he did, the doctors claimed it as a victory for Vietnamese medicine. But flu specialists worried that his case was actually part of a trend presaging pandemic.

Gina Samaan, a WHO epidemiologist, tracked the possible source of a fatal infection from Jakarta to an adjacent province and back again. Her examination of the victim's home showed no sign of contamination and that made Samaan suspicious.

promoted to professor, awarded his own laboratory, and installed in a corner office on the fifth floor of the Faculty of Medicine with a spectacular hillside view of the western approaches to Victoria Harbor and Lantau Island beyond. When he breaks for a cigarette, he stares out the floor-to-ceiling picture windows, watching the procession of freighters and the setting sun burning into the mist. He travels the world, lecturing and consulting. Shortly before my last visit to see him, he had flown to my home city of Washington for a meeting with U.S. health officials, sending me a message by BlackBerry when his airplane landed but departing again before we could meet. Just three days after he'd set out from Hong Kong, he was home again. "Fifty hours in transit," he told me afterward, somewhat bemused. The pace had left him little time for Jiangxi. Though he sent money home to support his brothers and sisters, he could only spare five days a year for his aging, melancholy mother. "I'm so sorry, mother," he told her. "For 360 days a year I belong to the world."

This globalized age offers untold firepower for fighting disease. It was via the Internet, for example, that some of the earliest rumors about SARS in southern China found their way to WHO, and the intense scrutiny of global media made it untenable for Beijing to keep the secret indefinitely. The agency's virtual lab network wired together the world's leading scientists as never before. Public health officials, even in poor, far-flung corners of the world like Mongolia and Vietnam, were quick to learn how to recognize, treat, and contain the disease with guidance from foreign reinforcements.

Yet for all the advances of this era, someone still has to grab the bird and swab its underside. There is no substitute for the grunt work of influenza field research.

Guan had continued to expand the sampling program that he launched with the lab at Shantou University Medical College. By 2005 he had about eighty people working for him, quietly collecting specimens every week from poultry and wild birds in seven provinces of southern China and Hong Kong. He had tapped into an old-boy network that dated as far back as Jiangxi, locally recruiting what he called his band of heroes. "They're very brave," he said. It wasn't just the health risk. Most had some background in veterinary studies or health care, so they knew how to take proper samples and protect themselves from infection. It was also politics. The Chinese government was wary of this outside meddling and at times tried to block it. But using the cultural smarts he'd developed as a boy, Guan helped win his staff access to poultry markets across the vast belly of China even as officials grew increasingly uncomfortable with the extent of infection his program was uncovering. He demurred when I pressed him for more details. "This is a kind of top-secret weapon, a top-secret system."

The logistics of maintaining this network were almost beyond Guan's ability. The financial burden of paying the staff was tremendous. "They are working for the good of China, working for the good of Hong Kong, working for the rest of the world," he kept telling himself. But the sampling of more than two hundred thousand birds over nearly a decade yielded an unrivalled library of ever-mutating influenza viruses. It came to represent the most comprehensive accounting of the pandemic threat, in essence an early-warning system for the world.

Guan told me in late 2007 that his research showed the virus was now smoldering in poultry across much of Asia, waiting to flare up. China had ordered a massive campaign to vaccinate chickens against bird flu, as had Vietnam and Indonesia. While this had helped curtail poultry outbreaks in many places and reduced the overall level of infection in birds, the practice had not eliminated the pathogen altogether. Birds were still spreading it but without overt symptoms. "The virus is covered up," he warned. "We're giving the virus a chance. Now the virus can travel freely and undetectably and easily be transmitted."

Many in government and media had mistaken silence for peace.

"Because we don't have a pandemic today," he said, "don't accuse of us of crying wolf."


Guan had been at his apartment watching television on Boxing Day 2003 when his wife called. Though the day after Christmas was a legal holiday in Hong Kong, she had gone in to work, where she'd heard a disturbing report. After a half-year hiatus, there was a new suspected case of SARS in China.

Guan was not surprised. The Chinese government had reopened the wild-game markets months earlier despite his objections. The world's concern over the disease had waned but not Guan's. He had continued sampling wild animals. He had even expanded his effort beyond Shenzhen to cover other markets across Guangdong province. His findings were alarming. Not only was he discovering the SARS *Coronavirus* in most of the civets he tested; he was also turning up evidence of infection in a wider range of species than before. When he learned in December that a Chinese television producer had been hospitalized with the disease and been put into isolation, Guan knew what he'd have to do.

A week later, he met with senior Guangdong health officials at a Guangzhou hotel to argue his case. The civets had to be slaughtered. Guan was emotional, perhaps too emotional. The officials were skeptical of his judgment and resisted such a radical recommendation. The trade in wild animals was worth at least $100 million a year to the provincial economy. But when Guan had them compare the genetic signature of the virus from the ailing journalist with the one he had isolated from civets, they were stunned to see that the two were practically identical.

Later that day, the governor of Guangdong ordered that all civets on the farms and in the markets of the province be culled. Though three more human cases would surface in Guangdong that month, the outbreak would be rapidly contained. WHO credited Guan for helping preempt a second SARS epidemic.

"Before it got into humans, I knew it was coming, but other people said I was crying wolf," he recounted. "After the first case, I said, 'Let's use direct scientific information to stop the outbreak.' So it was averted."

Guan now finds himself playing a prophetic role again. To anyone who listens, he says the moral of SARS is clear. The flu virus must be controlled in birds. Whatever it takes, the microbial agent must be extinguished before a readily transmissible flu strain jumps to people, because once it does, global spread is inevitable. There won't be time to stop it.

But he laments that his counsel is again being shunned. Only now, with the flu virus so widespread, it could be too late.

"I did my job," he said, rising to light another Mild Seven. "I can face God and say, 'OK, God, this mission I did. I gave all this advance warning. I provided evidence. I did everything a scientist could do. The remaining job is for governments and politicians. And each person must pay the price if they go against the laws of nature.'"

# CHAPTER SEVEN

# Cockfighting and Karma

The pair of Thai fighting cocks, long-legged and elegant, stalked each other around the dirt ring, feinting and probing for an opening. They puffed out their broad chests, flaunting their foot-tall physiques. Then they each settled into a brief crouch, face-to-face, beak to razor-sharp beak. As they spread the majestic plumage around their necks, electricity coursed through the arena with anticipation of first blood.

Generations of breeding had brought the prizefighters to this moment of steely, instinctive, hard-wired aggression, nurtured and shaped by hundreds of hours of training.

They attacked as one, lunging at each other through the air, colliding in midflight with the muffled thud of meat on meat and the frantic flapping of ruffled wings and tails.

Spectators leaped from their concrete-block bleachers, surging against the edge of the ring.

Feathers flew. Blood oozed from the wounded eye of one combatant, a lean, handsome rooster with rich black plumage and golden brown along the neck and back. Even more was flowing from the throat of his adversary, an equally graceful creature with a white body and black trim along the wings and tail. His neck was quickly staining red.

Cries swelled in the bleachers as the spectators doubled and redoubled their wagers like frenzied traders on the floor of a stock exchange. Phapart Thieuviharn, a lifelong cock breeder with intense brown eyes and straight black hair speckled with gray, shifted anxiously on the edge of his seat, clutching the notepad on which he had scribbled his bets. Hundreds and perhaps thousands of dollars in Thai baht would ultimately change hands once one of the roosters finally surrendered to its injuries.

But Phapart and 125 other spectators were wagering more than their banknotes. They were gambling with their lives. In the years since bird flu began racing across Southeast Asia, cockfighting has repeatedly been implicated as a killer. It has sickened cock breeders and enthusiasts from Thailand to Vietnam and spread the virus on to Malaysia and perhaps even to Indonesia through the smuggled exports of prized fighting birds.

Cockfighting has long been a prominent feature of rural Southeast Asia, intertwined with its history, a spectacle for kings and peasants alike. For centuries, it seemed to pose no human threat but was just one more tradition that wove together the lives of man and bird into the fiber of daily existence. Villagers shared their homes with their chickens, peddled poultry at live markets, and integrated wildfowl into religious rituals. But these traditions, benign for humans if not the birds, have lately acquired a sinister edge. They have proven largely impervious to the admonitions of public health officials, who have urgently warned that the practice could unlock flu's devastating potential.

In fundamental ways, modernity has recast this corner of the world, unleashing dramatic economic changes that have magnified the potential for a pandemic strain and weaving the region into a globalized planet now exposed as never before to viral threats born of Asia. Yet Asia's past could also be mankind's undoing if age-old conventions give the virus entrée into the human population. Time and again, the intimate contact between fighting cocks and their doting breeders has proven a fatal attraction. Even for spectators at the cockpit, the brew of rooster blood, breath, and mucus that sprayed the ringside could be lethal.

Yet flu seemed of little matter on this sultry Sunday when Phapart had agreed to take me to the fights. We had driven about forty miles from his home in the northern Thai province of Phayao, where the sport had been banned because of public health concerns, to neighboring Chiang Rai province, where it was still allowed. But gambling was not. So local villagers thought it prudent to build their arena away from the main roads, far from the inquisitive eyes and outstretched palms of law enforcement. We turned off the paved road and headed down a long, unmarked dirt track that stretched deep into the emerald rice paddies until we reached a clearing. Though barely midday, the dirt and grass lot was already filling up with dusty pickup trucks. A young man collected fifty cents from Phapart. By midafternoon, the attendant would net about two hundred dollars in parking fees.

We walked over to what Phapart called the stadium. It was actually a whole cockfighting complex, a cluster of open-sided sheds with thatch and corrugated metal roofs. The main events were held in the central arena, a twenty-foot-diameter pit with red padding along the sides surrounded by three rows of concrete bleachers and a fourth fashioned from bamboo. Side matches were staged in three smaller rings without any seating. Food stalls peddled Thai noodles, soup, and other simple dishes.

The matchmaking had already started. Several dozen men, looking for action, had carried over roosters in woven bamboo cages and set them down near the entrance. There they sized up the competition, judging the other birds for weight and size, the other owners for the depth of their pockets.

A middle-aged farmer in a plaid work shirt had struck a match with a teenager in a red soccer T-shirt, and they shook hands on the first bout. Their birds, the black rooster with golden brown patches and the white rooster with black trim, each weighed in at about five pounds. Their base wager, sure to escalate over the course of the bout, would open at thirty-three hundred baht, or slightly more than eighty dollars. The fight organizer wrote their names on a blackboard outside the main arena, Golf Chai versus Mae Yao, and dispatched them to their corners.

For fifteen minutes, they prepped their fighters. Like trainers at a boxing match, they massaged the roosters to loosen their muscles. They wiped down the birds with moist towels warmed on a portable gas stove. The white rooster had somehow lost a wing feather. So his teenage owner, determined that his bird be properly accoutred, produced a spare white plume and glued it in place.

The spectators, mainly men from surrounding districts, filed into the arena, claiming spots on the bleachers. Those on the far side were silhouetted against brilliant sunshine. But beneath the metal roof, the ring was shady and cool. The stadium workers distributed small notepads to the crowd so they could record their bets, and the scribbling began even before the referee barked the fight to a start. With a few words of whispered encouragement, the owners released their impatient, agitated birds into the ring.

The roosters strutted and stalked, and then they struck. Over and over they flung themselves at one another. They craned their lissome necks, red crowns high, and jousted with their beaks. They jabbed and kicked with the daggerlike spurs on their legs. Resolute and reckless, beautiful and brutal. First blood was just that, only the first. This fight was scheduled to go two rounds, twenty relentless minutes each, and toward the end of the opening round, their fine, well-groomed feathers were growing ragged and red from combat.

The spectators had crowded the lowest rows and were now hanging on every thrust and parry. Dozens leaned forward into the ring, their arms dangling at times within inches of the action. They scrutinized the rivals for a glimmer of doubt or weakness, a slight hesitation or momentary loss of heart that could presage final retreat sometime later on. The betting swelled, with the crowd barking out side bets across the ring almost as fast as they could jot them in their pads.

Two hundred baht. Five hundred baht! One thousand baht! Two thousand baht!!

When the referee called an end to the first round, the two owners rushed into the ring, swept their roosters up into their arms, and hustled them away. There was much to be done during the break.

On adjacent patches of dirt, the two owners followed the same, urgent regimen. First they scrubbed the blood from the birds. Clutching soggy rags in their bare hands, they firmly washed the roosters' faces, followed by their necks, stomachs, and legs, repeatedly wringing out the bloody cloths on the dirt. Next they slipped the birds painkillers to help get them through the final round, prying open their beaks and popping in pills with their fingers. Each owner then grabbed a spare feather and inserted it into his cock's mouth, twisting it in the throat to help clear blood and mucus. They withdrew the feather and ran it between their fingers, squeezing the slime onto the ground. Then they repeated the procedure.

As we joined the small circle of onlookers, Phapart explained that bruising and internal bleeding can become so painful that an owner must nick the swelling with a knife and suck out the blood with his mouth. Some owners have been known to remove excess mucus the same way. They do what it takes to keep the cocks in the game. "If the beak breaks loose from the mouth during a fight," he continued, "you can reattach it with a small net wrapped around the head and then begin fighting again. If a claw breaks off, you can bandage it. If the wing feathers are loose, you can glue them back on."

In the boxing matches of the West, prizefighters often rely on a "cut man" in their corner to help stanch bleeding from around the eyes so they can tough out another round. Since roosters are no better at battling blind, cockfighting has a similar craft. On this afternoon, the eyes of both birds were swelling shut. So in the final minutes of the break, the two men produced needles and thread and deftly stitched their roosters' eyes open.

The referee summoned the competitors for the second round. Their owners, now smeared with blood and mucus and bird droppings, returned the patched-up cocks to the ring. There, they resumed the brawl where they had left it.

When the climax came, toward the end of the round, it came quickly. The white rooster had been stripped of more and more feathers and ultimately of his confidence. The spectators immediately noticed this tentative turn, and the sound of cheers and jeers swelled in the bleachers. Those few who were still seated jumped to their feet.

The black-and-gold aggressor continued his pursuit, pressing his advantage, pushing his foe up against the side of the ring. Attacking over and over. The white rooster was broken, its spirit finally crushed. In a wholly unfamiliar act that betrayed his very nature, he scampered away in retreat. A holler rose from the crowd.

This unforgiving competition is an acute form of natural selection. Losers perish in the ring or become supper for their owners. Winners prevail to fight another day and, if they win enough, go on to father the next generation of fighters.

"This the best way to breed," said Apichai Ratanawaraha, an agriculture professor at Bangkok's Thammasat University and a scholar of this blood sport. "You get the best of the best. Because Thai people in the countryside have selected their birds this way for generations, the fighting cock breed in Thailand may be the best in the world."

Cockfighting spread centuries ago to Europe and onward to the Caribbean and Latin America. The fighting cock has transcended cultures as a symbol of virility and manhood. But the sport's roots are in Asia.

Apichai told me that the peoples of East Asia have been raising chickens as gamecocks for just as long as they have been raising them for food: about 7,500 years. It was in Southeast Asia that mankind first domesticated chicken, most likely in Thailand itself. The poultry found today on farms across the world are descended from the region's red jungle fowl, wild pheasants with golden bronze plumage draping the necks, wings, and backs and with black chests and tails that shimmer blue and green. The male of the species is much larger than the female, with a fleshy red wattle on his head and, during breeding season, an intense dislike for rival suitors. Early farmers found that pitting the males against one another made for a welcome diversion from the slog of subsistence agriculture, a way to unwind during the weeks after the harvest was finally in. "Cockfighting," Apichai claimed, "was the first sport for human beings."

Archaeologists around Southeast Asia have repeatedly uncovered relics portraying the pastime. Bronze artifacts discovered in Thailand's Kanchanaburi province near the Burmese border depict cockfighting from 1,700 years ago. The sport also appears on the sculpted walls of the magnificent Angkor Wat temple complex in Cambodia, one of the grandest finds of modern times. Along the exterior of Angkor's twelfth-century Bayon Temple, a three-tiered mountain of stone that rises at the heart of the complex, are extensive bas reliefs depicting the everyday life of the Khmer people nearly a millennium ago. Among the images of women peddling fish and giving birth, and of men hunting, kickboxing, and playing chess, is a scene of cockfighting that would be instantly recognizable anywhere in Southeast Asia today.

In Thailand, cockfighting assumed a special place in the national culture during the reign of King Naresuan, an accomplished military strategist and avid breeder of the late sixteenth century. When Naresuan was a still a boy, the armies of Burma overran the Thai kingdom of Ayuthaya and took him prisoner. They carried the nine-year-old prince off to Burma as a royal hostage to ensure the fealty of his father, the king, but allowed Naresuan to take his favorite rooster. In captivity, he pitted the bird against those of the Burmese prince and, as Thais tell the tale, vanquished them all. "Not only can this cock champion a money bet," Naresuan told his jailers, "it can also fight for kingdoms." The Burmese returned him to the vassal state of Ayuthaya at age sixteen in a prisoner exchange. During the following years, Naresuan became a renowned warrior, campaigning to drive the Burmese occupiers from Thai lands and declaring the restoration of the Ayuthaya dynasty. The Burmese dubbed him the Black Prince. On his father's death in 1590, Naresuan acceded to the throne and reigned for fifteen years, extending the Thai domain to unprecedented frontiers. He adorned his palace gates with images of the cock. Monuments to the king still depict him surrounded by his roosters. The fighting cock became a symbol of national resistance. Even today, one of the most sought-after breeds is the Gai Leung Hang Khao, a fierce black-feathered bird with gold around its throat like a necklace and a long white tail; a cock that traces its ancestry back to the one Naresuan had carried into exile.

After Naresuan, the sport took a firm hold on the Thai imagination. The woven bamboo baskets of fighting cocks became ubiquitous in the front yards of peasant villages across the country. And it remains a pastime for the elite, with the most coveted breeds selling for $10,000 or more and up to $250,000 in bets changing hands at top matches. Thai celebrities and entertainment moguls have rallied to the cause of cockfighting as the tradition came under fire from public health specialists. The country's most prominent devotee and outspoken partisan is a long-haired pop icon named Yuenyong Opakul, the godfather of Thai country rock 'n' roll. Yuenyong rose to the top of the charts penning edgy songs about social injustice and performing them as the singer and lead guitar player of his band, Carabao. He cashed in on his fame as the spokesman for a Thai beer company and then launched his own brand of energy drink, called Carabao Dang, which quickly claimed a significant share of the market. Then, after bird flu erupted in Thailand in late 2003, Yuenyong emerged as vigorous defender of cockfighting, clashing with the government over its demand that roosters in infected areas be culled, defying a ban on vaccinating the birds against the virus by immunizing his own. (Thai officials worried that any poultry vaccination could undercut the confidence of foreign markets in Thailand's massive chicken exports.) Ever the rebel, Yuenyong included the song "Vaccine for Life" in his CD *Big Mouth 5: Bird Flu*, which reportedly sold at least a hundred thousand copies.

Today Thai magazines devoted to cockfighting proliferate. Dog-eared copies are a common sight in farmyards and the front seats of pickups. For a few more dollars, glossy books with colorful plates detail the attributes and ancestry of different breeds, including some that trace their bloodlines even further back than Naresuan. The black-tailed Gai Pradu Hang Dam, for instance, is descended from birds raised by King Ram Khamheng of the Sukhothai dynasty, who six hundred years ago extended the Thai kingdom as far as modern-day Laos and Burma.

The Burmese have never accepted Thai claims of superiority, whether in politics or cockfighting. If Thai cocks are famed for their aggressiveness, then the Burmese retort that theirs are smarter and more stylistic. Many Thais in fact do not contest that. Like Thailand, Burma has traditionally been a major exporter of gamecocks to other Asian countries.

But passion for the pastime spread across the region centuries before these exports. In Indonesia, cockfighting has a long history on the main island of Java and the Hindu outpost of Bali. Despite Bali's mystical reputation as an oasis of transcendental peace, the island's villagers have long preferred spending their afternoons betting on blood sport than joining tourists to watch dance performances of the *Ramayana* epic. Remarkably, the Balinese have assimilated this sanguinary diversion into their spirituality, making the neighborhood temple a prime venue for cockfights. When officials in Bali tried to combat gambling by suggesting that cockfights should be limited to major festival days, villagers balked. "We believe that to purify our sacred temple, you should have cockfighting regularly," a Balinese matron named Made Narti told me when I spent an afternoon in her village. "The blood splattered by the cock will protect the temple and protect the whole village. If you go without cockfighting for a long time, our god Dewa gets unhappy. The bricks will fall out and the temple walls will collapse."


If fighting cocks are bred for valor, strength, and stamina, then Phapart Thieuviharn was bred to be a breeder. His grandfather had emigrated as a young man from China to northern Thailand at a time when thousands of other Chinese were making a similar migration to the lands of economic promise in Southeast Asia. He settled in the pleasant farming province of Phayao just outside the infamous Golden Triangle, where Thailand, Laos, and Burma all come together in a remote, hilly region that for years produced much of the world's opium. He found a piece of land in a valley nestled between high mountains, built a traditional wood home on stilts, and began raising chickens. His prize possessions were about ten cocks he bred beneath the house. He fed them on scraps from the family table.

Phapart's father substantially expanded the family's landholdings but always found time away from farming to pursue the family passion, raising cocks and training to become a cockfighting referee in local arenas. Later, around 1970, he bought a parcel of land in the province and built his own.

Phapart, who was forty-seven when I first met him at an empty fish restaurant beside the province's scenic Kwan Phayao Lake, said fighting cocks were a part of his life from birth. "Even before I can remember, I was already caring for them," he told me. When he was seven or eight, he would refuse to get a haircut unless he could take his favorite rooster with him to the barber. As a teenager, he rose hours before school to train his cocks and then pitted them against those of his teacher.

"When you raise fighting cocks, you see them from the moment you open your eyes in the morning. You can recognize the way each one coos," Phapart explained, pushing up the sleeves of his green work shirt and chomping on an ever-present piece of gum. He had recently recovered from heart surgery, and though he was fit and vigorous, cigarettes were no longer an option. "We give them more love than we would a baby. You see, you and your children can talk in the same language. They can tell you what they want. But fighting cocks can't. So you have to be even more attentive and give them even more care."

Over the years, his flock grew and grew. He raised dozens of cocks at his home in Phayao town and kept hundreds more at a family farm near Thailand's northern capital, Chiang Mai. Many of these cocks were bred as an investment and sold for a small fortune. Proven winners went for up to $2,500. Others he retained and personally groomed as champions. Their framed portraits now adorn Phapart's house. But none was more accomplished and lucrative than a beautiful bird named Lucky, who retired undefeated after twelve matches. A picture of Phapart embracing the champ occupies a place of honor in his living room.

"It wasn't about the money," Phapart stressed. "It's not like other gambling, like at a casino. It was about social status. It was about my pride as a winner."

Not long after Lucky called it quits, tragedy struck. In late 2003, bird flu erupted across Thailand, ravaging the country's poultry and outracing the government's ability to contain it. The epidemic reached Chiang Mai and within days had sickened the roosters at the family's farm. Birds bred for fierce character turned listless. "We saw the symptoms. We just killed them all," he recounted, too pained to say much more. He never notified the government, just slaughtered six hundred roosters himself and burned their bodies. The economic loss was staggering, at least $150,000, and the emotional loss was worse.

For good measure, he gave away ten other cocks he was grooming beneath a metal awning behind his house, fearing they might be next to catch the bug.

Phapart eventually restocked. He drove me over to his house to see his new flock. There were eight birds, still too young to fight but promising. They had good bloodlines, strong builds, and character. "I can tell they'll do well," he said with pride. He drew one of the roosters from its wire-mesh cage and cradled it in his arms. It was a handsome, frisky creature with a red head and black feathers tinged with brown.

"I love them, and I'm looking for more," he said. But he continued, "If the disease comes back again, I will do it again and cull them again."

He returned the bird to its cage and sat down on the edge of the practice ring he had built in his backyard. So far, the young cocks were healthy, he said. A government veterinarian had recently examined them and declared them free of bird flu. But Phapart quipped that he did not need some bureaucrat to tell him that.

"I know my birds. I check them every morning," he said, irritation flashing in his eyes. "You have a very close relationship with your fighting cocks, and the closer you are, the more confident you are about their health. You know their condition." Emphasizing each word somewhat defensively, he added, "That is why I am not afraid."


"The villagers around here and their fighting cocks communicate heart to heart," Phapart told me as we set off in his Isuzu 4x4 for a breeder's tour of Phayao town. "They share the same spirit and the same daily life." He had volunteered to give me a crash course on the care and conditioning of gamecocks. It would also be a chance for him to dispense advice to some of the townspeople who were planning to enter their birds at a few of the more competitive arenas. Phapart estimated that at least two-thirds of the local families raised fighting cocks. "Cockfighting connects people in the same community," he continued. "We're a farming country, and after we work hard in the fields, this is how we like to relax."

We drove though a leafy neighborhood and pulled up in front of a farmhouse set well back from the road. The front yard had been converted to a training camp. The air was heavy with the tart smell of poultry and the crowing of roosters. Two dozen metal cages were arrayed around the grounds.

In a makeshift ring fashioned from concrete block and plastic tarp, a hired trainer was teaching a young prospect how to feint and dart. The man, a dour fifty-six-year-old named Decha with a black shirt and green cargo pants coated with dander, had wrapped his bare right hand beneath the chest of an old, retired rooster, palming it like a basketball and rocking it back and forth toward the young cock. Decha lunged forward with the old bird and then jumped back. He thrust it in and pulled it out, then swung it from side to side. The young cock followed the moving target intensely, ducking and dodging, pecking and kicking. Decha looped the old bird over and behind his student, and the young prospect spun around furiously, feverishly trying to land a blow.

Decha had wrapped strips of black sponge around the spurs on the young cock's claws, both to keep them from cracking during practice and to protect the old rooster. But blood was still drawn, including the trainer's. His hands and arms were scarred and swollen from the errant attacks of his pupils.

When the exercise finished, Decha released a second young cock into the twenty-foot ring and let the two prospects spar for about five minutes. They puffed out the plumage around their necks and repeatedly pounced at each other in a whoosh of feathers and fluff.

"Train harder," Phapart counseled Decha. "They're not really strong enough."

Phapart explained that roosters must be exercised every day. They should be drilled on walking to build their leg muscles and drilled separately on kicking. At least once a month, he continued, they should be pitted against other cocks in full-length practice bouts, their beaks covered with little sacks much like boxers sparring with headgear. The birds would require three months of intensive training before they were ready for the arena.

Our next stop was the Khun Dej Camp on the edge of town, a

sprawling training facility that Phapart billed as a "gymnasium" but was really more of a boarding school for would-be contenders. In a long shed toward the rear of the property were ample quarters fashioned from wood and screen for promising candidates, with smaller cages for breeding hens. The owner, Sitthidej Sanrin, a solidly built forty-nine-year-old with a high forehead, thick mustache, and disarming smile, had been in the business for twenty-five years, following in the footsteps of his father and grandfather as a breeder and trainer. Many of the roosters were his own, but some belonged to clients who had entrusted him with the rearing and seasoning of their pricey investments.

When we drove up, Sitthidej was seated on the edge of a practice ring, clad in a red tank top and gray shorts. His legs were scratched and smeared with bird droppings, his feet bare. He had placed two roosters in the pit, an older one still inside a metal cage and a young prospect loose outside. The latter, an exquisite bird with yellow and white feathers around its neck, was circling the cage, stalking, and then lunging, trying to peck through the bars. The caged veteran watched warily, spinning on his claws, parrying the attacks. For half an hour this dance continued, the young cock exercising his leg muscles. He had already fought twice in the arena and won. A third bout in Chiang Rai was imminent.

Sitthidej lifted the lid of the cage, letting out the old bird, and the two roosters sparred for a few minutes.

"Your cock will do well in Chiang Rai," Phapart told his friend. "Over there, most of the cocks are very aggressive and like to fight up close. But yours likes to hang back and then kick. It's a good style."

Clearly pleased with this endorsement, Sitthidej scooped the young rooster from the ring and placed him in his lap. Then he reached over and grabbed a soft, moist towel, warmed it on a hot plate, and began to scrub the bird feather by feather, rubbing the muscles in the shoulders, back, and stomach. He gingerly held the rooster's red neck between his thumb and forefinger and leaned back, surveying his condition, and then resumed. It was bath, massage, and sauna rolled into one, and Sitthidej continued meticulously for twenty minutes.

As part of the strict regimen, Sitthidej served the rooster a lunch of champions: the grilled meat of a local mountain river fish called kang, a lean, brawny creature so tough that the villagers of northern Thailand claimed it can survive out of water for an hour. The meat was minced and mixed with honey and herbs, including garlic, pepper, and lemon grass, and then rolled into marble-size pellets. "This is our secret formula," Phapart offered. "It goes back generations. It makes them strong." Sitthidej nimbly slipped the food with his fingertips into the cock's mouth.

After a dessert of chopped banana, Sitthidej walked over to a wooden cupboard to grab the vitamins. The shelves were crammed with little bottles and containers, protein supplements, and various antibiotics. One jar contained yellow paste made from a local root, soaked and ground, used for special massages. Another contained facial cream to prevent skin disease and heal wounds. A third contained a red paste to be applied to the face before matches to toughen the skin. Beside it was a glass jar stuffed with replacement beaks and claws in case the bird's own cracked or snapped off. Beside that were a needle and thread to stitch wounds closed and eyes open.

Sitthidej returned the rooster to his cage and moved it into the tropical sunlight. It was time for the bird's daily sunbath before he would retire to his quarters. Phapart teased his friend that there was no music to serenade the bird, recalling that other trainers played Thai country songs while the roosters lounged in the sun. "In truth, the music is for the owners more than for the cocks," Phapart admitted. He smiled and the corners of his eyes crinkled. "But it makes the birds happy too. Sometimes they even try to sing along."

Komsan Fakhorm, an eighteen-year-old from the eastern province of Prachinburi, loved his fighting cocks, as Thai men had for generations. He would clear their throats by sucking out the blood, mucus, and spit from their beaks with his mouth. He would sometimes sleep with his favorite roosters.

On the final day of August 2004, the young cock breeder fell sick. He had a fever, a nasty cough, and difficulty breathing. Though Thai officials later faulted his family for waiting too long to get him medical care, by September 4, he had been admitted to the hospital. Three days later, he was dead.

There was no doubt this was bird flu. Thai health officials reported that thirty of Komsan's hundred roosters had died in previous weeks. But this was the first confirmed human case of the virus in Thailand for months and a jarring setback to Thai efforts at containing the disease. After a series of false starts and premature declarations of victory, senior Thai officials finally had seemed justified when they announced that summer they had turned the corner and quashed the epidemic.

Thailand would continue to struggle with bird flu over the coming months. But by the end of 2005, a massive campaign by health and agriculture officers coupled with thousands of local volunteers had again appeared to banish the virus. It would be déjà vu. In July 2006, after more than seven months without a case, a seventeen-year-old boy from a province north of Bangkok fell ill with a high fever, cough, and headache. He was hospitalized five days later, deteriorated rapidly, and died after another four.

Thai health officials concluded that he, too, had caught the virus from a fighting cock. He had been infected while burying roosters that had died of bird flu. "The victim failed to report the death of his fighting cock because he was afraid the authorities would slaughter his birds," Thai Prime Minister Thaksin Shinawatra told reporters angrily.

The livestock chief of the boy's home province, Pichit, alleged an even wider cover-up, saying villagers had declined to notify officials that some of their cocks were dying because the birds were so expensive. Tests by the national livestock laboratory ultimately confirmed bird flu in samples taken from the carcass of a dead cock. When authorities learned of the outbreak and ordered that the surviving roosters be culled, the owners resisted. The livestock chief himself was disciplined for failing to prevent the outbreak, and a complete ban was slapped on cockfighting in Pichit and neighboring Phitsanulok provinces.

Ever since the virus resurfaced in late 2003, cockfighting has played a role in spreading it around Southeast Asia. At least eight confirmed human cases were possibly caused by infected fighting cocks during 2004 alone.

These roosters have proven to be difficult targets for disease control efforts. Owners have frequently hidden their cocks when officials have ordered mass poultry culls. Others have smuggled the birds across provincial and even national lines to elude the dragnet. Each time they are moved, they risk introducing the virus to new flocks. When bird flu was confirmed for the first time in Malaysian poultry in August 2004, animal-health officers blamed illegal imports of fighting cocks from neighboring Thailand. In turn, several other countries banned imports of all Malaysian poultry, and an area of the northern Malaysian state of Kelantan, where the outbreak occurred, was put under quarantine to prevent further spread. The state's chief minister, Nik Aziz Nik Mat, who doubled as the spiritual head of Malaysia's Islamic opposition, slammed local cockfighters. He urged them to give up the sport and repent. While some Muslim clerics opposed the culling of cocks and other poultry as un-Islamic, Nik Aziz endorsed the draconian measure but asked that it be carried out away from villagers so as not to antagonize them.

Some animal health investigators have also suggested that cocks exported from Thailand were behind the far wider outbreak of bird flu in Indonesia. In 2002, the year before the virus swept the region, Thailand shipped nearly six thousand fighting cocks abroad, most of them to Indonesia, according to the Kasikorn Research Center in Bangkok. Later, Thailand may also have been on the receiving end. Health officials speculated that illegal cockfighting tours reintroduced the virus from Laos into the northeast of their country after a long hiatus. Thai cock owners, including a government livestock officer whose own roosters later died, had been stealing across the broad Mekong River to pit their birds against Laotian opponents despite an outbreak on the far side of the border.

The allegation by some Thai officials that cockfighting helped seed the regional epidemic leaves the sport's partisans seething. "The government is telling lies," Phapart retorted when I asked him about the claim. He insisted the poultry industry and not fighting cocks was to blame. (Some Thais have also objected to cockfighting on the grounds of animal cruelty, but their calls for prohibition have never gained traction.)

To regulate the movement of fighting cocks, Thai agriculture officials suggested a modern, digital fix for this ancient pastime. They recommended inserting microchips into the roosters. But this prescription was no better received than the diagnosis. "That's ridiculous," Phapart scoffed. "They move around. They don't stand still. How are you ever going to put a microchip into them?" He said a microchip could cramp their agility and leave them vulnerable. "When they fight, they get hit in every part of the body. It might create a weak spot. What if they get hit in the chip? They might run away and you would lose your thousand-dollar bet." Senior Thai officials ultimately agreed with this widely shared critique and dropped the microchip proposal.

They had only slightly better success with their plan for fighting-cock passports. Local veterinary departments began issuing travel documents for each rooster with its photograph on one side and a register of its movement on the other. Every time an owner planned to take his bird across district lines, he was required first to visit a government veterinarian, who would examine it, record the trip, and stamp the passport. These control measures were to be supplemented with random testing of fighting cocks. Phapart said he personally obeyed the regulations but the whole notion made him chuckle. "Many people don't use the passports," he explained. "Less sophisticated villagers don't care. They just keep breeding and pitting their cocks like they always did. They just tell their friends that they're going to meet up and hold a fight before the police come. They even have someone to look out for the police."

Phapart urged that cock owners be left to regulate themselves. In this age of cell phones, he said an outbreak in one village is instantly flashed across the district through text messages, and owners effectively quarantine the infected area themselves. No owner would want to see bird flu spread among his own prized roosters.

"The decision makers analyze the situation just on paper," he said, growing agitated again. "Their feet aren't on the ground. They don't really know how we treat the cocks and don't really share our feelings. We care more for the fighting cocks than the health officers do."

But for a growing minority in Thailand, the debate keeps coming back to the basic question of whether it is time to ban the pastime altogether. Phapart leaned forward intently and vowed, "They'll never be able to stop us from doing cockfighting."

In contrast to the brawny, exquisitely groomed gladiators of the cockpit, the vast majority of Asian chickens are scrawny, sorry creatures. In Indonesia they're known as *kampung* chicken, or village chicken. They root around in the dust and slime, often living off discarded rice, fallen leaves, morsels of overripe mango, and whatever else they happen across. So when Indonesians are given a choice between dining on these vagabonds of the backyard and on their plumper distant cousins raised commercially for the market, the response is unanimous, and surprising. "It's obvious," Ketut Wardana, a Balinese villager with a shock of dark hair and droopy mustache, confirmed for me. "Kampung chicken is healthier, tenderer, and much more delicious. Our kampung chicken is raised on natural food, not those chemicals like the broiler chickens in the market."

In Bali, as throughout the Indonesian archipelago and much of the region, a home is not a home without a chicken, or several dozen. They pretty much come and go as they like, sleeping as often in trees and under beds as they do in their own cages. The Indonesian government estimates that 30 million households raise poultry. It is their intimate presence in the lives of so many Asians coupled with the near-total absence of safeguards against contagion that makes backyard poultry farming what the U.S. Agency for International Development has called "the greatest single challenge to effective control of the spread of the virus."

Wardana raises twenty-five chickens behind the ornately sculpted walls of his family's traditional compound on Bali's lush east coast. When he invited me inside the courtyard, the air was fragrant with frangipani blossoms and the grounds tranquil, shaded from the sun's midafternoon rays by a stand of palms. At his feet, a black hen cackled. "It would be hard to imagine life without chickens," he said, nodding with a laugh. "Life would lose its flavor without chickens."

Yet culinary concerns are the least of it. Chickens are central to home economics, he continued, taking a seat on the wooden floor of his *bale dangin*, the raised pavilion at the heart of most every Balinese compound. They are a way out of poverty. If the family is hungry, they can always sup on chicken. If they're short of cash, they can hawk them in the market and earn a premium over the price for commercial poultry. When the new school year began and his brother needed shoes for his two children, Wardana sold several chickens to raise the money, and chicken paid the doctor's bill when his wife got sick. "They're our living wallet," he quipped.

Chickens also constitute part of the social contract binding communities together, providing what Wardana's matronly neighbor Made Narti described as the "solidarity of the centuries." Though her good fortune has translated into a flock of several hundred chicks, she recalls a time when she was hungry and had to turn to her fellow villagers of Tegal Tegu for chicken. She said she has reciprocated countless times. "For generations, chickens have lived very close in the lives of us Balinese," she recounted. The elderly raise chickens as a hobby. The devout raise them as a matter of faith. Four times a month, Narti slaughters a bird and carries it with a plate of fruit and flowers down the narrow, walled alley to a Hindu temple. It's an offering to the gods.

Some public health officials have urged an end to backyard farming. But it is so tightly stitched into the cultural fabric of Asian life that the prescription is sure to fail. "If you seriously proposed eradicating backyard poultry farming, you would get a lot of undesirable outcomes," said Jeffrey Mariner, a veterinary professor at Tufts University dispatched to work with the UN Food and Agriculture Organization in Indonesia. Senior FAO officials have cautioned that a ban would simply force poultry farming underground. This could also alienate villagers from other programs to control the virus, for instance notifying authorities of outbreaks, Mariner explained.

Mariner is not one to underestimate the threat of flu. In early

2006, he helped set up and train teams of inspectors to uncover out-breaks that had gone undocumented. "We thought at the time we'd find that bird flu is underreported. We never imagined the extent to which this is true," he said. They started with twelve districts on Java, then twenty-seven districts, then the outer islands. Everywhere the teams looked, they found the virus. Even on training exercises they found it. "It's very widespread, and it's difficult to address the disease, since it's in the backyard system."

In the two months before my visit to the village of Tegal Tegu, Mariner's teams had confirmed a dozen different outbreaks in Bali, including a pair just days earlier. Animal-health officials had burned more than a thousand chickens in a bid to contain the epidemic in one location on the resort island. Though Narti had heard about bird flu on television, she remained oblivious. "Bali is safe. There's no bird flu here," she assured me. Her warm eyes, full cheeks, and thick lips of-fered a mother's comforting smile. "It happened on Java. The chickens that got bird flu on Java had white feathers. My chickens mostly have green feathers."

Researchers elsewhere in Southeast Asia have found that villagers are widely cognizant of bird flu's perils yet continue to take risks with their own backyard birds. As they have for generations, they handle sick and dying fowl, butcher and eat birds that have died of illness, and even let their children play with infected livestock. The contradic-tion is not surprising. For years, long before the disease struck, they have seen their own relatives and neighbors engage in these practices and rarely come to harm. As a precaution, Narti volunteered she was in the habit of separating out any of her birds that seemed sick and fortifying the rest with vitamins, including a supplement to combat depression and stress. "But there's really nothing to worry about," she added. "I don't think it could happen here."


Asia's live poultry markets leave many Westerners queasy. Deep inside the cavernous Orussey Market in the Cambodian capital, Phnom Penh, a short walk from the fairy-tale spires of the royal palace, scores of live chickens and ducks are crammed together, legs bound, on wood pallets speckled with droppings. Shoppers stoop over to scrutinize the birds like customers examining cantaloupes in a Safeway, then hang the beleaguered creatures upside down from hooks to measure their weight. Some shoppers carry off their cackling purchases to finish them off at home. Others turn them over to butchers, who hunch on the muddy floor, slitting throats and plucking feathers. So, too, in the dim light of Jakarta's Jatinegara Market, one of more than thirteen thousand live poultry markets in Indonesia, a dozen boys squat amid stacks of pungent cages on a tile floor slick with death. With swift slices of the knife, these barehanded youths dismember the chickens and then tug out their entrails, heaping them up for waiting customers. Blood trickles down ruts in the floor and spills into the alley outside.

Many Asians swear by freshly killed chicken, duck, and goose, insisting they are tastier and more nutritious. But Robert Webster has vigorously argued that "wet markets" represent a perilous nexus where flu viruses can amplify, swap genetic material, and spread. WHO's expert committee on avian flu has endorsed this view. Each morning, live markets are restocked with birds, and with them new microbes. They can infect merchants, customers, and other animals, who by day's end may carry the contagion onward to new frontiers. Researchers in the early 1990s identified live markets as a "missing link" to explain flare-ups of a low-pathogenic strain among poultry in the United States. After the 1997 human outbreak in Hong Kong, investigators ultimately traced six of the deaths back to wet markets. The city instituted new safeguards, including the screening of poultry from mainland China and a ban on the sale of live ducks and geese. But the virus nonetheless returned to Hong Kong's markets five years later. On the mainland, where an even wider array of flu strains has continued to circulate amid the poultry stalls, Chinese researchers concluded that six city dwellers who came down with bird flu had likely caught it during recent market visits. These patients had no other known exposure to sick poultry.

Yet despite these scientific warnings, Asian governments have been hard-pressed to break people of their longtime passion for freshly butchered meat. Some countries, notably Vietnam, have begun phasing out wet markets and building modern slaughterhouses. But Webster counseled that, as with backyard farming, a complete ban on wet markets would simply drive this commerce underground. Demand would remain strong and prices high while monitoring for disease would become far more difficult.

Among Southeast Asians, it is Vietnamese who take fresh furthest, with a delicacy called *tiet canh vit*. This popular pudding is traditionally prepared from raw duck blood and served at meals to mark the anniversary of a death in the family, the celebration of Tet, the lunar New Year, and other special occasions. It is typically washed down with rice wine. *Tiet canh vit* is also sold widely in the market. Health investigators suspect that at least five people from two families in northern Vietnam contracted bird flu after feasting on the dish. After hearing this, I told a Vietnamese friend in Hanoi that I simply had to have the recipe. She e-mailed the following.

1. Cut a small incision in the duck/chicken to get the blood in a bowl. Pour some drops of lime into the blood bowl.
2. In a separate bowl, mix chopped bowel and stomach together.
3. Mix the blood liquid (the first bowl) and the stock (the second bowl) with the ratio of one spoon blood to two spoons stock. You will also put some fish sauce in, as much as desired.
4. Set aside for about half an hour. The mixture will form a texture like pudding cake.

For many in Asia, birds are an essential element of everyday life, synonymous with sustenance, commerce, companionship, and even national identity. Moreover, for some, they are also linked to aspirations not just for this life but for the life to come.

Over the centuries, Buddhists across much of Asia have released the sorrows born of sickness, hunger, and war through the simple, cathartic act of buying caged birds and setting them free. In front of the shimmering gold pagoda of Wat Phnom, erected on the wooded knoll that lent Phnom Penh its name, Cambodian devotees reach inside the metal and wire mesh cages, draw out sparrows, swallows, munias, and weavers, often in pairs, and raise them in cupped palms to their lips. The adherents mumble a prayer and, often with a kiss, set them free into the warm, still air. But this tradition, in which devotees seek blessings for this life and the next, could now prove to be a curse. The lethal flu strain has been isolated from some of the wild species most commonly peddled outside the shrines of Buddhist Asia from Thailand to Taiwan. The hazards posed by the collection and release of these so-called merit birds is akin to that of live poultry markets.

Kong Phalla has been selling merit birds from the cobblestone sidewalk at the base of Wat Phnom since she was eight. A slight woman in her twenties with small brown eyes, she had the familiar look of those who trade their childhood for the hustle of the street: a thin veneer of smarts overlaid on innocence. She approached me with a lotus stem in one hand and a cage crammed with birds in the other. She said the birds had been shipped into the capital overnight by riverboat. She had already sold nearly three dozen to worshipers. "They want to free their depression, free their sadness and illness with the birds," Kong Phalla explained. Her dark hair was tucked under a red knit cap despite the day's gathering heat. She rested her load in the shade beside a table of incense sticks and flashed a weak smile, saying she had brought five cages to the pagoda that morning and was confident all one thousand birds would be sold by nightfall. The birds went for about fifty cents each, good money in Cambodia, though Kong Phalla got to keep only a tiny fraction. On holidays like Cambodia's New Year, when business was especially brisk, she said, prices could triple.

Bird flu was of no concern, Kong Phalla continued, patting the cage. It's only the foreign tourists who fret. She snickered. "They'll only open the doors of the cages and ask me to release the birds myself so they don't have to touch them," she said, adding with a boast, "Bird flu has never happened to me."

Kong Phalla spied one of her frequent Cambodian customers drive up to the curb in a new Toyota sedan and get out. She instantly abandoned her thought, grabbed the cage, and gave chase. She followed him up the long brick staircase, past the statues of lions and pink balustrades of mythical serpents and beyond the stone stupas above, beseeching him at each step to purchase some of her birds. He acceded just before vanishing into the sanctuary on the crest of the hill. Kong Phalla put down her cage on a stone bench beside those of other peddlers and waited for her next chance.

To understand this Buddhist custom, I sought out a monk named Khy Sovanratana. I found him at his monastery in the center of Phnom Penh, a once romantic city of French colonial villas still trying to collect its thoughts three decades after Pol Pot's reign of terror. The Khmer Rouge had abolished religion, decimating the country's Buddhist institutions. Since then, Buddhism has revived, monks bearing alms bowls have returned in large numbers to the early morning streets, and Khy Sovanratana has emerged as a commentator on morality and social issues. Though his close-cropped hair was still black with youth when I met him, his learning had already elevated him into the ranks of senior clergy. When he received me, he was seated cross-legged on a thin cushion, his orange monk's robe draped over his left shoulder.

The monk started by recounting a legend of Prince Siddhartha, the Indian nobleman who would later attain enlightenment and become the Supreme Buddha. The young prince and his cousin were walking through the woods when they spotted a swan. The cousin drew his bow and shot the swan with an arrow. Siddhartha raced to the injured bird, refusing to relinquish it. His cousin grew furious. But Siddhartha caressed the swan, eventually nursing it back to health before setting it free.

"This kind of conduct has had a big impact on Buddhist practices," the monk said softly. "Giving life is very much extolled in Buddhism." He explained that the simple gesture of releasing birds is rich in significance, and he slowly explicated the different layers of meaning. First, by giving life, a devotee follows in the footsteps of the Buddha. Second, the act of releasing the bird helps to cast off the "torments and tortures" of everyday life. And third, the act of liberating a living creature earns devotees religious merit toward reincarnation into a better life. For a person with financial means, the only limit on the number of birds to be released is his kindness. Sometimes, the monk said, adherents have been known to free not only birds but fish, turtles, and even cows and buffalo that are tied up awaiting slaughter.

But setting aside the sublime, Khy Sovanratana acknowledged that believers should not be blind to the dangers of this tradition. "There's no point if you don't get benefits but instead catch a virus," he counseled. "Monks should be given this kind of awareness and pass it on to devotees when preaching."

That's a tall order in Cambodia, where this tradition is intertwined not only with religion but national identity. The king himself frees doves, pigeons, and other wildfowl about four times a month—in especially generous numbers to mark royal birthdays—and this has complicated efforts to regulate the practice. Its adherents rarely comment on the contradiction of trapping birds only to set them free, an irony compounded by the success of some boys in catching fowl moments after their release so they can be sold yet again. Not long before my audience with the monk, an environmental group based in the United States had tried to curtail the practice on the grounds that the sale of merit birds represented illegal trade in wildlife. The organization, WildAid, had established a rapid-response unit that included Cambodian military police and forestry officials and carried out several raids on bird peddlers. The campaign culminated in the confiscation of birds sold at Wat Phnom and elsewhere. But this provoked a religious and political backlash. The government suspended further raids.

Even in Hong Kong, which so successfully overcame public opposition in its decisive response to the initial bird flu outbreaks, officials have been reluctant to tackle this revered ritual. Nearly ten years after the virus first jumped to humans, fears of a new outbreak in Hong Kong surged when several dead birds recovered from city streets tested positive for the lethal strain. Among these were munias, which are not native to urban Hong Kong but imported by the tens of thousands from mainland China each year for Buddhist rites. The discoveries prompted Richard Corlett, an ecology professor at the University of Hong Kong, to publicly warn that bird releases posed the principal threat of reinfection in the city. Agriculture officials urged people to refrain from freeing captive birds and asked religious organizations to make a similar appeal to their members. But while the government ultimately suspended trading at Hong Kong's famous Bird Garden market after an infected starling was discovered there, a similar ban was not imposed on merit-bird releases. The cultural sensitivities were too great.

By the banks of Phnom Penh's Tonle Sap River stands an ornate, carnival-colored shrine called Preah Ang Dang Ker. Under its steeply pitched roof rests a likeness of the Buddha gazing across the broad gray waters. Around the outside linger peddlers surrounded by cooing and chirping. "I have no concern about getting sick with bird flu, and the buyers have no concern," offered Srey Leap, a stocky woman in a sweat-stained shirt keeping vigil from the shade of an umbrella. "They never worry about this. It is our Cambodian tradition." When a family approached, Srey Leap and the other hawkers converged. The five visitors paused to haggle, then purchased an entire cage frenetic with the flapping of about a hundred pairs of wings. They carried it to a low stone wall above the water's edge. They pulled the birds two by two from behind the mesh and, with the occasional whisper of a prayer, set them loose, casting a line of silhouettes down the ancient river until the entire contents of the cage and whatever contagions it concealed had disappeared along the banks.

# CHAPTER EIGHT

# Sitting on Fire

To thwart a gathering pandemic, the perimeter must hold. Once it is breached, there's no turning back. This precarious frontier, the first and last line of defense separating the pathogen's animal hosts from the human race, runs through thousands of remote Asian villages. These outposts are vulnerable and often unsuspecting, like the Javanese hamlets that scale the lush, terraced slopes of the Mount Lawi volcano. There, an Indonesian animal-health officer who goes only by the name Suparno had been drafted into keeping the virus in check before it crossed to people. But the day I met Suparno, he preferred to go to lunch.

It was late one morning in May 2005 when this lanky, good-humored veterinarian arrived at an elderly woman's farmhouse partway up the slopes. Clad in the tan uniform of a civil servant, Suparno announced he'd come to inoculate her chickens against bird flu. While a human vaccine had so far proven elusive, workable poultry vaccines were already in production, and several Asian countries, including Indonesia, had made them the centerpiece of their efforts to contain the virus. Suparno knew the woman kept some chickens. Nearly every family in her village did.

"How many do you have?" he asked her.

"Twenty-five," she answered. The woman motioned initially toward a low, concrete barn out back where she kept some of them. Then she swept her right arm in front of her, indicating the rest were wherever he might find them.

Suparno led his team around the side of the house into the cramped backyard. Crouching on the dirt, he set down the small, pink pail that held his gear. He took out a plastic bottle of vaccine, then slowly drew the fluid through a tube into an automatic needle. His colleagues produced five black hens from the barn, one by one, and clasped their wings and legs tightly while Suparno injected half a milliliter of vaccine into their breast muscle.

After only a few moments, he rose to his feet and got ready to leave.

"What about the rest of the birds?" I asked him.

"Too hard to catch," he responded. They might be hiding in the trees or in the crawl space beneath the house.

Then, changing the subject, Suparno and his fellow officers agreed it was time to eat. He invited me to join them. With no irony intended, they suggested a local joint specializing in chicken.

I had come to the province of Central Java to spend several days observing Indonesia's much-publicized effort at fighting the infection that had been coursing through the country's flocks for more than a year. Central Java, as its name implies, is at the center of Java island, which, in turn, is home to the majority of Indonesians and has always dominated the country's politics. My base would be the old royal city of Solo, host to one of Java's two main sultanates. Solo remains the premier seat of Javanese culture and tradition. So I'd figured, given the political, cultural, and geographic centrality of the city, that the surrounding countryside would be at the forefront of the national campaign to root out the disease.

At first I was encouraged. The chief livestock officer in one nearby district, Sragen, told me how she'd set up a twenty-four-hour bird flu command center. Sri Hardiati, a gregarious yet autocratic woman with a stylish haircut and piercing dark eyes, described how her office monitored poultry outbreaks and even had a small diagnostic lab for dissecting stricken birds. But as I toured the countryside with my assistant, we discovered that containment efforts were just public relations. We had asked to see the vaccination campaign at work. Yet in district after district, livestock officials declined. They said they had none to show us. Finally, after some pestering on our part, Hardiati asked us to accompany her chief vet, Suparno. He made only one stop, pausing long enough to vaccinate the woman's five black hens. When he bypassed all the other homes in the village, I realized the outing had been simply for my benefit, little more than a photo op.

Over the coming days, we would learn the extent of the ruse. Indonesia's central government was claiming it provided millions of free vaccine doses for small and midsize Javanese farms and that 98 percent of these had already been used. But local officials and peasants told us this was fiction. "Maybe the farmers get the vaccine. The percentage who use it is small," said the chief livestock officer in neighboring Karanganyar district. In Boyalali district, the chief livestock officer told me he had a hundred thousand doses in a refrigerator, but no one had asked for any in months. He was content to let them sit there.

As we continued to drive the narrow byways of the Javanese countryside, we were also starting to learn from villagers and local veterinary officers that die-offs among chickens had been occurring much longer than we'd believed. Indonesia had officially confirmed its first poultry outbreak in January 2004, not long after Vietnam and Thailand initially reported theirs. But the local accounts we were hearing contradicted the version we'd been provided back in the Indonesian capital, Jakarta. We were fast realizing that Indonesia's central government had covered up the mounting epidemic for almost half a year, since mid-2003, until it was too late to reverse the tide.

Now, as we explored Central Java in May 2005, Indonesia had still to confirm its first human cases. But that too would change within months when death struck a suburb of Jakarta and Indonesia joined the growing list of countries with casualties. It wouldn't be long before the death toll in Indonesia outstripped that of anywhere else on Earth.

Yet Indonesia wasn't alone in concealing the disease. I would come to learn that every Asian country with major outbreaks in livestock had hidden them from view, for months or even longer. The fatal strain's progress across East Asia had been a journey veiled in secrecy and blessed with neglect. This microbial killer, born in the deep south of China, had repeatedly slipped across international borders over the previous decade, evolving and increasing its virulence until the toll on both people and poultry could no longer be denied.

But even then, when it became untenable for governments to keep up the lie, they often chose to discount the danger rather than mount a serious campaign to defeat it. Instead of attacking the virus, they too often went after the scientists, journalists, and other whistle-blowers who tried to reveal the threat.

The virus exploited this opportunity to put down roots. It became entrenched in Asia's poultry, thus posing a long-term menace to humanity. While some Asian governments eventually intensified their efforts to contain the disease, total eradication was now a distant prospect at best. Not a single one of these frontline countries—China, Indonesia, Thailand, and Vietnam—had adopted the most powerful disease-fighting weapons: truth and transparency.

Later, when I arrived on my first trip to Geneva, a senior WHO official gave me a piece of advice. He counseled me that influenza is all about politics. And those antiquated politics have proven every bit as intractable as the virus itself.

This tale of death and deception begins in the coastal Chinese province of Guangdong, close to where the first human cases were confirmed in Hong Kong back in 1997. The H5N1 flu strain, which went on to ravage farms on at least three continents, infect hundreds of people, and pose the most serious threat of pandemic in a generation, was first isolated in a sample taken from a sick goose during a Guangdong outbreak in the summer and early fall of 1996. That was more than seven years before China first acknowledged any infection in its flocks. By the time Chinese officials went public in early 2004 and stepped up efforts to contain it, the virus had already seeded outbreaks in the country's neighbors.

Molecular biologists were later able to identify the Guangdong pathogen as the common ancestor of all subsequent H5N1 viruses by analyzing the eight segments of RNA that all flu viruses contain. Each

of the segments in a single virus has its own signature, a specific sequence of basic building blocks called nucleotides that make up
the RNA. As viruses evolve, these segments mutate. They can even
be completely replaced as the promiscuous flu strains swap genetic
material. In the lab, genetic genealogists determine the pedigree of
viruses by looking for similarities in their RNA. Isolates that share the
same pattern are often related and descended from the same specific
virus.

A combined team of researchers from the U.S. CDC and China's
official National Influenza Center reported in 1999 that the virus that
had killed people in Hong Kong two years earlier was related to the
Guangdong goose isolate. The specific H5N1 subtype was identified
at the Chinese influenza center. At least three other academic papers
written in Chinese by Chinese researchers also reported in 1998 and
1999 that H5N1 had been isolated in Guangdong in 1996, disclosing
in one instance that up to 40 percent of the geese on a farm stricken
by the outbreak had died. These findings had been reported by Chinese government researchers in government publications. Yet for years,
senior Chinese officials continued to deny publicly that Guangdong
had been struck by bird flu in 1996 or that it had spawned the wider
epidemic.

Over the coming years, Chinese and foreign scientists continued
to report periodic outbreaks in southern China, including a large poultry die-off in Guangdong just months before the Hong Kong cases in
1997. A series of research articles published between 2000 and 2002
further documented that H5N1 viruses were continuing to circulate
in southern China, for instance in geese and ducks exported from
Guangdong to Hong Kong. The virus was also isolated in a specimen
taken from frozen duck meat exported from Shanghai to South Korea
in 2001. A team composed primarily of mainland Chinese researchers
later reported that tests on ducks from southern China between 1999
and 2002 had repeatedly come back positive.

The most damning evidence came in a study published in 2006 by
scientists from China, Indonesia, Malaysia, Vietnam, and the United
States, which identified China as the wellspring of the international
epidemic. "We have shown that H5N1 virus has persisted in its

birthplace, southern China, for almost 10 years and has been repeatedly introduced into neighboring (e.g., Vietnam) and distant (e.g., Indonesia) regions, establishing 'colonies' of H5N1 virus throughout Asia that directly exacerbate the pandemic threat," the researchers wrote. They concluded that addressing the "pandemic threat requires that the source of the virus in southern China be contained."

Chinese initiatives to tackle the disease may have only fueled its spread and honed its lethality. Just days after the government first disclosed in January 2004 that it had detected the disease, *New Scientist* reported that Chinese poultry producers had been vaccinating their flocks against it for years. But the vaccination campaign may have been mishandled, obscuring the usual symptoms without eradicating the virus itself. "The intensive vaccination schemes in south China may have allowed the virus to spread widely without being spotted," the magazine alleged.

Animal-health experts later told me about an even riskier strategy that China had adopted to suppress the spreading virus years before officials publicly disclosed its presence. Acting with the approval and encouragement of the government, Chinese farmers had tried to douse major outbreaks among chickens with amantadine, an antiviral drug meant for humans. As a result, international researchers concluded that this drug might no longer protect people in case of a flu pandemic. The H5N1 subtype circulating in Vietnam and Thailand had become resistant to the drug, which is one of two types of medication for treating human influenza, though another viral subtype found elsewhere was still sensitive to it.

China's use of amantadine violated international livestock regulations. It had long been barred in the United States and many other countries. But veterinarians and executives at Chinese pharmaceutical companies said farmers had been using the drug to contain the virus since the late 1990s. "Amantadine is widely used in the entire country," confirmed Zhang Libin, head of the veterinary medicine division of Northeast General Pharmaceutical Factory in Shenyang. He added, "Many pharmaceutical factories around China produce amantadine, and farmers can buy it easily in veterinary medicine stores." Zhang and other animal health experts said the drug was used by small private farms and larger commercial ones. China's agriculture ministry had approved the production and sale of the drug, and local government veterinary stations instructed Chinese farmers on how to use it by adding it to the chickens' drinking water—and at times even supplied it.

One veterinarian steered me toward a popular Chinese handbook, titled *Medicine Pamphlet for Animals and Poultry,* which provided farmers and livestock officials with specific prescriptions for amantadine use to treat chickens and ferrets with respiratory viruses. The manual, written by a professor at the People's Liberation Army Agriculture and Husbandry University and issued by a military-owned publishing company, prescribed 0.025 grams of amantadine for each kilogram of chicken body weight. Farmers also used the drug to prevent healthy chickens from catching bird flu, giving it to their poultry at least once a month and often mixing it with Chinese herbs, vitamins, and other medicine.

After China's misuse of amantadine was first reported in the *Washington Post* in June 2005, the Chinese government angrily denied the account. "This is groundless and isn't in accordance with the truth," the agriculture ministry said in a statement. "The Chinese government has never permitted farmers to use amantadine to treat bird flu or other virus-related disease." But the government-owned *China Daily* acknowledged in a front-page article that some farmers had in fact used the drug in poultry. The newspaper said the agriculture ministry planned to send inspection teams across the country to halt the practice. I was never able to determine whether it was indeed brought to an end nor the extent to which it had been part of a concerted Chinese effort to conceal the outbreaks. In private, officials at WHO had also been asking the Chinese government about the improper use of antiviral drugs in livestock. They, too, were never given a satisfactory answer.

From China, the virus had ventured south. It stole across the border to Vietnam and began infecting poultry there no later than 2001, more than two years before the children in Hanoi's National Pediatric

Hospital started succumbing to the mysterious outbreak of respiratory disease in late 2003.

Vietnamese scientists knew the truth about the early poultry infections. These researchers had participated in a study that the CDC's intrepid investigator Tim Uyeki had helped put together sampling various species of poultry in the live bird markets of Hanoi. Uyeki had suspected that the novel flu strain was still circulating in China even after Hong Kong's mass cull of 1997 had expunged the virus from that city. But mainland China was, for all practical purposes, off-limits for the kind of scientific study he envisioned. So he tried the next-best thing: peering in from a distance by sampling birds south of the Vietnamese border. In the fall of 2001, the researchers discovered two geese with H5N1 in a pair of Hanoi markets. This strain was closely related to the isolate from the Guangdong goose. Lab tests revealed that this pathogen was devastating for chickens, killing every single bird inoculated with the agent within forty-eight hours.

Sometime in 2002 or the first half of 2003, a new variant of the H5N1 virus again slipped across the border from China. This one was later traced to China's Yunnan province, just north of Vietnam. It was even more virulent. Veterinary officers working in the rural districts outside Hanoi later recounted how it decimated poultry in the province of Vinh Phuc half a year before the government first announced Vietnam's outbreaks in January 2004.

Among the first farms stricken was a commercial operation run by Japfa Comfeed, an Indonesian conglomerate with operations in four Asian countries and whose president commissioner, a retired general, would later become Indonesia's director of national intelligence. The die-off was never publicized. But Japfa's annual corporate filings later disclosed that bird flu had been discovered in June 2003 and severely affected the company's Vietnamese operations.

The first outbreak killed twenty thousand of Japfa's chickens. The company suspected from the symptoms that it was bird flu and sent samples to the agriculture ministry's veterinary department for testing. Government officials hushed up their findings, informing the company that the deaths were due to an "unknown agent." But Van Dang Ky, an epidemiologist in the veterinary department, later admitted that "the first signs of an epidemic" were found in Vinh Phuc in July 2003. The timing was inconvenient. "Vietnam was preparing activity for the Twenty-second Southeast Asian Games, and we did not announce it for political and economic reasons," Ky said.

The Southeast Asian Games came and went that fall, with the prime minister opening the sporting competition at Hanoi's new national stadium and Vietnam racking up the mother lode of gold medals, and still there was no announcement of the outbreak. Vietnam wanted no rerun of the SARS drama earlier that year, which had tripped up the country's humming economy. Senior Vietnamese officials hoped to quietly contain this latest epidemic before it delivered a new blow to tourism and trade. But the virus refused to cooperate. It quickly spread to other commercial farms in Vinh Phuc and then to neighboring Ha Tay province, the heart of Vietnam's poultry industry.

Early on, disease also struck the Vietnamese flocks of the Thai conglomerate Charoen Pokphand. The first outbreak eventually confirmed by the government was a massive die-off of chickens at one of the company's industrial farms in Ha Tay. CP, as it's commonly known, is Thailand's largest business enterprise. Started by a family of poor Chinese immigrants, it grew from a small Bangkok seed company into a sprawling multinational, with operations in twenty countries ranging from agribusiness and retailing to telecommunications. The family's connections to Thailand's senior ministers are legend, prompting allegations from critics in Thailand that CP used these to cover up flu outbreaks on its farms. (By November 2003, the virus had begun devastating flocks in Thailand itself, though the government in Bangkok concealed this until late January 2004.) The company's chief executive denied that CP had helped spread the disease across Southeast Asia or cover it up.

During one of several trips to Ha Tay, I heard an alternate explanation. Nguyen Xuan Vui, the deputy director of a local animal-health office, described how tens of thousands of chickens were hauled every day from China into the markets of northern Vietnam by small-time traders. Family farmers also drove up to China and brought back nutrients to fatten up their flocks. These imports could easily have been contaminated. Vui's account appeared to dovetail with the findings of microbiologists, who analyzed the genetic fingerprints of the virus circulating in northern Vietnam and concluded that it had been introduced from China on at least three separate occasions between 2001 and 2005. A single, even massive introduction by a multinational company would have left a different signature. But although the culprits were minor-league merchants, the impact was tremendous. Vui told me that 2 million chickens had been culled in his province at more than 175 different locations over the course of 2003 before the central government ever got around to acknowledging the first outbreak in January 2004.

For a brief period when the initial human cases were confirmed in early 2004, Vietnam opened its doors to dozens of WHO disease specialists. But after the first wave of human cases ebbed, Vietnam closed back up. This was a country that for years had depended mostly on its own wits and was finally discovering a modicum of prosperity after more than a generation of war and deprivation. Senior officials were deeply suspicious of foreign meddling. Dr. Hans Troedsson, WHO's chief representative in Vietnam, publicly complained in the summer of 2004 that the government had not responded to a new offer of outside expertise nor provided virus samples for analysis. In private, he fretted that his telephone calls and urgent faxes to the health ministry were going unanswered.

WHO officials in Asia and Geneva acknowledged they were completely in the dark about the details of Vietnamese cases, how they arose, and even how many there were. "So basically, bugger all, we still don't know what the numbers are," wrote one agency official. The frustration would continue to mount through early 2005 as the rate of human infection in Vietnam accelerated and a pandemic seemed to grow closer. WHO's office in Hanoi repeatedly tried to raise its "grave concerns" with the government but was again met with silence.

In Geneva, senior officials debated whether to admit in public that they were flying blind. They had neither accurate information from Vietnam nor the freedom to conduct their own field investigations. But WHO did not want to jeopardize its relationship with Vietnam's health ministry, since they collaborated on a raft of public health issues. "We cannot openly blame Vietnam (at least not yet) but cannot also let people believe that we have access to all the data we need," one senior WHO official said at the time. "At some point in time we may have to make a radical choice." A second senior agency official warned that nothing less than the world's ability to contain a global epidemic was at stake. "I am concerned," he stressed, "that we are pretending to our Member States and also internally in WHO that we can assess the situation, are capable of detecting the emergence of a pandemic virus and initiating early interventions." Without transparency, there could be no hope of containment.


Duc Trung and Hoai Nam aspired to be Vietnam's Woodward and Bernstein. Duc Trung, at age thirty-four, was the senior member of the team. He was a university journalism graduate who had been working at the *Thanh Nien* newspaper in Ho Chi Minh City for five years when I met him. He had been attracted by the paper's reputation for investigative reporting. Lanky, with large, soft eyes and smooth skin, his tastes ran to pastel designer shirts, teal on the day we were introduced. He sported a gold watch, and his shoes were impeccably polished. He was well spoken and did most of the talking for the pair, choosing his words carefully.

Though a year his junior, Hoai Nam looked far older. He never reached university, instead spending five years in the Vietnamese army, retiring as a sergeant major. He was short and skinny with an uneven haircut, white sleeves rolled up to his elbows, shoes old and scuffed. He was quick to smile, a broad disarming smile, and when he did, the crow's-feet would deepen around his eyes and the creases multiply on his cheeks.

"This guy looks like a chicken trader," Duc Trung told me, patting his partner on the back. "I don't look the part but he does. That's why he was chosen for the project."

A few months earlier, in the summer of 2005, the pair had swapped their office clothes for the soiled shirts and shorts of poultry traders and loitered around the livestock markets until their disguises soaked up the odor. Duc Trung poked a hole in his shirt, near the waist, affixing a tiny video camera barely an inch wide to the inside. Hoai Nam did the same. Posing as novice chicken sellers, the two men repeatedly filmed the illicit, predawn commerce at a Ho Chi Minh City slaughterhouse where merchants bought and sold forged health documents certifying that their birds were free of avian flu. Through these nearly invisible punctures, the reporters offered Vietnamese an education in the corruption and governmental misconduct that would continue to bedevil the country's battle with the disease.

By the fall of 2005, it looked as though Vietnam had turned a corner. After a long internal debate and repeated delays, the agriculture ministry launched a campaign to vaccinate about 250 million chickens and ducks in all sixty-four provinces. New restrictions were imposed on the breeding, movement, and sale of poultry. To win more cooperation from farmers, the government promised to pay them greater compensation when their birds were culled. As the months passed, poultry outbreaks and human cases declined. Vietnam was widely praised by UN agencies. The U.S. Agency for International Development called its performance "remarkable."

Yet if not an illusion, this judgment was certainly premature. The virus remained entrenched in the Vietnamese countryside. And every time Vietnam thought it had finally put out the fire, new outbreaks flared, first in ducks and then chickens in the Mekong Delta. "The situation is alarming," reported Hoang Van Nam, a senior agriculture official. Soon the disease returned to farms across the country and eventually Vietnamese again fell sick. Hoang Van Nam attributed the resurgence to the failure of local officials to carry out the vaccination campaign and enforce a ban on hatching ducks.

Even during the brief lull in outbreaks, Duc Trung and Hoai Nam were documenting the rot inside Vietnam's control efforts. Their reporting, which illuminated the disconnect between national policy and officials on the front lines, left little doubt the virus would resurface.

Their editors at *Thanh Nien* had decided to tackle bird flu soon after the government announced its new drive against the virus in 2005. Though the country's television stations and newspapers, including *Thanh Nien*, were state-controlled, reporters had increasing latitude to dig into corruption, from crooked traffic cops to soccer referees fixing matches. Duc Trung and Hoai Nam had been tipped by sources that the government's effort to keep sick poultry out of Ho Chi Minh City was a sham.

The pair headed to a livestock market near the city's edge. There they asked around until they found a veteran chicken hawker who could tutor them on the poultry business. The merchant coached them on how to behave, what to wear, and what to say. They bought their cheap disguises and borrowed a cage for hauling chickens.

In the hours before dawn, in a teeming quarter of the city, they approached the Manh Thang slaughterhouse, cited months earlier by the city for health violations. Barefoot, like other retailers, they entered the building. It was chaos, chickens screeching, traders shouting over the roar of the plucking machine. "It was risky for us," Duc Trung recalled. "The people at the market and the people at the slaughterhouse would threaten us if they knew what we were doing. And we were exposed to all those chickens and their virus."

On a table, they spied two baskets stacked with health certificates, called quarantine papers, which were already stamped and signed. A staff veterinarian told the reporters they could buy the papers for just twenty cents. There was no need to examine their chickens. The document would allow them to claim their birds were free of disease, and it could be used over and over. Four more nights, the pair returned to the slaughterhouse. Each time, the reporters obtained government-issued certificates without any question.

They also turned their lenses on three of the city's main roadside inspection stations, this time using a video camera concealed inside a plastic bag. The stations were little more than shacks erected along the highways entering the city. The infection had been widespread in the surrounding countryside. At each post, inspectors were to check that truckers hauling live animals had health certificates for all their livestock and that the number of animals matched the figure on the license.

But the reporters discovered that officials were at best taking a cursory glance at the paperwork, never inspecting the trucks themselves. "At least they should check the papers thoroughly," Duc Trung said. "But whenever we observed them, the inspection activities were so neglectful." At the post on the Hanoi highway, the reporters found that the officer serving on the front line of defense for the city, and potentially for the world beyond, was actually asleep in the yard behind the station.

Their revelations hit the streets under the headline, QUARANTINE PAPERS ARE SOLD LIKE VEGETABLES! Immediately, the owner of the slaughterhouse stormed into the newspaper's offices, demanding a meeting with the reporters and their editor. He was furious and menacing. "We showed him our evidence," Duc Trung recounted. "We know we'll be threatened when we start an investigation and we're ready to face the threats."

*Thanh Nien's* findings were also publicly challenged by the city's veterinary agency. But members of the People's Council, essentially the city council, were troubled enough by what they read that they ordered officials to look into the reported irregularities. The article had concluded ominously: "Facing the risk that bird flu might break out, the authorities of this crowded city should have the feeling they were sitting on fire. But according to how the work of inspections is carried out, the prospect of bird flu breaking out is unavoidable."


At almost the very moment the virus had struck the farms of northern Vietnam in the middle of 2003, Indonesian chickens began to die two thousand miles to the south. And Indonesian officials, like their counterparts in Vietnam, would wait half a year to confirm the arrival of the virus. That delay allowed the scourge to spread to nearly one-third of Indonesia's provinces without any official resistance and become entrenched. Human deaths were only a matter of time.

The Indonesian virus, like that in Vietnam, had its origin in China. But while scientists traced the Vietnamese subtype of H5N1 directly back to China's southern province of Yunnan, the provenance of the Indonesian strain was Hunan province, farther to the east.

At first Indonesian poultry experts were stumped by the abrupt die-off of nearly ten thousand chickens at a commercial farm in Pekalongan, a town in Central Java known mostly for making batik. Soon after, another outbreak burned through a poultry farm hundreds of miles away in a suburb of Jakarta. The National Commission for Eradicating Poultry Disease was divided over whether the cause was bird flu or a less virulent ailment, Newcastle disease. So they called in Chairul A. Nidom, an Indonesian microbiologist at Airlannga University.

Nidom reported that the carcasses showed unmistakable signs of avian flu. Their combs had gone blue, their legs tattooed with red stripes. Those suspicions were reinforced by the separate findings of a pathologist, who cut open a stricken bird soon after the initial outbreak and discovered abnormal brain tissue consistent with bird flu. Within two months, Nidom's lab research determined it was indeed that virus and moreover a strain genetically related to the H5N1 pathogen identified in southern China seven years earlier. He urged an aggressive response. "When the outbreak began in Pekalongan, if the government had acted to stamp it out and closed the case and did it properly, it would not be going like now. It would be finished," Nidom told me about two years later as the virus continued to rage.

But the country's poultry industry blocked the disclosure. Indonesia's national director of animal health later said that poultry company owners, who had personal ties to senior agriculture ministry officials, had insisted that any containment efforts be pursued in secret. Eight farming conglomerates in Indonesia account for 60 percent of the country's poultry, and they feared the publicity would harm sales. The owners even lobbied Indonesia's president, Megawati Sukarnoputri. "They said, 'It's better to do it with confidentiality. Do a hidden, silent operation,'" recounted Tri Satya Putri Naipospos, Indonesia's animal health director at the time. "I said, 'It won't work if you do a silent operation. This is a disease that can't be hidden. It's too risky.'"

Yet through January 2004, the government maintained the deception. To allay growing suspicions, the agriculture minister and several of his lieutenants summoned the media and feasted on chicken satay. "As of now, there are no findings in the field that can confirm the spread of the disease in Indonesia," insisted the director of animal husbandry in an interview with the *Republika* newspaper. "For the moment we are still free from bird flu."

The very day that interview was published, Nidom broke ranks and announced his findings to the competing *Kompas* newspaper. He said 10 million chickens had succumbed to the disease over the previous three months. He said he had forwarded at least a hundred samples from infected birds to the central government as proof.

A day later, the agriculture ministry publicly confirmed the outbreak. But already the disease had spread across Java and on to the islands of Bali, Borneo, and Sumatra. The plague was bleeding from the commercial sector into backyard holdings, infecting tens of millions of free-range chickens that had been left unprotected. "It was too late. The virus was everywhere," Nidom recalled.

Though scientists concluded that the virus had been introduced into Indonesia on only a single occasion, the disease would go on to infect at least thirty-one of the country's thirty-three provinces, transmitted by the trade in poultry and poultry products. The virus would eventually leap to people, and by the end of 2005, Indonesia was registering more human cases than any other country. WHO and other international agencies would grow ever more exasperated with Indonesia's continuing negligence. "It is important for the Indonesian government, in its interest and the interest of the international community, to take the necessary political decision" to tackle the virus, urged Bernard Vallat, head of the World Organization for Animal Health, in comments he would make nearly three years after the outbreak started. "Indonesia is a time-bomb for the region." He could have added, "for the world."

As soon as Nidom went to the media with his findings, the national poultry commission fired him from his advisory post. But he continued to press. Nidom grew increasingly nervous about the prospect of the epidemic spreading to people in Indonesia, a country with an impoverished health-care system and the largest population in the region. He arranged a conference in late 2004 at his university to discuss the disease, inviting four of the world's premier influenza researchers, from the United States, Japan, Hong Kong, and mainland China. Yet shortly before its scheduled date, he told me, a senior agriculture official contacted the head of Nidom's institute and ordered that foreign participants be barred. Officials threatened to have police break up the conference if it went ahead as planned. Nidom canceled the program altogether.

The Indonesian government also turned its ire on foreigners, including WHO staff, who spoke out of turn. The agency's team leader for avian flu in Indonesia told the media in 2006 that human cases would continue as long as the disease was circulating widely among birds. This was not only WHO's official position but basic science. But the government subsequently expelled him from the country. Though the health ministry never supplied a formal explanation, some WHO officials concluded that his remarks had contributed to his ouster.

A similar fate befell Naipospos, the country's animal-health director. Commonly known as Dr. Tata, she was passionate and opinionated, with dark eyes that seemed both probing and vulnerable behind her thick glasses. She was a rare professional in the ranks of the agriculture ministry. She had earned a master's degree from England, a doctorate in veterinary epidemiology from New Zealand, and widespread respect from disease experts at WHO and other international agencies. Naipospos first disclosed the government's cover-up in an interview she gave me in early 2005 for the *Washington Post*. She repeated her allegations five months later but this time in Indonesian, in an interview with the *Kompas* newspaper. A day after the article was published, the agriculture ministry fired her.

Agriculture Minister Anton Apriyantono told me he dismissed Naipospos because he was not happy with her handling of bird flu and her working relationship with top ministry officials. This explanation outraged Naipospos. She countered that she had been sacrificed by the ministry not only because of her candor but because of party politics. With the minister's upstart party trying to cast itself as a force for government reform, Apriyantono sought to tar her with the failures of his department, even accusing her of corruption. That charge was never pursued by prosecutors. And UN officials publicly criticized the government for ousting its most respected animal-health expert at the height of a crisis.

Naipospos alleged that bird flu had never been a priority in the agriculture ministry. Agriculture officials had not even tapped available emergency funds to pay for disease control. "I talked to the

minister about it many times," she recalled. "He said a disease outbreak is not a national emergency, not a disaster."

Naipospos was ultimately vindicated when she was named to a
new presidential commission established in 2006 to oversee the
government's avian flu policies. The body's primary charge was to coordinate the efforts of rival ministries, in particular health and agriculture. But it, too, proved impotent. The commission received little
support from the ministries, and even its own director was just a part-
time appointee. Indonesia would continue to be singled out by senior
UN officials for "the lack of a national strategy, the lack of political
involvement." Still, the avian flu commission made a small contribution to setting the record straight. In a press release three years after
the disease erupted, the commission formally acknowledged that the
virus had indeed first made landfall in Indonesia in the middle of
2003, many months before the government had admitted.

The trip that took Margaret Chan to Beijing's Great Hall of the People
was a delicate mission. Just three weeks before, in early November
2006, the former Hong Kong health director had been elected WHO's
new director general. She was to replace Lee Jong Wook, who had
died suddenly after brain surgery. Chan's election had been hotly contested, with the former Hong Kong public health director besting ten
other candidates. China had sponsored Chan's candidacy, lobbying
hard for her, and saw the contest as a measure of the country's growing clout on the world stage. When months of quiet pressure and
diplomatic horse trading were finished, the Chinese government had
lined up far more than enough votes, including crucial backing from
the United States, and Chan became the first Chinese national ever
to hold such a high post at a United Nations agency. Now she was
visiting China's monumental seat of government to personally thank
President Hu Jintao. "I will remember the support given to me by the
country in my heart forever," Chan told him.

But she also had more sensitive matters to raise. Beijing had
stopped supplying flu samples to WHO's affiliated labs. No human
specimens had been shared since the spring, and the most recent was

from a year-old case. The newest samples from infected birds were also a year old. Flu viruses are notoriously mercurial, and China's defiance of repeated international appeals was keeping WHO from staying abreast of the virus's twists and turns. This was undercutting international scientific efforts to understand the behavior of this unusual virus and anticipate its further evolution. At stake was the timely development of vaccines and drugs tailored to the prevailing strains.

In the days after Chan was elected, reporters had pressed her on whether she would stand up to the Chinese government. One commentator in her hometown of Hong Kong even asked whether she'd be simply a "foot soldier for Beijing." Chan insisted that her allegiance was to WHO. "First and foremost, now that I have been elected as director-general, I will no longer wear my nationality on my sleeve. I'll leave it behind," she replied. As she met with Hu, skeptics were watching for a sign of such independence.

Chan told the president it was vital for China to disclose its outbreaks and share samples. She urged him to speed up the process of providing specimens to WHO. She pressed the same points with Premier Wen Jiabao.

"The president and premier have a very good understanding of the potential impacts, on the health of the people, and also on the economy," Chan told me later. "They understand the importance of being transparent." In meetings with the leaders, she had made sure to praise China for strengthening its disease monitoring system since SARS. But she also reminded them that the country still had a hard time staying on top of emerging threats in the provinces. "China has a challenge, being a vast country," Chan continued. "The importance is to make sure that policies go down to the lowest possible level so that the implementation is not impeded."

That point was not new for either Hu or Wen. "They understand the challenge," she put it to me. "But you know . . ." Her voice trailed off. It was a tacit acknowledgment of her own doubts about the potential for change.

After she exited the meeting in the Great Hall of the People and was asked by waiting reporters about the prospects for cooperation between China and her agency, Chan offered a diplomat's assessment:

"We will have to look at the actual situations, but we all agree with it in principle."

In one of Chan's first speeches as the world's top health official, she outlined in 2007 what she called an unwritten code of conduct requiring governments to tell the truth about infectious disease. "No nation has the right to conceal an outbreak within its territory," she declared. Chan didn't mention China by name. But Beijing's handling of SARS had clearly been the most flagrant breach.

Now China was again a cause for concern. It wasn't just that China was still failing to provide virus samples. People were continuing to get sick from bird flu, but Chinese authorities were not confirming any related outbreaks in the birds. Investigators were being robbed of crucial details about how people were contracting the virus.

In an e-mail to WHO headquarters in early 2006, one of the infectious-disease experts in the agency's Beijing office noted a series of recent human cases that had all occurred in the absence of any reported poultry outbreaks. "What on earth is going on with the animals and the virus?" she wrote. The health ministry, though repeatedly pressed for an explanation, offered none. By the middle of 2007, cases in China had reached twenty-five, and only one could be explained by a related outbreak in poultry.

One possibility was that China's hugely ambitious campaign to vaccinate its entire population of chickens, ducks, and geese was hiding outbreaks by keeping birds from getting sick, without fully disarming the virus. A second explanation, put to me by influenza researcher Robert Webster, was that the poultry might be receiving a kind of natural inoculation from another, less lethal flu strain that was prevalent among the birds. He suggested that exposure to this second strain, H9N2, might offer limited protection to poultry when infected by H5N1. The birds would carry the virus and spread it but not show symptoms.

Yet in some instances, birds were indeed dying and agriculture officials were not reporting the fact. China's agriculture ministry has long denied details of livestock diseases, even to officials at China's own health ministry, claiming that animals are none of their business. When the two ministries clash, as they have repeatedly since the 1990s, the agriculture ministry inevitably prevails. It is far more powerful and prestigious because China's senior leaders place their top priority on development, and agriculture is a central part of that, according to Yanzhong Huang, a professor at Seton Hall University specializing in the politics of China's public health system.

Huang explained that China's mishandling of the SARS outbreak had transformed the country's health sector. Chastened, China invested heavily in disease surveillance and laboratories and lifted the ban on disclosing infectious diseases, which had been considered state secrets. But these advances did not extend to the agriculture ministry, where the prevailing view was that any candid discussion of animal diseases would only undercut productivity.

Equally vexing is the divide between China's central government and local officials. A week after China confirmed its first outbreaks in birds, Vice Premier Hui Liangyu admitted to WHO that the central government wasn't sure what was transpiring in the provinces. For local officials, career advancement hinges on success in promoting economic development. Better to keep quiet about infectious diseases that could deter investment and scare off tourists. So it's little surprise that Chinese local officials look unkindly on journalists and whistleblowers who publicize these outbreaks, even arresting and expelling them.

Qiao Songju was a simple farmer in Jiangsu, a coastal province just north of Shanghai, when the young man heard a piece of gossip that changed his life. Qiao's father told him that more than two hundred geese had died a mysterious death at a friend's farm in the next province. Afraid that local officials were covering up a bird flu outbreak, Qiao called all the way to Beijing and notified the chief of animal husbandry at the agriculture ministry. That in turn prompted a formal investigation, which soon found that at least 2,100 geese and chickens in the village could have the virus. The suspect birds were all slaughtered. Chinese media dubbed Qiao the "farmer hero." China's state-owned television nominated him to receive its award for economic figure of the year.

A month later, Qiao blew the whistle again, this time reporting a suspected outbreak in his own home county of Gaoyou in Jiangsu. The police came for him a day later. They arrived at midnight and asked him to accompany them to the station for a "chat." They made the arrest formal on the following day. Qiao was charged with blackmail and extortion. The accusation was that he had wrung thousands of dollars out of veterinary institutes by threatening to report them for manufacturing bogus flu vaccines. His lawyer called the charges fabricated. His family and supporters, including scores who recorded their outrage over the Internet, questioned the timing of his detention.

But few sympathizers were found among his fellow farmers. The price of eggs in Gaoyou had fallen by half after he'd raised the alarm. The price of chicken meat had tumbled even more. "Qiao Songju is a sinner to all Gaoyou farmers," said Chen Linxiang, an official in Gaoyou's agriculture and forestry bureau.

Five months after he was detained, Qiao went on trial in the Intermediate People's Court of Gaoyou. His lawyer complained that not one of the agriculture ministry officials who could have testified in his defense had chosen to do so. After another three months, in the summer of 2006, the court handed down its verdict and convicted Qiao on six counts. He was fined nearly four thousand dollars and sentenced to three and a half years in prison.


Deep in the interior of China, more than a thousand miles from Gaoyou, in the midst of the great green grasslands of the Tibetan Plateau, is a vast body of salt water called Qinghai Lake. Many in the region consider the lake, China's largest, to be sacred, and pilgrims still circumambulate its 220-mile shoreline. Sheep and yaks graze on its banks, distant mountains reflected in its azure waters. In the northwest corner is Bird Island. Though this rocky outcropping is technically more of a peninsula than an island, the first part of the name is apt. Each spring, thousands of geese, swans, cormorants, and other wildfowl from 189 species congregate here, migrating over the Himalayas and from Southeast Asia to lay their eggs. By the time the rapeseed of summer has turned the pastures a brilliant yellow, the birds have continued on their way.

In late April 2005, something stunning occurred at Qinghai Lake. The birds began to stagger around like drunks. They became paralyzed, their necks trembling, contorted. Then they would die. This was not supposed to happen. Sure, birds were dying elsewhere in Asia, but those were chickens and other domestic poultry. These victims were wild birds, mostly bar-headed geese and some gulls. For millennia, migratory waterfowl had been nature's reservoir for flu viruses, meaning these birds could carry the pathogen without getting sick themselves. But now they were dropping at a rate of one hundred or more a day. By the time the full numbers were tallied, at least five thousand had perished from what Chinese authorities confirmed was bird flu. The novel strain had abruptly turned its fury on its own natural hosts. Equally alarming was the fear that the birds at this major migratory hub might now carry the infection on with them, speeding its spread westward through the network of overlapping flyways.

WHO officials had been clamoring for details about the Qinghai outbreak since they first heard about it. They were appalled to learn that Chinese authorities had sampled only twelve of the sick birds at the lake and checked no healthy-looking ones for signs of infection. Though China finally allowed a team of investigators from WHO and FAO to visit Qinghai, the government continued to refuse them access to samples, test results, and the sites of related outbreaks. "They're doing all they can to block information from us," WHO's Beijing office reported to headquarters.

Yi Guan, the maverick microbiologist in Hong Kong who had been amassing bird flu samples across southern China, was not to be denied. One of his former students headed to Qinghai, surreptitiously collected nearly a hundred specimens, and sent them on to Guan for analysis. Barely two months after the die-off on Bird Island began, Guan's team had published its findings in *Nature*. By comparing the genetic material in these samples with others taken in live poultry markets of southern China, the scientists were able to establish a similarity between the Qinghai virus and two others previously isolated in Shantou, a city on the Guangdong coast. "This indicates the virus causing the outbreak at Qinghai Lake was a single introduction, most probably from poultry in southern China," the team concluded.

The Chinese government was infuriated by the claim that the virus had spread from southern China to Qinghai. This made China the source of the global threat rather than an innocent victim of someone else's birds. No sooner had the *Nature* article appeared online than China's official Xinhua news agency fired back, quoting Jia Youling, the ministry's animal-health director. No bird flu has broken out in southern China since the previous year, Jia claimed.

Government officials had been monitoring Guan's research for more than a year. His genetic detective work was steadily unraveling their deceit. "They were very unhappy," he recalled. "This is supposed to be a black box. Nobody is supposed to know what is going on. Now I'm opening the door and I have a strong case."

The article about Qinghai was too much for the government. Jia went on the attack. He announced that Guan's team had fabricated the data, saying the researchers had lied about taking samples at Qinghai. Soon after, Jia suggested that the team's lab was so poorly equipped that the tests could have been contaminated and the results meaningless. He further alleged that the scientists had failed to apply for government approval to conduct their research and that the Joint Influenza Research Center at Shantou University Medical College, which Guan helped set up four years earlier, lacked adequate safeguards for doing the tests.

The government ordered the Shantou research center shuttered immediately. Virus samples were to be destroyed or turned over to the ministry's official animal-flu lab in Harbin, one of only three institutes in China that would now be allowed to conduct this research. Jia later denied any political motive, saying the Shantou lab was one of four around the country that had been closed because it failed standard inspections. But to be sure Guan understood how severely they viewed his activities, agriculture officials accused him of the cardinal sin of disclosing state secrets.

"It was a lot of pressure," Guan admitted, baring his still-fresh wounds. "Doing the science is simple. The big problem is that people try to stop you from writing." As he recounted the incident, his voice rose an octave and he sputtered, almost spitting out his grievance as he struggled for the right words in English. "They tell me, 'You're a human doctor and this is an animal matter. Don't interfere in my animal issue. It's none of your business.'" He paused and caught his breath. Then, filling his lungs with the smoke of a freshly lit cigarette, he continued, "Working on this, I'm not sure how far I can go, how safely I can go. They kept saying I'm leaking state secrets."

Yet Guan kept going, and his network kept collecting samples even after Shantou's front doors were closed. That fall, his team published another piece of disturbing research. It documented a new wave of disease that had appeared in China's poultry and had already spread elsewhere in the region. After sampling more than 53,000 birds at live poultry markets in six southern Chinese provinces, the scientists discovered that a new H5N1 subtype, which they labeled "Fujian-like" because of its similarity to an earlier isolate from Fujian province, had rapidly squeezed out its predecessors. In just a year, the rate of infection in China's poultry had nearly tripled. This new wave of transmission had renewed poultry outbreaks in at least three Southeast Asian countries and sickened people not only in China but also in Thailand. The scientists said China's massive campaign to vaccinate all its poultry was possibly to blame for stimulating the emergence of this resistant subtype.

The government responded by calling a press conference to belittle the findings and impugn Guan's ethics. "The data cited in the article was unauthentic and the research methodology was not based on science," Jia said. "In fact, there is no such thing as a new 'Fujian-like' virus variant at all. It is utterly groundless to assert that the outbreak of bird flu in Southeast Asian countries was caused by avian influenza in China and there would be a new outbreak wave in the world."

A month later, Guan shut down his research network in mainland China and told his army of sample takers they would have to find other work. Over the previous eight years, Guan and his "thankless heroes" had sampled more than two hundred thousand birds. Many of the virus samples available to researchers and vaccine developers around the world had come from work done by Guan and his Hong Kong University colleague Malik Peiris. Time and again, they had alerted the world to crucial turns in the behavior of the virus.

But now he was switching off the radar, pulling the plug on the world's early-warning system. "Let Dr. Jia Youling come and find a solution," Guan told me in late 2007. "Let him clarify that I didn't make a mistake or make up the data so I can recover my honor. Why should I sacrifice my honor?"

"Malik and I did a lot of work for the world," he continued, his voice rising again. "Who continued on a weekly basis to do sampling for eight years? Anybody else? No. We are second to none in the world. What more can we do?"

# PART THREE

PART THREE

# CHAPTER NINE

# The Secret Call

Nguyen Sy Tuan was conscious but could barely talk. His wasted frame was tucked beneath a white sheet on a metal hospital cot, arms spindly and useless by his sides. Never a husky man, he had lost more than a third of his weight since the virus set upon him nearly two months earlier. Tuan had withered to eighty-four pounds. His head was propped up slightly on a thin pillow. His face seemed frozen in horror, cheeks sunken and lips agape. His bulging eyes were fixed, staring for hours at the ceiling as if the young man was still haunted by a specter that the doctors had predicted would surely claim him. Never before had the specialists in the intensive care unit of Hanoi's Bach Mai Hospital seen anyone survive such a massive attack.

"We thought there was no way Mr. Tuan could make it," confided Dr. Nguyen Hong Ha, head of the ICU, as he stood just outside the doorway. "No patient who we've put on a mechanical ventilator has ever survived."

Yet when I visited Tuan, the twenty-one-year-old had already cheated death. He no longer needed the machine to help him breathe. He could even stomach a little rice on his own.

The small, white-tiled hospital room was silent except for Tuan's occasional dry cough and the muted sound of distant car horns wafting through the second-floor window. To his right was another cot. Not long ago, his teenage sister lay there beside him, burning with a fever of 105, gasping for air just like her brother. Now the cot was empty. Somehow, she too had eluded death. She had already returned to school in the village, where her classmates, much to her consternation, had nicknamed her Miss H5.

Some doctors on the wards claimed these two cases as a triumph for Vietnamese medicine. But flu specialists nervously monitoring the virus in the spring of 2005 knew better. This unexpected turn of events was no reason to celebrate. The survival of Tuan and that of his sister, ironically, were part of a deeply disturbing trend.

These two siblings, the young seaweed harvester and the mischievous schoolgirl, were at the epicenter of a renewed outbreak in northern Vietnam that signaled to some of the world's leading virologists and field investigators that the virus had mutated. It wasn't just the increasing number of cases. It was the pattern. They were coming in larger family clusters, and the overall mortality rate had dropped substantially in a matter of months, suggesting the virus was edging toward pandemic. It may seem counterintuitive, but an astronomically high kill rate can be bad strategy for a prospective epidemic. After all, a virus that swiftly dispatches most everyone it infects gives itself little chance to spread. The 1918 flu virus, by contrast, settled on a far more modest fatality rate, claiming fewer than 5 percent of those infected. Yet it was ultimately able to kill at least 50 million people and perhaps many more.

Over the following months of that spring, new laboratory findings would emerge from northern Vietnam that apparently explained the shifting pattern, confirming that the field observations were no coincidence. Hard science seemed to show that the virus had crossed another threshold. A year earlier, in 2004, this novel strain had demonstrated conclusively that it could pass from one person to another, though widespread transmission had still been elusive. Now, in 2005, that fateful barrier appeared to be falling. Some in the know even concluded that the pandemic had already broken loose. But disease specialists at WHO never publicly disclosed their fears. Instead, they sweated in private, secretly weighing whether to sound a global pandemic alert.

If they did so, the economic fallout could be tremendous. Though the blow would fall hardest on Vietnam, decimating tourism and trade, the whole region could suffer. Multinational companies might suspend their operations. Foreign governments might evacuate their nationals. Airlines might cancel routes, leaving countries isolated and visitors marooned. Stock markets would plunge. These reverberations would be felt worldwide. Yet the danger of waiting to sound the alarm might be catastrophic.

The quandary was compounded by gaps in the evidence. The scientific data were incomplete and contradictory in places. So the flu hunters were forced to make pivotal decisions with only a partial view of the truth. In battling this virus, science has time and again failed to provide the solid answers needed to decipher the pathogen and keep it in the box. Since the last flu pandemic in 1968, the revolutionary field of microbiology has indeed succeeded in breaking the genetic code of the microbes that menace us. But laboratory science has still failed to unlock the secrets of how this mercurial agent evolves and mutates, how it strikes its human prey and when.

This presents a different kind of challenge than those that stem from the Asian landscape. The limits of current science in understanding and disarming the disease are largely independent of the realities on the ground, whether there or elsewhere along the expanding frontier of viral spread.

Nor are scientific constraints the only ones. Both sides of the man-versus-microbes equation pose difficulties. On one side, global efforts to contain flu are hamstrung because WHO and other human health agencies focus on the people afflicted by the disease, at times to the exclusion of the animals that are the source. In addition, money is tight. The resources that frontline states need to identify, contain, and ultimately eradicate the disease among both people and livestock are running short. On the other side of the equation, the essence of the virus itself often eludes disease investigators, whether in the lab or the field.

So on a Friday afternoon in June 2005, WHO's flu team secretly convened in the agency's underground command center in Geneva, linked by a dedicated communications network with some of the world's most elite medical specialists from Atlanta and London to Tokyo, Manila, and Canberra, and prepared to gamble.

Something odd was happening outside of Hanoi. Within a few weeks of one another, three separate clusters of bird flu cases had appeared in a single province southeast of the capital. One of the largest included Tuan and his sister, their grandfather, and a local nurse.

Thai Binh province, where Tuan grew up, is mostly a flat plain of lakes and emerald paddies, part of Vietnam's rice basket. After Tuan had finished his schooling, he left Thai Binh to look for work in the seaport of Haiphong. Many in the West know Haiphong because of President Richard Nixon's decision to mine its harbor during the Vietnam War. But today this port city at the mouth of the Red River Delta flourishes as northern Vietnam's premier industrial center, and there Tuan found a job collecting seaweed for producing agar, a gelatin used in local cuisine.

In early February 2005, all Vietnam took a breather for the Tet holiday. Across the country, Vietnamese bought new clothes, cleaned, repaired, and even repainted their homes, and decorated them with small kumquat trees, pink peach blossoms, and yellow apricot blooms. They stocked up on *banh chung,* or pork cakes, and on candied fruit and other traditional delicacies. Then they invited the spirits of their ancestors to join them in marking the lunar New Year. Sons and daughters who had moved to the cities crammed trains and buses, streaming home to celebrate this extended festival with their relatives. Tuan joined this mass migration. He headed back to the remote village in Thai Binh he had left more than a year earlier and ambled down the dirt alley to his family home, a one-room brick dwelling with a cement floor built beside a creek. Just outside the front gate, ducks paddled in the murky water as they had since his childhood. On the opposing bank, a verdant field of tobacco stretched into the distance. For the reunion, his family bought a chicken in the local market and butchered it in the yard. Tuan's fourteen-year-old sister, Nguyen Thi Ngoan, clasped the bird's wings and legs. Tuan slit its throat. The chicken was likely infected. Soon the siblings were, too.

Tuan broke into a fever about four days later, his wizened father told me over a cup of tea. When I arrived, Nguyen Sy Nham, the family patriarch, was visibly exhausted. For weeks he had been commuting by bus to the Hanoi hospital seventy-five miles away, keeping vigil for hours at a time on a plastic stool at the foot of his son's cot. Yet Nham offered me a carved wooden chair at his table, turned down the volume on the television, and, between puffs on his traditional *dieu bat* bowl pipe, softly shared his family's ordeal.

Tuan's fever had lasted for about two days and then subsided, his father recounted. Tuan took some aspirin, had a bath, and felt better. But the fever soon returned and spiked at 104 degrees. His head throbbed. His chest ached. He started coughing and had trouble breathing. A village medic was summoned, and Tuan was taken to the local health center, where X-rays showed a white smudge in his left lung. The center's deputy director suspected it was severe pneumonia. Because bird flu had previously been identified in the area, Tuan was transferred to a larger hospital in the provincial capital after less than a day. There the doctors concluded he had indeed contracted bird flu and immediately rushed him to the tropical disease institute at Bach Mai Hospital.

By the time he made it to Hanoi, the X-rays showed the white smudge had clouded the entire lung. Soon it consumed the other one also. "Just from the morning to the evening and from one day to the next day, it spread very quickly," recalled Dr. Nguyen Thi Tuong Van, deputy director of the Bach Mai Hospital ICU. The doctors gave him oxygen to ease his breathing, but it continued to grow more labored. After ten days they inserted a tube down Tuan's throat and hooked him to a mechanical ventilator. The infection marched on, damaging his kidneys and liver. The pain was excruciating. "We thought it was very likely the bird flu would kill him. We were very pessimistic," Van continued. "Then, when it seemed the situation couldn't get much worse, it started to get better. Two weeks later, when he didn't die, I thought maybe we could cure him."

Tuan's kid sister, Ngoan, helped care for him during the early days of his sickness. Researchers who later studied the genetic signature of the pathogen concluded that Ngoan may have caught the bug from her brother, noting that some genes in her virus were practically identical to his. By the time I met Ngoan at her home, she was fully recovered. A tall, somewhat gangly teen with a mane of black hair falling to the small of her back, she bubbled over with nervous energy. She told me she loved badminton, chess, and drawing. Her eyes were narrow, but in rascally moments or when feigning surprise, she'd open them wide, big and black. When her father hesitated in retelling the tale, she'd prod him with a playful slap on the leg or she'd interject with details of her own.

Ngoan said she fell ill several days after her brother. "I felt some pain in my legs and some chills," she recounted. "I started coughing a little." At the district health center, X-rays revealed her lungs were clear but a subsequent test came back positive for bird flu. She was quickly transferred to the Hanoi hospital, where her fever ascended to searing levels. "I felt so tired because I had so many injections and I couldn't sleep much," she recalled. Hospital staff moved her to the same room as her brother and the pair bantered as always to keep up their spirits, until Tuan could no longer speak. Ngoan's older sister brought her a pad and a pen. Ngoan, who would always draw people when she was feeling down, sketched the doctors and nurses to ease her mind.

After four days, her fever broke. It returned to normal within two weeks. Barely a month after she got sick, she was back in school as something of a local celebrity.

Though the grandfather also contracted the virus, he objected that he had never felt sick and indeed had shown no symptoms. Vietnamese health officials happened across the old man's infection while testing all the family members. To flu specialists, this was more evidence that the virus was experimenting with new, deceptive paths of infection. Though the man might not feel sick, he could possibly be contagious and, if so, how would anyone know to steer clear of him?

The nurse was the most disquieting case of all. Nguyen Duc Tinh was a tall, skinny twenty-six-year-old with an earnest manner and a whisper of a mustache. He was on duty at the local health center when Tuan was brought in. During the brief, overnight stay, Tinh took

Tuan's blood and temperature, gave him injections, and helped him walk. Within a week, Tinh had developed severe muscle aches, eye pain, and a high fever, symptoms of what he believed was ordinary flu. But when the fever subsided only to return two days later, he grew alarmed.

"Then I suspected I had bird flu," he recalled, his brown eyes widening. "I was really, really afraid of dying." But just two weeks after being reunited with Tuan in the Hanoi hospital, Tinh was discharged. "I had lost hope when the fever came a second time. When I returned to my hometown, I felt as if I were born again."

Vietnamese officials were loath to admit he might have caught the disease by caring for Tuan. This could be an admission that a more dangerous strain was taking hold. They offered a raft of possible explanations for how Tinh got sick: He had sick chickens in his village. There were sick chickens at his girlfriend's home. He'd eaten a sick chicken. He had eaten it at his girlfriend's home.

Tinh dismissed them all with a snicker. "I haven't had any direct contact with poultry or eaten chicken or duck in a long time," he countered. Nor had he dined at his girlfriend's house for weeks before he fell ill, and, thank you very much, all her chickens were healthy. "But," he added, "I was the one in the health center who had the closest contact with Mr. Tuan."


A rare voice of candor among Vietnamese officials was Dr. Nguyen Tran Hien, the astute and able head of the National Institute of Hygiene and Epidemiology (NIHE), the country's CDC. He told me in April of that year that his researchers had identified a significant number of mild and even asymptomatic cases, like Tuan's grandfather. "The symptoms are not as severe as before. Also, the transmission may be faster and easier," he reported. "We are concerned that if the virus is changing, maybe a new virus is coming in the future." Hien wanted more data. And he wanted outside expertise.

Keiji Fukuda returned to Hanoi in mid-April 2005 as part of a special WHO mission. Fukuda was still with the CDC in Atlanta and would not officially transfer to WHO for several more months. But his standing and experience in Vietnam made him a natural for the assignment. Joining him was Dr. Aileen Plant, a fellow epidemiologist from Curtin University of Technology in western Australia known for her passion and wicked sense of humor. Plant had headed WHO's highly successful response to SARS in Vietnam two years earlier. The third member was Dr. Lance Jennings, a virologist from the Canterbury Health Laboratories in New Zealand and his country's acknowledged authority on flu.

For about a week, they shuttled from the WHO offices on Tran Hung Dao Street in downtown Hanoi to the handsome French colonial edifice that houses NIHE about a mile away. That grand old building, with its warm, mustard-colored facade and gray shuttered windows, is located on a street named for a foreign disease detective, Alexandre Yersin, the French-Swiss bacteriologist who discovered the pathogen causing bubonic plague a century ago and later adopted Vietnam as his home. There, down the elegant, high-ceilinged corridors of NIHE, Fukuda and his team huddled with their Vietnamese counterparts, reviewing their findings.

"We were looking at everything we could look at," Fukuda recalled. "Was there an increase in cases? Any differences of patterns in cases? Anything different about the people who were getting infected themselves, and so on." The evidence wasn't conclusive. "But on the other hand, it began to appear there were some differences from earlier patterns in Vietnam."

During the previous year, bird flu had killed nearly three-quarters of those Vietnamese it infected. Yet over the first few months of 2005, the mortality rate had dropped by more than half, suggesting that the virus was edging closer to a pandemic strain. The shift was especially conspicuous because it took place only in the north of the country. The cases were also increasingly coming in family clusters, and often the time that elapsed between cases within the clusters was growing, making it ever more likely that relatives were passing the disease to one another rather than all catching it from the same source. Nine cases were from Thai Binh alone, including Tuan's cluster.

"We thought that it appeared there really could be some changes going on," Fukuda said. "We were in that sort of gray area of understanding. Even if you have enough data to suggest, you don't have enough data to tell you really what's going on, and you need help interpreting."

Fukuda and his colleagues pressed for a wider review of the evidence, and WHO scheduled a private conference for the first week of May, just ten days after the mission left Hanoi. It was to be held in Manila, the Philippine capital and home of the agency's regional headquarters. WHO summoned staff from Geneva and across Asia. Senior government health officials from Vietnam, Thailand, and Cambodia were pulled in, as were outside specialists from the United States, Japan, Britain, and Australia. Leading laboratories, in particular the CDC in Atlanta and the National Institute of Infectious Diseases in Tokyo, were also asked to prepare genetic analyses of H5N1 specimens so these could be compared with the pattern of cases in the field.

"We called for the consultation knowing that it was a lot of trouble to bring a lot of people in rather quickly, but on the other hand these weren't academic questions," Fukuda recounted. "If there really was a change going on, we really wanted to try to come to grips with that as quickly as possible."

For two days, the experts cloistered in Manila and sifted the evidence. Afterward the agency issued a report that cited the shifting patterns of infection in northern Vietnam, including a wider age range of victims, more and larger clusters involving cases over a longer period of time, cases without symptoms, and a declining mortality rate. The document said this was all consistent with the possibility of human transmission and greater infectiousness.

It also detailed genetic changes in viruses isolated in northern Vietnam. One mutation involved the place on the virus where it binds to either human or animal cells and could make it easier for the pathogen to infect people. Another change was near a site related to the lethality of the virus. The report also revealed that a sample from Nguyen Thi Ngoan, the mischievous teen from Thai Binh, showed a mutation that could cause resistance to the antiviral drug Tamiflu. If that change became widespread, it could rob doctors of a vital weapon.

"While the implications of these epidemiological and virological findings are not fully clear," the report concluded, "they demonstrate that the viruses are continuing to evolve and pose a continuing and potentially growing pandemic threat."

In the United States concern was mounting. Just three days after the Manila conference, the Central Intelligence Agency sponsored an exercise to model the global impact of a pandemic strain erupting out of an unnamed Southeast Asian country. Participants were drawn from five federal departments, including Defense and Commerce. The conclusions were sobering: economic downturns, international tension, and political instability.

On May 26, two weeks after Nguyen Sy Tuan was finally discharged from Bach Mai Hospital, WHO's senior communicable disease officer in East Asia, Hitoshi Oshitani, got an alarming e-mail. It was from an epidemiologist in the agency's Hanoi office. Vietnamese researchers at NIHE had been testing specimens taken randomly at health-care facilities in Thai Binh province. The sampling had not specifically targeted suspected bird flu cases. But 10 percent of the 170 specimens had come back positive for the virus, an exceptionally high proportion.

The results seemed to underscore the frightening scenario mooted in Manila. Even worse, the data lent credibility to separate tests conducted by Canadian scientists in Vietnam, which Oshitani had been hearing about.

Without a word to WHO headquarters in Geneva, he flew to Hanoi to see the Canadian microbiologist responsible for the research, Dr. Yan Li. WHO's flu hunters in Asia were trying to keep the startling information from leaking out prematurely. "We didn't want a huge panic with unverified information," explained Peter Horby, the agency's lead flu investigator in Vietnam.

Based on their briefings in Hanoi, Oshitani and Horby drafted a confidential report and on Tuesday, June 7, shared its contents with

Geneva. They reported that Li, a Beijing-born scientist based at the Canadian health department's National Microbiology Laboratory in Winnipeg, had begun a project earlier in the spring to help train Vietnamese scientists responsible for flu research in testing and laboratory techniques. As part of the work, the Canadians had sent in their own mobile lab. They began testing nearly two hundred samples previously collected by the Vietnamese. These were blood samples, or more accurately serum, the clear liquid that remains in blood once red and white cells and platelets are removed. The Canadians were using a technique called Western blot that could detect the antibodies that the human immune system produces in response to a bird flu infection. Though the Western blot technique was not entirely reliable, it did not require advanced lab safeguards like other antibody tests and could be done under local conditions in Vietnam.

According to the confidential report, the researchers tested 86 specimens from people with suspected cases of bird flu. About two-thirds came back positive for the telltale antibodies, indicating the patients had caught the bug. Another 101 samples were from people who had had contact with confirmed cases or infected birds. Nearly as many of these, about three-fifths, were also positive.

Separately, the Vietnamese had run tests using a different technique on the samples from the Thai Binh health facilities. Scientists at NIHE had established that 10 percent were positive by using a method that looked for genetic evidence of the virus itself rather than for antibodies. This technique, called polymerase chain reaction or PCR for short, uses special strands of highly sensitive genetic material called primers. Scientists would combine these with the sample and, if they matched, the primers would cause the virus's own genetic material to rapidly reproduce until there was enough of it to identify.

Finally, the Canadians and their Vietnamese counterparts had conducted an analysis of thirty-eight samples and found that many had specific mutations in the surface proteins of the virus, strongly suggesting it was becoming less deadly. These mutations could help explain some of the milder and asymptomatic cases in Thai Binh and elsewhere in northern Vietnam, such as those of Nguyen Sy Tuan's sister and grandfather.

The report concluded that the disease could be spreading among people more readily than anyone had thought. Moreover, if most cases were mild or lacked symptoms altogether, identifying those who were infected would prove nearly impossible. Even in hospitals, it would be challenging to recognize bird flu patients and segregate them from others. "Extinguishing a pandemic strain by early identification and targeted use of anti-viral [drugs] and public health measures is not going to be successful," the document warned.

Klaus Stohr, the influenza chief, was taken aback. But at the same time, there was something about the results that struck him as not quite right. If the virus was already racing across Vietnam, shouldn't the hospitals be flooded with patients? They weren't. "It should stick out like a sore thumb," he thought.

Calling his staff into his fourth-floor office at WHO headquarters on the morning of Thursday, June 9, Stohr said he planned to urgently convene an outside panel of experts to evaluate the information. "We'll never have perfect data," responded one of his lieutenants, but added, "We have data sufficient to consider raising the pandemic alert level."

Stohr began drafting a memo to Lee Jong Wook, the agency's director general, outlining the arguments pro and con for sounding the global alarm. Raising the alert level would immediately activate steps to contain the outbreak. Stockpiles of antiviral drugs could be rushed to Vietnam and the surrounding region. A warning against travel to Vietnam and nearby countries might follow. Every day mattered. Any delay could hand the disease an even larger head start, potentially costing the lives of untold masses of people.

But what if it's a false alarm? "If you raise the level and you're wrong," Stohr thought, "you'll be blamed." The Vietnamese would be stigmatized. Their economy damaged. The move would spark waves of unnecessary panic worldwide, and WHO's own credibility would suffer. Future warnings might be ignored.

"You make the decision based on the data you have in a responsible way," he later explained. "You need to get ready to defend it. You need to get ready to take the blame."

At that moment in mid-2005, the alert level was at level three, meaning the virus had succeeded in achieving no more than very limited human transmission. Based on the new information, WHO could hike the level to four or five, signifying greater human transmission and alerting the world that a full-blown pandemic, level six, was imminent.

An internal document written that same day suggested the situation might be even graver: "If the results are correct . . . this could be the signal that an influenza pandemic has begun."

"Good morning, good afternoon, and good evening to everyone," Stohr said as he opened the conference call at ten minutes past noon on Friday, June 10. His greeting was familiar to those who had sat in on his previous calls but the setting was not. He had summoned his staff to the WHO bunker, and they gathered around the large, circular conference table in the mezzanine overlooking the main floor of the SHOC. The command center offered a sophisticated communications network that could handle the large call while its secure doors assured that access to the session was kept strictly limited.

On the call were WHO officers from the Hanoi office and Manila regional headquarters. Also invited to participate were Dr. Nancy Cox, chief of CDC Atlanta's influenza division, Dr. Roy Anderson, a senior epidemiologist at London's Imperial College and chief science advisor to the British defense ministry, Dr. John Horvath, the Australian government's chief medical officer, Dr. Masato Tashiro, head of virology at Japan's National Institute of Infectious Diseases, and Dr. Kiyosu Taniguchi, chief of infectious-disease intelligence at NIID. Dr. Yan Li, the Canadian scientist, was also on the line.

They had all been supplied copies of the report detailing the test results from Vietnam. Now Stohr wanted their feedback. He said they would go in alphabetical order.

Anderson would have started, but he was running late. So Cox, the tough-minded virologist from the CDC, was first out of the box. Wasting no time, she went on the attack, calling the tests into question.

"The results I'm most concerned about," Cox said, "are the anti-body test."

She and her colleagues at the CDC had previously done extensive work developing various tests for antibodies and had discovered that some of these techniques picked up false positives for H5N1. Some even purported to show evidence of the virus in specimens from U.S. blood donors who had never been within thousands of miles of an actual H5N1 outbreak, whether in people or birds. In particular, Cox was skeptical about the reliability of Western blot testing. To avoid misleading findings, the sensitivity and precision of this kind of analysis had to be calibrated using the results of other tests.

On the call, she grilled Li about his techniques in performing Western blot and interrogated him about whether he had used proper scientific controls to gauge the accuracy of his findings.

"Very limited controls," Li acknowledged. He was already on his heels. "The number of controls is very low."

Cox pressed on. She said the CDC had tested similar serum samples from Vietnam, and they'd all come back negative for virus antibodies. She called the rate of positives in Li's test, or assay, "exceedingly high."

"I really wonder if your assay is picking up false positives," she continued. She urged that the samples be retested using a more respected technique for detecting antibodies, called microneutralization. "The serology is very much in doubt and must be repeated with another test in another lab," she insisted.

"I understand the risks of Western blot," Li responded defensively. He stressed that his findings were only preliminary. He admitted they might have overestimated the extent of infection in Vietnam.

Cox wanted everyone on the call to take a deep breath and think hard before rushing to conclude there was evidence of an emerging pandemic. So she continued to pound away, questioning the overall reliability of Li's approach.

"The results are in question because they were obtained with an assay that hasn't been validated," she argued.

"I agree," Li conceded, wishing his results had never been passed to the agency. "I didn't even want that circulated."

None of the other experts had yet weighed in, and already they could feel Li squirming on the other end of the line.

For a moment, he seemed to win a respite. The call shifted to a discussion of the genetic changes Li had detected and whether these were making the virus less lethal. He said that the mutations had been found in most of the human samples that his team analyzed. He suggested that either the virus was evolving into a less deadly form, or an entirely new flu virus was now circulating in Vietnam.

But he promptly came under fire again, this time from a new quarter. Anderson, the British epidemiologist, had joined the call and pushed Li on whether those mutations alone would determine the lethality of the virus. Anderson personally didn't think so. He also noted that the analysis had examined only an "exceedingly small" number of samples.

Cox joined back in, suggesting that a whole series of genetic changes would be needed to alter the lethality, not just one. And even if these changes did make the virus less deadly in animals, as some scientists suspected, she was unconvinced these changes would do the same in people. She wasn't even sure the mutations had occurred at all.

"The results need to be verified," she said, repeating her refrain. "It's quite possible they are true, but they need to be verified."

Horvath, the Australian chief medical officer, now entered the fray. He wondered whether the researchers had tried to make sense of their highly unusual lab findings by comparing them with the actual experience of patients in the hospital.

"It's difficult to understand, naturally," Li offered meekly.

Then, abruptly, the tone changed. Having roughed up Li, the expert panel shifted gears and began to ponder what it would mean if he actually proved to be correct.

"The problem is serious, very urgent," Anderson conceded. "We need access to information. It's urgent that an independent lab confirms the changes." But Vietnam might not share that sense of urgency, and Hanoi's record of cooperation was poor. "It may be necessary to elevate the pandemic level to get the information," Anderson suggested.

Stohr pressed the issue of whether to in fact raise the level. "How concerned should we be?" he queried.

"On face value, it's a very serious report," Horvath answered. "But it's not sufficient to take action. I'd be concerned with raising the level of pandemic because it might shut Vietnam down."

"We'd have to weigh that risk," Anderson said. The move could be highly unpopular in Southeast Asia. He said the agency would have to weigh "the urgency to understand more versus the danger of upsetting the political balance."

With time on the call running down, the experts turned to the third and final set of findings. These were the results of the PCR tests on the samples from the Thai Binh health-care facilities, which had revealed a 10 percent rate of infection. They were the most troubling findings. PCR was a more reliable test than Western blot.

"Ten percent is extremely high," Cox noted. During a seasonal flu outbreak, far fewer than 10 percent of people tested under similar circumstances would have been positive. "But if this is a pandemic beginning, it's not impossible. It's urgent that the results be verified."

Stohr concurred. "There's an urgent need to verify the accuracies of the findings," he concluded. WHO would rapidly assemble a new team of experienced virologists and epidemiologists, he decided, and dispatch them to Hanoi. WHO's chief representative to Vietnam would hope to meet early the following week with senior Vietnamese officials and pressure them to let the team in. The Vietnamese government would be urged to turn over serum and other samples and related material for further testing.

A decision on sounding the pandemic alarm, he said, would be temporarily postponed.

At two o'clock Geneva time, Stohr adjourned the call.


Uncertainty is what often separates public health from laboratory science. Lab researchers select questions they believe they can answer and bypass those they can't. They design experiments to produce definitive proof. Success means generating results that can be reproduced by other scientists in other labs. Public health, by contrast, rarely gets to choose the questions it must answer. People get sick. Doctors try to diagnose. They prescribe treatment and hope it will work. Symptoms, family history, and even test results are suggestive but often not definitive. The whole purpose of getting a second opinion on medical care is not to reproduce the conclusions of the initial physician but to discover if there are different ones. If the affliction is the common cold, uncertainty can be annoying. If it's cancer, it can be tragic. If it's flu, it can be catastrophic.

Medical science as a modern discipline is little more than a century old. Molecular biology goes back barely two generations. In that short period, the field has reshaped human knowledge and answered some of the basic questions about our very essence. Yet in confronting pandemic influenza, the greatest human killer in history, science still comes up short.

The difficulties begin with diagnosing the virus. Scientists have had trouble coming up with a fast way of determining whether someone has H5N1, and this lack of an accurate rapid test has repeatedly fostered confusion and even panic, most notably when the disease came knocking on Europe's door. In the final days of 2005, villagers in the frigid mountains of eastern Turkey began falling sick with a mysterious respiratory ailment. Flu specialists suspected the novel strain. But initial tests by Turkish health officials came up negative. Days later, they reversed themselves, announcing that further tests had confirmed bird flu in two patients. More would follow. Of eight patients ultimately confirmed with bird flu and treated at a hospital in the regional center of Van, rapid flu tests came back negative every time. So did a separate enzyme-linked immunosorbent assay, or ELISA, test. Even the initial PCR test for four of the cases failed to identify the virus, though they were already very sick. Only after additional samples were taken did they test positive.

Then, as fear spread, Turks from around the country flooded hospitals at the first sniffle or cough. Day after day, hundreds of test samples poured in and overwhelmed the country's national lab, which was equipped to complete at most a few dozen each day. Soon, instead of missing cases, Turkey was reporting false positives. The crest of the epidemic suddenly seemed to rise far higher than anyone had seen before. The government would eventually announce more than twenty human cases. But follow-up testing by WHO partner labs outside the country could confirm only a dozen.

Researchers in Indonesia reported similar problems with their initial flu tests. So did doctors in Thailand, who said in 2006 that these tests were actually becoming less able to detect the virus over time. Though rapid tests for seasonal influenza are commercially available and widely used, WHO says they are not sensitive enough to be used for bird flu. False negatives are common, and even positive results cannot distinguish between H5N1 and other influenza strains. Both the U.S. government and European Union have funded efforts to come up with a fast diagnostic test for bird flu. In spring 2009, the U.S. Food and Drug Administration approved the initial marketing of a test that can take less than forty minutes to work.

PCR tests that check for genetic evidence of the virus are more accurate but take several hours, longer than some victims can afford. And should the world face an emerging pandemic, even a short delay in detecting the virus could scuttle hopes of slowing its progress. Moreover, these are expensive tests that are frequently ill suited to much of Asia and elsewhere in the developing world. They require precise sampling, highly trained technicians, and the proper primers. As the virus strain mutates, which it does with haste, existing primers may no longer match and the tests will fail. And poorly matched primers can incorrectly detect other genetic material in the sample and yield a false result. Tests for antibodies to the virus can be even more accurate, especially established techniques like microneutralization. But these often demand labs with sophisticated safeguards. And they take much longer. Since antibodies to a virus appear only after the infection, these tests are of little use for responding to the disease in real time. A pandemic could be well on its way before antibody tests confirmed it. The "gold standard" for laboratory science is growing the live virus itself from a sample, traditionally in eggs, and using this to identify the pathogen. But this too is a long process. And for the many countries that lack high-security labs, the risk of the live virus escaping makes this technique far too risky.

The uncertainty goes well beyond testing. Flu specialists have been asking themselves for more than a decade why they know so little about the novel strain. Not long after it first crossed to people in 1997, researchers had already ascertained that the disease could be contracted through exposure to infected birds. But as Fukuda explained, exposure could mean many things. "Do you have to touch it? Do you have to rub your face after you touch it to get the virus near your mouth or your nose? If your face is close enough to infected birds, does that mean you can really breathe it in?" Maybe you don't have to touch it at all, he suggested. "If you mean proximity, do you mean two feet, five feet, ten feet? Bird feces the same as having contact with birds? Is touching the wall in an area where birds were in the last twenty-four hours, does that count?" These are not idle distinctions. This information would help inform a strategy for stemming the spread of the virus and in turn reduce chances of a sinister mutation.

Research has been slow to offer answers, in large part because it is hard to conduct. Sick people don't wander into studies. Those who are exposed and infected are often too ill to report how they got that way. Nor are time-consuming field studies involving large numbers of people a priority for countries reeling from these outbreaks. "Most of the countries where the first cases have occurred do not have traditions of analytic field investigation and the high profile of 'bird flu' does not encourage immediate openness," wrote Dr. Angus Nicoll of the European Center for Disease Prevention and Control.

Dr. Michael Perdue, an American microbiologist assigned to WHO's influenza program for several years, said it is hard to get standardized reporting from countries because each health ministry has its own way of doing things and few are anxious to publicize their disease outbreaks. "When it hits their country, they shut down," he lamented. "WHO has to walk a fine line between demanding we have to have this information and shutting down the communications with the country, where they say, 'Whoa, they're coming after us. WHO is coming after us.' You don't get as much information as you like. There's sort of a fine political dance that you're doing."

Perhaps more surprising is the lack of autopsies. Of the first two hundred confirmed deaths from the virus, fewer than a dozen victims underwent these postmortem exams. "This has posed some major problems in understanding the biology of the disease or fully understanding the virus underlying it," explained influenza researcher Malik Peiris from the University of Hong Kong. Autopsies could examine how the agent wages its assault on the lungs and what other organs might also come under attack. These studies could help explain why some people get infected and not others, why the disease is so severe, and how well antiviral drugs work. "These are all pretty fundamental questions," Peiris put it to me.

Why so few autopsies? The conventional response from leading researchers and senior officials is that local cultures bar the practice. In Chinese tradition it is said a corpse must be buried intact. Similar strictures are thought to apply in the Buddhist cultures of Thailand and Vietnam and the Muslim culture of Indonesia. "You have to get permission from the family to get that done," explained Dr. Triono Soendoro, head of laboratory research at Indonesia's health ministry. "People here are not used to having postmortem procedures." Had Indonesian officials actually asked families for permission? "No, we haven't tried," Soendoro added.

When I broached the subject of autopsies with victims' relatives, I got surprising responses.

Rini Dina Prasetyaningsih was among the first Indonesians known to succumb to the disease. Months later, I visited her small but comfortable middle-class home in South Jakarta and raised the delicate subject with her husband, Agus Mardeo. "I wouldn't object to a postmortem as long as it was in the hospital and done by a doctor, as long they didn't dig up the body after it was buried," he said. Mardeo's long, thin face turned somber. "Personally, I really wanted to know what happened to her. I would have been more than happy to have an autopsy." But he added, "I'm just a common man. I don't know about these medical procedures. No one suggested the idea to me."

In a poorer suburb just outside the capital, a woman named Ibu Samida offered the same answer. She received me in her dimly lit living room, where she sat on a tattered couch surrounded by her four children and two grandchildren. The only one missing was her middle daughter, nineteen-year-old Ina, who had died not long before of bird flu. "I would have allowed an autopsy for the benefit of my neighbors and my family and everyone else concerned about the disease," Ibu Samida said firmly. But again the refrain: "No one asked me. If they had asked, I would have given my permission."

Yet even with autopsies, field studies, and fast, accurate testing, the answer to the most critical question could remain elusive: What will it take for the novel strain to become a pandemic strain? In other words, what specific genetic changes are required for the disease to become easily passed among people, and what is the sequence of biological events in animals and humans that will foster those changes? Scientists suspect these could involve mutations that make it easier for the virus to bind to human cells. But that's likely not the whole story. Though modern science has delivered an intimate view of the pathogen's inner machinery, the world still has little insight into how this fateful transition would work.

David Heymann, who was WHO's chief of communicable diseases before retiring in 2009, conceded we're not much further along in that regard than we were in 1918. "We're handicapped by a lack of knowledge about the risk factors that cause the virus to change," he said. "There are a lot of things we just don't know."

Scientists have little comparative data to go on because there have been only a few flu pandemics in modern times. Despite groundbreaking work on the genetic makeup of the Spanish flu strain, Fukuda said there's no way to be sure whether the 1918 experience is relevant to the avian flu virus. And without knowing how a pandemic strain would develop, it is nearly impossible to stop it. What steps should be taken to interrupt the evolution of a pandemic? What practices are speeding it on its way? When do changes in the pathogen represent a real and present danger? When are they a red herring? What to do about them?

"Sometimes the decision is simply 'Let's wait.' We don't have enough to go on. Let's see how things evolve," Fukuda said. "Or, it's 'We don't have as much information as we like but we're worried enough that if we're missing something, it's worth the price of jumping now.'"

For five days, WHO had been trying to get a meeting with senior Vietnamese officials. Finally, on the afternoon of Tuesday, June 14, Hans Troedsson's delegation filed into a conference room at Vietnam's Ministry of Agriculture and Rural Development, an unremarkable building around the corner from the monumental granite mausoleum, modeled after Lenin's Tomb, where Ho Chi Minh, father of Communist Vietnam, still lies in repose.

Troedsson was no influenza specialist. He was a Danish doctor who had previously been responsible for children's health issues at WHO. But as the agency's chief representative in Vietnam, Troedsson essentially held the rank of ambassador. Flanking him were six other foreigners, including senior figures from the United Nations, World Bank, and European Commission.

The Vietnamese side was led by Deputy Prime Minister Nguyen Tan Dung. Considered a reformer within Communist Party ranks, Dung would be promoted a year later to become his country's youngest prime minister since Vietnam was reunified. He was joined at the table by the agriculture minister and his deputy, the deputy health minister, and five other top officials. Such audiences with senior dignitaries would often merit a few minutes on Vietnam's nightly news and a mention in the morning papers. But this one was to be strictly confidential.

After opening formalities, Troedsson explained to Deputy Prime Minister Dung the significance of the recent lab findings. They indicated that bird flu infections in Vietnam were more common than previously thought.

"Vietnam could be on the verge of experiencing a nationwide epidemic, which could spread globally and cause a pandemic," Troedsson continued. If an epidemic broke out, it could do serious harm to Vietnam's economy and the health of its people. If it went global, it could kill as many as 40 million people. So, he warned, WHO might have to elevate the pandemic alert level to either four or five if the test results were accurate. Several UN agencies stood ready to help Vietnam fight the virus and prepare for a pandemic. But the troubling lab results had to be checked by international experts, he insisted. This was urgent.

Troedsson closed with a veiled threat. If the results could not be checked, WHO might have to raise the alert level as a "precautionary measure."

Then, one by one, Troedsson's colleagues tightened the vise. Jordan Ryan, the UN's ranking official in the country, told the Vietnamese it was no longer a question of *if* a pandemic was coming but *when*, and he urged Dung to do the right thing. Ryan's counsel carried weight. A dozen UN agencies were active in Vietnam, engaged on multiple fronts, ranging from agriculture to childhood nutrition.

Next came Klaus Rohland, who ran the World Bank office in Vietnam. Over the years, Vietnam had received more than $5 billion in foreign aid for development, and the World Bank was in charge of coordinating that. Rohland told Dung that his country now faced a situation that could be "disastrous." He strongly advised him to accept WHO's offer of assistance.

Finally, Ambassador Markus Cornaro, head of the European Commission's delegation in Hanoi, warned that foreign donors to Vietnam were counting on the government's full cooperation. How the government responded to the "offer of international support" would indicate the degree of trust and cooperation between Hanoi and the donors, Cornaro told him.

The words were diplomatic, the threat unmistakable.

Dung listened intently to his visitors. Then he listened to presentations from his own health officials outlining the lab findings in greater detail.

When everyone else had finished, Dung vowed that Vietnam was concerned for the welfare of the whole world, not just itself, and wished to be proactive, neither complacent nor hasty. And so, Dung said, he would propose inviting an international team of experts to work with Vietnamese scientists in checking the recent lab findings and exploring the wider threat. The team could come immediately, but on the condition that the lab findings remained secret until the review was done.

It seemed WHO had gotten its way. But when Troedsson reported back by telephone an hour later, Stohr and his colleagues in Geneva doubted Vietnam's intentions. Hanoi hadn't said anything about actually turning over the original lab samples for independent testing, and nothing less would do.

"It's a trap," warned one senior official in Geneva.

Troedsson didn't want to dally. The deputy prime minister had offered full cooperation and Troedsson took him at his word. What Troedsson didn't say was that some of his staff in Hanoi believed time was already up and the pandemic was under way.

"It's urgent here," he pressed. "We want to have a team here next week. Klaus, don't let us down on this one."

When the conference call was over, Stohr and his colleagues in Geneva were agitated. "We didn't get the message across to Vietnam," Stohr grumbled.

But could they afford to continue bickering with the Vietnamese?

"We have to get a team in there as fast as possible," urged Tom Grein, the veteran investigator who helped coordinate the global response to Vietnam's early outbreaks a year before. "Six hours on the ground, ask for the samples." If Vietnam balks, Grein added, "WHO has to have the guts to pull the team out."


That week, Yan Li, the Canadian scientist at the center of the storm, sent an e-mail to many of those who had participated in the secret conference call the previous Friday. He said he agreed with their criticisms of his Western blot study and wanted to explain why he had proceeded with this technique.

Li said it would have been difficult to analyze the serum samples using a more reliable test because the Vietnamese didn't have the proper lab safeguards. "Therefore, it would be better for us to first use the Western blot method for initial testing and provide on-site training to Vietnamese scientists to use this technique in their regular laboratory," he said. "But the prerequisite was that all, or at least those serum samples initially screened positive by Western blot, be sent back to us so we can further test them." Li's own lab in Winnipeg could perform the verifying tests.

After the initial tests came back positive, Li wrote that he had repeatedly told officials from NIHE and WHO that "these were very preliminary results" that could reflect contaminated samples and would have to be retested using a more established method. "It was clear to us," he said, "that the Western blot method was only for initial screening and training but not for the final decision-making test."

Dr. Frank Plummer, scientific director general of the National Microbiology Laboratory in Canada where Li worked, later provided me with a similar account of the Canadian collaboration with the Vietnamese, adding that the results of the Western blot tests were supposed to have remained private. "This interim information was neither meant for distribution nor for decision making," Plummer said. He called it unfortunate that the information was shared with officials, prompting public health concerns.

To some at WHO, Li's defense rang hollow. They'd heard something similar from the Canadians before, two years earlier, to be precise. That's when a baffling respiratory outbreak had raced through a nursing home in the suburbs southeast of Vancouver, Canada, and Li's lab had weighed in with an alarming and ultimately misleading diagnosis.

In the first week of July 2003, scores of elderly residents at the 144-bed Kinsmen Place Lodge in Surrey, British Columbia, started coming down with the sniffles. So did dozens more of the staff. By August about a hundred residents and at least forty-five staff members had developed what looked like a summer cold. Seven residents had died during that period, not an unusually high number for the home. But pneumonia was implicated in three of the deaths, and that was worrisome. Initial tests at several Canadian labs failed to identify the cause of the outbreak. Then, Plummer announced to the media that his scientists had discovered it was SARS.

The revelation rattled Canada. The country was still on edge after the global SARS epidemic earlier in the year. Toronto, Canada's premier city, had experienced the largest outbreak outside Asia and 250 people had fallen sick. Forty-four had died. Li and his colleagues at the National Microbiology Lab had been deeply involved in researching that epidemic.

The new cases at Kinsmen Place Lodge certainly didn't look like the same SARS. Most were mild and nearly everyone recovered. But Li's lab was now reporting that its genetic analysis had revealed an almost perfect match with SARS. Separate tests conducted by the lab on samples from the nursing home had also found antibodies to the SARS *Coronavirus*. The virus could be returning in a new, unpredictable form. WHO urgently dispatched a SARS expert from Geneva to investigate.

In suburban Surrey, the six-story brick nursing home was placed under quarantine. Ill patients were put into isolation. A sign on the front door greeted visitors: STOP. RESPIRATORY OUTBREAK. STRONGLY RECOMMENDED YOU DO NOT ENTER THE BUILDING. For days, residents could only wave through the windows to relatives on the sidewalk below. At Surrey Memorial Hospital, nineteen health-care workers who had contact with a patient from the nursing home were restricted to home quarantine. Then another nursing home in western Canada reported a similar outbreak.

As time passed, however, tests at other labs, including CDC Atlanta, failed to turn up SARS in the samples. Instead they isolated another *Coronavirus* called OC43, known to cause common colds. Both WHO and Canadian officials ruled out a new SARS outbreak, and the quarantines were lifted. The misdiagnosis by the National Microbiology Lab was ultimately blamed on a combination of contaminated lab samples and overly sensitive tests that generated false positives. The lab's reputation took a drubbing.

Li later told the *Wall Street Journal* that his lab's results had only been preliminary and were faxed out to the public without his permission.

So as a crack team of investigators secretly set out for Hanoi in June 2005, they suspected that Li's Western blot results on the samples from northern Vietnam were flawed also. That thought offered a measure of comfort. But there were still those disturbing results from the separate PCR tests conducted by the Vietnamese, and the Vietnamese had a good track record.

Maria Cheng was in China when she took the call asking her to head for Hanoi. Her colleagues in Geneva told her that something was going on in Vietnam but they weren't sure what. They were more nervous than she'd ever seen them. "They were really scared about what was going on. They thought the pandemic was starting," she recalled.

Cheng was a bright, young Canadian journalist who had written from Asia for several publications before WHO hired her in the middle of the SARS epidemic to help handle the flood of press calls. She impressed her seniors in Geneva and, after that crisis faded, stayed on, expanding her communications portfolio to include bird flu, polio, and other diseases, splitting her time between headquarters and the field. She was an old hand at flu by the middle of 2005. Yet she didn't realize how serious the situation might be in Vietnam until Dr. Julie Hall, an influenza expert in WHO's Beijing office, sat her down.

"If there is a pandemic and you get infected, you're stuck there," Hall told her. "Are you really going to do this?"

Cheng said yes.

The Beijing staff showed her how to properly wear a mask fitted to the contours of her face and sent her off.

"It was freaky," she recounted.

The Vietnamese government had decided, after all, to cooperate with the special mission and provide the required lab samples. By coincidence, Prime Minister Phan Van Khai was due that same week to make the first trip ever to the United States by a top Vietnamese Communist leader, and this was no time to pick a fight. In an interview with the *Washington Post* on the eve of his trip, Khai signaled his government's good intentions. "Due to the limited capacity and conditions in scientific research facilities, therefore, we are working closely with the international community," he said. "A sample has been provided for international experts and foreign countries to carry out joint research."

In Geneva, Margaret Chan was planning for the worst. Just days earlier, she had been brought in as WHO's new director of communicable diseases and special representative for pandemic flu. She instructed Stohr to draw up a new list of experts who could advise the director general, Dr. Lee, about whether to sound a pandemic alert once the special team reported back. She recognized this would be as much a political determination as a scientific one.

"They should be from different backgrounds," she told Stohr. "We need policymakers or advisors to ministries of health, not pure scientists."

Chan was also concerned that news of the mission would leak. She called Hitoshi Oshitani in the regional office, and they agreed that the agency would say as little as necessary.

In Vietnam, Maria Cheng would have the unenviable task of trying to keep the press in the dark without actually lying to them. It would be hard to miss the foreign scientists that began descending on Hanoi on Monday, June 20. The elite team included Masato Tashiro, the chief virologist from Japan's NIID; Angus Nicoll, an epidemiologist who was then head of communicable diseases for the British Health Protection Agency; and Oshitani. Also joining the mission was Nancy Cox, the CDC Atlanta's influenza chief, who would finally get to make sure that the questionable lab findings were properly verified.

The mood in Hanoi was eerie. The team members were on edge. There was a sense, Cheng recalled, that something really bad could be happening.

They dropped their belongings at the Melia Hotel in downtown Hanoi and headed around the corner to the WHO office for a briefing with Hans Troedsson and Peter Horby. "They were convinced this was the beginning of a pandemic outbreak," Cheng said. "Peter told me it was definitely happening. We just didn't know how bad it was." There was no doubt that the alert level would be raised. The only question, Horby told Cheng, was whether it would be hiked to level four, which signifies increased human transmission of the virus, or five, which means widespread transmission.

Horby said he had a high degree of confidence in the Vietnamese lab technicians. They were careful, and their results had always turned out to be right. That's why he was so worried.

Wilina Lim and her assistant were the last to arrive. Oshitani had turned to Lim for help in retesting Vietnam's worrisome samples. As the head of Hong Kong's public health laboratories, she had been intimately familiar with H5N1 longer than nearly anyone else in her field, having identified some of the earliest cases in 1997. She had been reluctant at first to join the mission, asking instead that Vietnam ship the samples to Hong Kong, where she could do the analysis in her own lab. But the Vietnamese refused.

Lim had never been to NIHE. She had no idea what equipment they had or whether they knew how to use it properly. Even the slightest miscue could contaminate the samples, making gibberish of the results. Lim wasn't taking any chances. She and her assistant started preparing everything they'd need to run the tests, and that meant everything. In one large cardboard box, they packed their own pipettes and pipette tips, their own primers, and even containers of water for testing. They packed their PCR machine, called a thermocycler, into another box. They also crammed a suitcase full with gloves, sanitary wipes, paper towels, and other assorted supplies. If the samples could not come out to Hong Kong, then Lim would recreate a miniature version of her own Hong Kong lab right inside NIHE.

While Lim was packing, the other team members were already in Hanoi at work, scrutinizing the lab techniques the Vietnamese had used, evaluating the test results in light of information from the field. The days started early, not long after dawn, and went until evening. The team members compared notes over late dinners, night after night, and then went to bed so they could begin anew. Everyone was stressed.

Lim caught an early morning flight to Hanoi on Thursday, June 23, and by midmorning was on the ground. She and Cox consulted on how best to run the new test and check the Vietnamese findings. Lim was anxious to set up her equipment and get started. But there was no room at NIHE. The institute was busy with its routine work. She had to wait. Finally, at 7:00 P.M., the lab staff cleared out, and Lim and her assistant began testing the first of thirty samples. They continued until midnight, then returned early the next morning, using their PCR machine to amplify genetic material in the samples. By that afternoon, they were seeing a clear pattern emerge.

Several colleagues were waiting anxiously in the WHO office when Lim returned with her findings.

Lim reported that her tests were coming back negative, every last one. There was no trace of genetic material from the virus. The patients sampled in Thai Binh had not been infected after all.

The earlier results had been flawed. The positives had all been false positives. The Vietnamese, it turns out, had been using a set of primers they'd been given by the Canadians, and these were detecting rogue bits of genetic material. The new tests using Lim's primers had all come back clean.

In the office, the relief was tremendous, the swing of mood extreme.

Nicoll had been studying the field results when the pair reported back that there was no evidence the virus was becoming more infectious. He immediately put aside his papers. "What problem?" he suddenly thought. "It all goes away. It's like water disappearing into sand. You think you have a cluster and you don't."

Horby, hugely comforted, packed up his belongings and left. Twenty minutes later he was bound for a previously scheduled vacation in Britain.

"We dodged a bullet there," Cox said.

And with that, she and Nicoll decided to go shopping. Nicoll loaded up on souvenirs in the markets of Hanoi's old city near St. Joseph's, the city's neo-Gothic cathedral. Cox bought an *ao dai*, the traditional, form-fitting Vietnamese gown with a high collar and slits up both sides. A day later, the pair headed north to the haunting waters of Halong Bay, one of Vietnam's most popular tourist destinations, for a cruise amid its sculpted limestone islets.

When officials from Geneva called Hanoi for an update, they were informed that Cox and Nicoll were on a boat and unavailable.

The scare would remain secret. The world would never be told how close WHO had come to putting it on a war footing. Just about the only hint provided by the agency came one day after the team departed Vietnam. Citing WHO, the French press agency reported that an international team of virologists and epidemiologists had left the country after assessing that the threat posed by the bird flu virus was less than they'd suspected. Few other details were offered. "The most important thing," Troedsson was quoted as saying, "is that we could rule out that there was an immediate, imminent pandemic." It was meant to be reassuring. But his comments acknowledged more about the agency's fears than officials ever had before.

# CHAPTER TEN

# Let's Go Save the World

Her maroon minivan had just edged into Friday morning traffic and already Gina Samaan was troubled. The case just didn't make sense.

Seated in the backseat, Samaan flipped through the file yet again. This much was for sure. A twenty-nine-year-old Indonesian woman from the east side of Jakarta had died two days earlier. She had suffered from acute pneumonia. She had been ill for at least a week. The victim's doctors suspected it was bird flu, and indeed her samples tested positive for the virus at a government laboratory.

The specimens also came back positive from another, secretive lab run by the U.S. Navy in downtown Jakarta. This was one of three overseas labs established by the navy to specialize in infectious diseases confronting U.S. forces on foreign terrain. Recently it had been forced to lower its public profile in Indonesia because of rising anti-Americanism fueled by the wars in Afghanistan and Iraq. Still, the lab continued to operate, perhaps ironically, right across from the city's main prison, where Indonesia had jailed some of its most notorious terrorism suspects. The navy facility was more sophisticated than anything the Indonesians could muster, quietly supporting the government's efforts to contain the spreading bird flu outbreak. If the lab confirmed the victim had died from the virus, it was sure to be so. But that scientific fact by itself revealed precious little about the case at hand.

Samaan's van inched through the traffic. Outside, another equatorial morning had settled on the Indonesian capital, air so thick and languorous that breathing was a chore, the sky depressingly gray from the smog hugging Java's swampy coast. Commuter buses muscled through some of the third world's worst gridlock while motorbikes swarmed like mosquitoes. Though the van's front and rear windows were emblazoned with the shield of the United Nations, few took notice of the vehicle or the young WHO investigator inside. Even fewer made way.

At twenty-nine, Samaan was the same age as the victim. Dark eyes earnest and intent behind rimless glasses, brown hair tied back with a pink hair band in a practical ponytail, Samaan pursed her lips as she continued to review the case. The local media were reporting that chickens in the victim's neighborhood had recently fallen sick and died. But Samaan had access to test results on samples collected by the city's veterinary department. These seemed to show that the local poultry were healthy. "That makes me a bit worried," she admitted to me in a broad Australian accent. "Is there anything different here?" she wondered. "If so, what's different?" If the source of infection wasn't chickens, could it be another person? If so, it might mean the virus had mutated into a form more easily passed among humans.

That prospect seemed to grow over the following days as this influenza gumshoe followed a trail of clues leading unexpectedly into a neighboring province and then back again. In its own way, the stakes of her investigation were every bit as high as that of the special WHO mission dispatched to Hanoi seven months earlier to verify whether the virus was perilously mutating. Northern Vietnam had been a source of intense anxiety because of flawed test results that appeared to show that the novel strain had broken the pandemic code. But in the months before Samaan set out on this morning in January 2006, more people had died of the disease in Indonesia than anywhere else. And if a global epidemic were to erupt, there was no place more likely for it to start.

Samaan and the Indonesian health ministry had both agreed to let me accompany her on her investigation as long I respected the victim's medical privacy by keeping her identity confidential. This afforded me

an intimate view, not only of the hunt but of the challenges and frus-
trations epidemiologists face as they try to run this killer to ground.

Samaan wasn't taking any chances. She had stashed a bag with
masks and plastic shoe coverings in the back of the van. In her bulky
brown satchel of a handbag, buried beneath a cell phone, digital camera, and BlackBerry, she kept a small bottle of pink antiseptic hand
lotion and a cheap thermometer. She had been taking her temperature
twice a day since she had arrived in Jakarta eight months earlier, dispatched by WHO from the Australian health ministry. "It's important
to know your baseline," she explained, adding that hers was a cool 97.5
degrees. She had also carefully considered her footwear, donning
simple shoes with covered tops so her feet would be protected against
any contamination on the ground. They had flat bottoms so they were
less likely to get cruddy with dirt, chicken droppings, and other possible
sources of infection. And because she was constantly removing her
shoes according to Indonesian etiquette before entering someone's
home, she settled on a pair that was easy to slip on and off, so she
didn't have to bend down near the ground. After each outing, she
washed them in the sink and, as a result, was going through a pair
every few months.

Yet for all her preparations, Samaan wasn't expecting what awaited
her when we finally pulled up in the victim's neighborhood. It was as
if the whole community had turned out and was now sitting in rows
of folding chairs on the block captain's grassy front yard, anxiously
anticipating an explanation for their tragedy. Samaan waded through
the crowd along with her translator, a stocky Indonesian with thick
sideburns, wispy beard, and crisp American accent acquired during a
childhood in Pittsburgh, and a middle-aged woman in batik named
Ibu Eni from the health ministry's national lab. When they announced
that blood samples were to be taken, dozens crowded around, offering
their arms.

This was not the way things were done back home in Australia,
where Samaan had recently graduated from an elite program in epidemiology at the Australian National University in Canberra. She would
have preferred to proceed calmly, methodically. She would have
wanted to interview the victim's relatives and neighbors in their own

homes, carefully taking stock and observing them, noting their answers, calibrating their responses, piecing together the timeline, then drawing samples only from those who had immediate contact with the dead woman. Instead this was verging on chaos.

Ibu Eni took a seat behind a wooden school desk set out in the yard. She snapped on a pair of rubber gloves and began accepting one thrusting arm after another. After the first man had given blood, Samaan buttonholed the graying old-timer and tried to clarify the conflicting reports about sick poultry.

"Have there been any bird deaths in the *kampung*, or neighborhood?" she asked.

"Nothing," he demurred. "Everything's been ordinary."

A second man in a black Mercedes cap interjected, correcting him, "Yes, there *were* two chicken deaths, about thirty meters away from the victim's home."

"How many were dead?" Samaan asked, confirming the number.

"There were two," the second man repeated.

Then a third man, with a bushy mustache, stepped forward, weighing in, "Yes, there were two dead chickens a month ago."

Samaan listened intently, producing a large, spiral notebook from her handbag and jotting down the details. "What were the symptoms?"

"It was all of a sudden. We didn't see any symptoms," answered the mustachioed neighbor.

"What made you think the birds were sick?" she pressed.

"My wife found them," he answered. "My mother-in-law poured kerosene on them and burned them. I just cleaned up the remains."

The three men started buzzing among themselves. The translator fell silent, trying to follow the conversation. Samaan waited. Perhaps the neighbors were trying to recall the details. Or perhaps, like many other Indonesians, they were reluctant to disclose them, fearing that local veterinary officers would raid the *kampung* and cull the remainder of their birds.

"How's it going?" Samaan prodded after a few minutes.

"We didn't see any chickens with symptoms," insisted the man with the mustache. "They were just kind of dead. We still don't know whose chickens they were."

Samaan rolled his remarks over in her mind, examining them for any elusive clue. Two dead birds, no symptoms, no samples. It was hard to conclude they were responsible for the death. She excused herself. Then she went to locate the victim's relatives in the crowd.

The victim's mother was a handsome woman in a coarse, orange dress with thick brown hair pulled back in a bun. Samaan ushered her to the shade of a stately jackfruit tree in a corner of the yard. The great green fruit dangled from the canopy. Another neighbor dragged over two folding chairs so the pair could sit down.

"When did your daughter become sick?" Samaan asked, this time pulling a large calendar out of her handbag.

"Nine days ago." That made it the first week of January. The timing could eventually prove crucial to cracking the case.

"Did she have a temperature?"

The mother furrowed her brow trying to summon the specifics. "She was very hot. She had a terrible cough. She had a bad infection in the lungs."

Samaan inquired about the rest of the family, where they lived, and whether the other members were healthy. Everyone else was fine. She asked about the family's recent travels. She asked who did the shopping and the cooking in the family.

"Do you own chickens?"

"No, but sometimes the neighbors' chickens come around," the mother responded, squinting as she thought about the answer.

"Did any of those chickens die?"

"We didn't see any sick chickens. But most of the time we weren't home. We're a working family," she said, lines deepening across her dark brow. Then she added, "My daughter's a nurse at a hospital."

This last disclosure landed like a bombshell. Though Samaan's expression remained placid, her heart fell. If the victim had been a nurse, she might have been infected by another flu patient, signaling that the novel strain had mastered the ability to jump from person to person. Or she might have passed the disease to one of the many people whom she treated on the wards. Was the hospital itself breeding an epidemic? Had it already burst beyond the walls into the teeming quarters of Jakarta? The possibilities were frightening. The prospect of pandemic instantly seemed closer. But Samaan refused to let her anxieties betray her. She pressed on.

"Which hospital did she work in? Where is it located?" Samaan probed.

"West Java."

"Which department did she work in? ICU? Pediatrics?"

"I don't know," the mother said. "She's a regular nurse."

As Samaan wandered down the narrow concrete alley to the victim's home, her mind was humming with questions. First, what was the culprit? She wanted to uncover the precise source of her victim's infection to determine how the sickness spread and whether others were at risk. Had the mercurial virus changed its modus operandi? If so, this might reflect a fateful mutation, and the world would want to know. Samaan also needed to find anyone who'd been in contact with the deceased to see if they required medical care. Most urgent of all, was this extraordinarily lethal disease now being relayed from person to person? Was it time to sound the alarm?

The answers lay in the quotidian details of *kampung* life. Samaan intended to sift through the family's habits, peering under their prosaic routines for the tracks of a killer.

That meant delving into the family's medical history, whom they worked with and lived with, how they prepared their meals, and where they slept. She would need to know if they raised animals and how they kept them. She would learn where the family shopped and who butchered the poultry, whether they took traditional medicine or had a taste for liver, eggs, or raw meat, whether they were hunters, farmers, or regulars at the city zoo. To solve this whodunit, Samaan would also reconstruct the victim's final weeks and days. She would weave together the account in the hospital records with the recollections of relatives and coworkers whose memories were now buffeted by fear, confusion, and grief.

Epidemiology is often likened to detective work. But the metaphor is misleading. In reality, disease investigators all too rarely enjoy the satisfaction of collaring their suspect. Clues are few and easily overlooked: a smudge of bird feces on an eggshell, mucus on a feather. Trails go cold quickly. Infected birds may have flown away; victims may have been diagnosed too late and died before investigators could question them. Witnesses forget innocuous but crucial details: a friend's songbird, fertilizer in the garden, or a quick stop at the market. Even when a death coincides with an obvious outbreak in a neighbor's flock, the evidence of infection is often circumstantial. When several family members contract the virus, it's often impossible to conclude whether they all got it from the same sick chicken or shared it among themselves.

If a nascent pandemic strain were circulating in an Indonesian hospital, Samaan would find this out. If it wasn't, she would offer the world a measure of reassurance. But she also realized she might never be able to stamp "Case Solved" on her file. A study of Indonesia's first 127 confirmed cases found that that investigators had failed to come up with any possible explanation for one-fifth of them. At best, Samaan might only be able to suggest a plausible hypothesis for her victim's death. Yet this would leave much of the flu's mystery unrevealed and possible strategies for defeating it obscured.

"Sometimes you can only go so far and stop," she told me. "Sometimes you have to draw the line."

The victim's neighborhood was not as poor as Samaan had expected. Small, tidy dwellings lined the alley with laundry hanging limply out front. Sewage trickled down a ditch beside the alley.

After two minutes, Samaan reached the family home, a simple, middle-class house at the top of the alley. The victim had lived there with her parents, husband, and four-year-old son. Samaan crossed the porch and removed her shoes at the doorway. She joined half a dozen Indonesian officials inside. The air was close, tangy with sweat. They sat cross-legged on a woven mat in the front room, unfurnished except for a large wooden cabinet with glass doors, the walls unadorned but for family portraits. The mother reached up, took down a photograph in a carved wood frame, and passed it around. It was a formal portrait of the deceased and her young boy. The woman was fleshy but attractive with wavy brown hair and thick red lipstick. For the photograph, she had donned a green tunic over a red batik sarong. She was standing with her hands folded before her. At her side was the boy, smart in a traditional Javanese sarong and black *pece* cap.

From outside on the porch came the staccato of a youngster coughing. Samaan looked quizzically at her Indonesian colleagues. None of them were wearing any protection, no mask, no gloves. Samaan asked whether the boy might be infected. Unlikely, the Indonesians assured her. He had been coughing for weeks, long before the victim fell ill, and otherwise seemed healthy. As if to prove the point, the youngster suddenly appeared in the doorway on a bicycle. Besides, the Indonesians explained, he had already tested negative for the virus.

Samaan scribbled a few notes. Then she returned her attention to the victim's mother.

"In the ten days before she became ill, did she go to any parks? Any farms?"

"Nothing," the mother answered.

"A traditional market?"

"Yes, she went to a traditional market. A couple days beforehand, she bought us chicken."

Samaan jotted down the details.

"Did she kill it herself?" Samaan continued, wondering whether the victim had been infected while butchering a sick or contaminated bird.

"It was already killed."

"Feathers or no feathers?"

"It was already cut up and ready to cook."

"When did she go to the market?"

The mother shrugged and turned toward a bushy-haired man sitting quietly in the far corner. It was the victim's husband.

"I don't know," he said softly.

"Before New Year's?"

"Yes, before New Year's," he confirmed.

Rising to her feet, Samaan asked to see the rest of the house. She entered a long room in the rear of the house. There was a refrigerator. She opened the door and peered inside. It was mostly empty. Next she went into the cluttered kitchen, where she instantly spotted a metal wire basket suspended about six feet off the ground. She pulled a pair of rubber gloves from her handbag and tugged them on. Then she craned her neck back, reached up, and gingerly unhooked the basket, placing it on a table. Inside were about a dozen brown eggs. She plucked them out one by one and scrutinized them, looking for traces of chicken droppings that could have contained the fatal dose. Nothing.

"Very clean," she acknowledged. "No sign of feces."

Samaan replaced the basket and snapped a picture of it with her digital camera.

The general lack of filth was becoming ever more disturbing, the absence of livestock unnerving. None of the neighbors seemed to have the wooden cages common in most Indonesian yards for raising chickens. In fact, except for a pair of black chickens lingering in a bush near the block captain's home, there was practically no trace of poultry anywhere in the neighborhood.

"It's perplexing and it makes me a bit worried," Samaan confided to me. "She was a health-care worker. If there's no sign of infection here . . ." Her voice trailed away.


Inside WHO, Tom Grein is a legend. His exploits as an epidemiologist are unsurpassed in his generation. He has ventured to some of the world's most remote corners and confronted some of its most horrifying diseases. When bird flu initially exploded in Vietnam, he was among the first from WHO on the ground in January 2004. When the virus recorded its single largest cluster in the highlands of Sumatra in May 2006, again he was there. Yet Grein readily admits the limits of his calling. "Epidemiology never proves. It only highly suggests," he put it to me. "As frustrating as that might be, not having a smoking gun itself is not failure."

A lanky German with still, blue eyes, a strong chin, and closely cropped brown hair tinged with gray, Grein has a reserved, methodical manner. His colleagues say he can be dour when he's stuck in Geneva and this demeanor has earned him the moniker Gray Cloud. Nothing makes him happier than when he gets to leave the office.

In 2005 he left and headed for southern Africa, responding to the deadliest outbreak of Marburg hemorrhagic fever ever recorded. He drove for ten hours from the Angolan capital, Luanda, into the country's hilly interior, where he discovered villagers dying by the scores. Victims would develop a high fever, crippling headache, diarrhea, and vomiting. Then, as one of the most lethal pathogens of all turned its fury on the liver, spleen, and other organs, blood would often gush into the body's cavities before emptying out through the nose, mouth, eyes, rectum, and other orifices. Barely one in ten survived. As if that weren't daunting enough, Angola was just coming out of a twenty-seven-year civil war and remained unstable and violent. The landscape was still littered with land mines, the few concrete buildings scarred by gunfire. "There was a lot of hands-on work to do in a very difficult climate and very difficult environment," Grein told me. "It took a long time to contain the outbreak."

For nearly a month he and his team remained in the Angolan province of Uige. They went hut to hut carrying out inspections, working inside low, dark dwellings, the air thick with death. They disinfected bodies and buried those who had perished. Their protective suits, double gloves, and face shields were all that separated them from a similar fate. "There's a perception that you're putting yourself quite at risk. Though you take appropriate measures, the brain plays its games," he admitted. All the time, he feared the disease would spin out of control. The international team would contain it in one place, believing they had broken the chain of transmission, and it would erupt somewhere else. They were never able to identify the original source.

Twice before, he had been dispatched to face down Marburg in Africa. He had investigated and ultimately helped control outbreaks in the Democratic Republic of Congo, formerly known as Zaire. In 1998 he had responded to a flare-up in southern Sudan of what health officials suspected was also hemorrhagic fever. Ebola, an equally cruel and contagious cousin of Marburg, had first been identified a decade earlier in an equatorial province of Sudan. Now that same pathogen was believed to have struck a quarter of the people in four isolated farming villages high in the savannah. Two Sudanese medical investigators who first reached the area had gotten ill, and one died.

To get there, Grein and his entourage hiked from dawn until the middle of the night, fording four rivers along the way. "It was like Dr. Livingston with high grass left and right," he recounted. "It was something out of an old movie with people in single file carrying things on the top of their heads." The affliction turned out to be a particularly nasty respiratory tract infection. But not long after, he would confront Ebola itself in the Republic of Congo.

Then there was the time he was airlifted into the mountains of Afghanistan. Once again, hemorrhagic fever was suspected. An outbreak had pervaded three-quarters of the homes in a remote hamlet of northern Afghanistan, which was then controlled by the forces of the storied Tajik militia leader Ahmed Shah Masood. During the summer, it would take a week by mule to reach the village. But the disease broke out during the height of winter and the outpost was cut off by snow. Grein and a WHO colleague flew in by helicopter from the neighboring country of Tajikistan. They treated about twenty patients with antibiotics and collected specimens for further testing. The outbreak was ultimately blamed on an "influenza-like" respiratory infection.

I had known of Grein's reputation long before I met him. When I finally caught up with him in Yogyakarta, a historic royal city on Indonesia's Java island, he was exhausted. Six days earlier, a powerful earthquake had struck just south of the city, killing nearly six thousand people and displacing 1.5 million others. Now the neighboring Mount Merapi volcano was threatening to blow. Grein, who had coincidentally been in Indonesia for several weeks investigating the Sumatra bird flu cluster, was asked by WHO to extend his stay and set up a system for monitoring possible outbreaks of cholera and other disease in the quake's aftermath. He had done the same a year earlier in the Indonesian province of Aceh after it was struck by the massive South Asian tsunami.

"Epidemiology is a different world than lab work," he explained to me over strong coffee. "Most field epidemiologists only understand the basics of what happens in the lab. Laboratory science has advanced so much that it's impossible for field epidemiologists to keep up. But laboratory results without epidemiological information are limited. If you send in lab samples from the field without epidemiological background, they throw a fit."

We were sitting in the restaurant of the Santika Hotel, one of the few hotels in town still functioning and a base for humanitarian operations. Around us, relief workers in blue-and-white vests mingled at the buffet with Japanese soldiers in camouflage. Grein was clad informally in a short-sleeve blue shirt open at the collar, brown jeans, and heavy black shoes. The fine lines across his forehead deepened as he tried to regain his train of thought.

"You can't go by the first piece of information you have," he resumed. "You have to be able to listen and understand, look for the small details, be a little obsessive, and not give up. You have to keep trying to track down the disease." He paused to let the point sink in, then continued. "You also have to understand the cultural sensitivities. You are dealing with a lot of people who are grieving, who have lost family members. You have to strike a balance between pushing your agenda and not upsetting the local population."

During the Marburg investigation in Angola, villagers had stoned his team's vehicles, forcing WHO to temporarily suspend its efforts. The foreigners had initially descended on the community in the back of pickup trucks wearing what appeared to be strange "astronaut suits" and carried off the victims' bodies. The locals were prevented from showing their ritual respect to the dead by washing and embracing the corpses. They suspected the foreigners of stealing their loved ones. Grein said the WHO team ultimately defused the confrontation by putting on their protective gear in front of the villagers and taking the time to explain what they were doing. "It takes trust, transparency, and openness. Without it, it's an uphill battle."

Despite the potential for cultural miscues, Grein said he found the third world an easier place to work. Developed countries often keep WHO epidemiologists at a bureaucratic distance, preferring to run their own investigations. In developing countries short on resources, he added, "we get to work on the ground."

In Asia he stalked bird flu through the tropical jungles of Indonesia, the rice paddies and river deltas of Vietnam, and the deep mountain snows of southeastern Turkey. Other WHO epidemiologists had picked up death's trail elsewhere. Yet for all the months and miles, even the most seasoned investigators had failed to crack the mysteries of the novel strain. "Since it began in Vietnam in 2004, every case was followed up on one way or another but very little is known," he admitted. The precise incubation period remained unclear because investigators were rarely able to determine the source of exposure and thus the exact moment the victim was infected. But that knowledge would be vital for crafting public health policies to contain an emerging epidemic. Nor had they been able to detail how the virus infected people and why some were susceptible while others were not. These elusive specifics could be crucial for limiting human exposure to the disease and developing drugs to combat it.

Though Grein is the best at what he does, even his own flu investigations have often yielded partial success at most. Frequently victims take their personal histories with them to the grave. It's simple to blame poultry for the infection since they're everywhere in Asia, he noted. What's harder is ruling out the possibility that the source was actually another person. "That's always going to be a question. You can't really prove it," he explained.

When investigators manage to break open a case, the novelty makes it striking. Late in the winter of 2006, they'd hit a wall after the virus had suddenly erupted in Azerbaijan, a former Soviet republic situated on the cusp between eastern Europe and western Asia. Of the eight confirmed cases, seven were from one remote farming settlement so poor that it lacked electricity. All but one of these cases were from the same extended family, mostly girls and all between ages ten and twenty. They'd fallen sick within two weeks of one another. It was the largest family cluster of cases yet recorded, and the intriguing pattern promised new insights into the disease if only it could be deciphered. WHO investigators and their Azeri counterparts trekked to the village over and over, yet the family refused to cooperate, even denying the sickness was bird flu. "They wouldn't give us any information about how the patients got infected," recounted Dr. Caroline Brown, a WHO virologist who led the team. "There was no way they were going to talk to us. That's for sure. Even the young children wouldn't talk to us."

The family kept some chickens, but they looked healthy. Other poultry had died in the community, but veterinary officials were claiming it was not bird flu. Brown and her colleagues, however, had heard of a large die-off of wild birds in Azerbaijan a month earlier that was attributed to the flu virus. Next they learned from villagers that dead swans had been found at a lake on a nature reserve near the settlement. At least one member of the stricken family was a hunter, and his neighbors confided that he'd brought home some of the swans. It was an open secret that locals could earn good money by plucking the feathers for making pillows. That was usually the job of teenage girls and young women. But hunting swans was a crime, and the penalty for poaching could run into hundreds of dollars. The family adamantly denied any contact with wild birds, much less infected ones.

As Brown was leaving the family's house after another frustrating interview, she looked around the garden and spied at least three large, white pillows hanging from a clothesline. "You couldn't miss them," she said. "You could see the feathers sticking out." It was the smoking gun. Later, the mother of one victim later confirmed that, yes, all those afflicted had been plucking contraband swan feathers. This marked the first time that investigators anywhere had proven a link between wild birds and human cases of the virus. Yet the breakthrough, like the Azeri outbreak itself, remains exceptional.

"We don't know much," Grein told me with heightened urgency. "All of us, we really need to get our act together to enhance our knowledge and get better tools to fight this disease."


The clock was ticking on Samaan's investigation. Somewhere on the wards of a private hospital in neighboring West Java province, the novel strain could be incubating, amplifying, provisioning for an inexorable march toward the capital, toward the international airport only a ninety-minute drive away, toward the regional hub of Singapore another ninety minutes by air. By the time Samaan made arrangements with Indonesian officials to visit the hospital, it was already Saturday morning, and she hadn't had a chance to line up the minivan. She flagged down a white city taxi on the street near her downtown apartment building. Her translator was off, so she called an

Indonesian colleague from WHO and rousted her with a playful, early-morning plea, "C'mon, let's go save the world."

Samaan's career had been building toward this moment. Born to Iraqi parents, she was raised in Kuwait until the political upheaval wrought by Saddam Hussein's 1990 invasion sent her fleeing as a teenager to Australia. There she went on to study clinical psychology. For a year she worked with refugees detained on the remote South Pacific island of Nauru, a tiny outpost of little more than eight square miles. Many of the migrants, held in camps while applying for amnesty in Australia, were suffering from depression and other maladies. The experience piqued Samaan's interest in broader issues of public health and led her back to the university to take up epidemiology. During her first stint with WHO, she was assigned to Manila just as bird flu was sweeping East Asia. But she was kept primarily in the office. At her desk by 6:30 A.M. each morning, she was tasked to troll the Internet, international media, and the agency's own network of informants for rumors and emerging reports of flu outbreaks. A year later she rejoined WHO and headed into the field.

Now, as the taxi finally escaped the congested streets and pulled up beside the pink walls of the five-story hospital, Samaan was hoping to do better. This was a new facility with a clean, white-tiled lobby that contrasted with the filth and disorder just beyond the gates. No sign of panic or plague, just a few parents waiting to have a doctor examine their children.

The hospital's personnel director, a woman in a pink blazer, led Samaan through the breezy corridors to a conference room. The hospital's vice president, a woman in a Muslim headscarf and brown batik dress, and three other officials joined them around an oblong table. Samaan briefed them on her investigation. She shared the reports of poultry deaths in the victim's neighborhood but noted they remained unconfirmed. "So we cannot discount other possible sources of infection," she told them without revealing her suspicions about the hospital. Then she asked about the victim's recent work history. The personnel director recounted that the victim had reported for the afternoon shift on New Year's Day complaining of fever and chills. She had been initially diagnosed with dengue fever or possible typhoid. "We let her

go home early," the director said. The ailing woman had never returned. Her disappearance, the director said, had left the maternity
ward short staffed.

So the victim, it turns out, had been a midwife. That was a welcome detail. Because the maternity department tended to see healthy
patients, this reduced the chances that the woman had contracted the
illness in the hospital. It would also be easier to spot anyone else in
the ward she might have infected. But Samaan was not yet sanguine.

"Can you look at the hospital records?" she asked. "If you concen-
trate on the ten days before she felt sick, would she have had contact
with anyone who had an influenza-like virus?"

The hospital's vice president said that was an easy one to answer.
She referred to a report in front of her. The victim hadn't come to work
at all in the week before she became ill. She had been on vacation.

"That's not what the family told us," Samaan objected politely.
"They said she was working."

"We'll double-check our records. But maybe she was working in
another hospital?" the vice president responded, suggesting a new,
disquieting possibility.

The personnel director produced a stack of schedules and time
sheets. Samaan and several others huddled around. The documents
showed, in fact, that the family was right: The woman had indeed
reported for work four times during the final week of December. Two
dates in particular drew Samaan's interest. The woman had worked
the overnight shifts on December 27 and 28, just when she would
have likely contracted the virus. Samaan made a mental note.

Glancing up from the time reports, she asked whether any of the
women who had given birth in the maternity ward had gotten sick.
Nothing unusual, she was told. She inquired after the other midwives.
All healthy. She requested to meet a few.

The maternity department was spotless. It had twenty-three beds
in a series of rooms, mostly vacant at the moment. Four newborns
slumbered in small, glass-sided cribs. The private waiting room was
large, with padded chairs, a sofa, and a television on the counter. An
air conditioner hummed in the wall.

The personnel director showed in a pair of midwives dressed in pink uniforms. The younger one, a slight woman with short brown hair and lively eyes, named Swarni, was the chattier of the pair. Samaan asked about the last time she saw her dead colleague.

"I wasn't working the same shift as her that day," Swarni began. "I was leaving after the morning shift, and she was arriving for the afternoon shift."

Samaan nodded.

"I asked her why she was wearing a jacket," Swarni continued. "She said she wasn't feeling well. The next time I saw her, she was in the intensive care unit."

Samaan asked the midwives how they were feeling. They reported their health was good, further allaying Samaan's concern that the hospital was the source of infection. She shifted her line of questioning, asking about the woman's final days.

"Did she do anything before she got sick that comes to mind? Any activities?"

"She mentioned she had been coughing for a while. That's all."

"Did she talk about going to a poultry market?" Samaan pressed, following up on the tip from the previous day.

"Sometimes she would go after work," Swarni recalled. The young midwife mentioned a particular market in East Jakarta and then giggled softly, covering her mouth. Samaan waited for the rest of the answer, puzzled. "She would go buy chicken feet," Swarni added. "That was one of her favorite foods."


Epidemiologists are often faced with two complementary questions. The first is like the one that was stumping Samaan: Why does someone fall sick? The second question is less obvious, though the answer can be even more revealing: Why *doesn't* someone fall sick?

For each victim laid low by bird flu, there are hundreds, perhaps thousands, who should have been. These are the cullers, the armies of peasants, soldiers, veterinary officers, and day laborers who have slaughtered several hundred million birds across Asia, Africa, and Europe since 2003 in an orgy of bloodletting aimed at exterminating infected flocks and stemming the spread of the virus. They often did so with minimal protection, lacking masks, goggles, and even gloves. Yet a full decade passed after H5N1 claimed its first confirmed victim in 1997 before a single culler ever got seriously sick—a Pakistani man who died shortly after helping carry out a two-day poultry slaughter in October 2007. In Vietnam alone, more than ten thousand people participated in the great massacre without a reported case. This mystery has defied explanation, posing one of the great riddles confronting flu investigators.

Field sampling has revealed that some cullers were in fact exposed to the pathogen. Five South Koreans who helped stamp out infected flocks eventually tested positive for antibodies though none ever became ill, and a lone culler in Indonesia also showed elevated antibodies without any outward sign of illness. So, too, nine workers who helped carry out Hong Kong's mass slaughter in 1997 tested positive, with only one displaying mild symptoms. These findings suggested that bird flu may have spread farther than the flu hunters suspected but that many human cases were asymptomatic. If so, this would not be welcome news. It would mean a greater chance for the strain to stumble across the genetic mutations required to unlock a pandemic. Field investigators turned up other worrisome evidence. Blood taken from two elderly but otherwise healthy relatives of Vietnamese victims in 2005 had antibodies to the virus. A year later, a pair of young brothers in Turkey were also positive despite showing no symptoms.

But these instances proved to be rare exceptions. Efforts to uncover a rash of asymptomatic cases have found none. In one telling study, researchers canvassed nearly a hundred homes in a Cambodian village where a man had contracted the virus days earlier. Poultry outbreaks were widespread. If the virus were adept at causing cryptic cases, certainly some would turn up here. But not a single one of the 351 villagers tested had antibodies. A subsequent study of 674 people in two Cambodian villages where there had been both human and poultry outbreaks revealed that 7 of the villagers, or about 1 percent, had antibodies to the virus, indicating they'd been exposed even if they hadn't had symptoms. For some reason, they were all younger than eighteen. Though this last study suggested some mild cases were being overlooked by health officials, the results hardly signaled a silent epidemic. The enigma of the cullers endured.

At times, epidemiologists have also been stumped by the converse scenario: sick humans without sick birds. Indonesia detected its first human case in mid-2005 and over the following year confirmed that fifty-four Indonesians had the virus. Yet Samaan and her fellow investigators were unable to identify a possible source of infection in a quarter of those cases. During 2007 nearly one-third of the Indonesian cases were ruled inconclusive because there'd been no direct contact between the victims and poultry of any sort. Tjahjani Widjastuti, head of the agriculture ministry's bird-control unit, called the behavior of the virus in Indonesia "mysterious."

One morning in July 2005, just days after the country's first human case was announced, I drove west out of Jakarta and two hours later found myself unexpectedly in a California-style subdivision of palm-lined streets and middle-class bungalows. Health officials had reported earlier in the week that three residents of the town had succumbed to bird flu. The deaths of Iwan Siswara Rafei, a government auditor, and his daughters marked Indonesia's first confirmed fatalities from the virus. I had expected to find a typical Javanese village, with serpentine alleys, ramshackle homes fashioned partly from bamboo, and scrawny hens scavenging in the mud. Instead Villa Melati Mas was a gated bedroom community of bankers, businessmen, and doctors. Neighbors were anxiously trading rumors across the metal fences separating their neatly landscaped yards. Mothers were keeping their children off the quiet streets, and some families were considering whether to pack up their belongings in their SUVs and abandon their homes.

"We've really got a panic attack," said Kresentia Widyanto, a mother of three in a floral housedress. "People have been asking, 'Do we need to evacuate and go somewhere else, to vacate this place?'" For fifteen years, Wiydanto and her husband, a physician, had lived around the corner from Rafei's brown cottage with its pitched terra-cotta roof and sweet purple flowers out front. Widyanto's son was eight years old, the same age as Rafei's older daughter, Sabrina. When the girl was hospitalized a month earlier with a high fever, diarrhea, and cough, word spread quickly. Rafei's second daughter, one-year-old Thalita, developed symptoms days later, followed by Rafei. "I'm wondering why this happened," Widyanto told me while she finished her business with a street peddler selling broccoli and cauliflower. "Can we get it? We're trying to be calm."

But despite a six-week investigation by Samaan and her colleagues from WHO, the Indonesian health ministry, and the U.S. Navy lab, they remained utterly stumped. Even Grein was called in to help. Around the subdivision, a few parrots and other pet birds twittered from cages hanging on porches and balconies. But neighbors did not raise their own poultry, instead shopping at a Western-style supermarket. A solitary pet bird in the neighborhood tested positive for the virus, but Rafei's home was free of contagion. Tests conducted on him and his daughter, coupled with the timing of their deaths, suggested that the virus might have been passed among family members. Yet this was never confirmed, and the original source of infection was never discovered.

Rafei's wife and mother both told me they had no idea how he and the daughters got sick. Rafei was a busy professional who set out early every morning on his long commute to Jakarta's downtown financial district and returned late in the evening, leaving little time for side trips to farms or chicken markets. His wife, Lin Rosalina, eyes red from crying, said she was also certain her children had not come into contact with live poultry. "I'm very sure," she added, switching from Indonesian to English to make the point.

Her family's tragedy also highlighted yet another mystery. In clusters of cases, the virus has targeted blood relatives almost without exception. The disease that struck down Rafei passed over the remaining members of his household: his wife, two housekeepers, and his son. All but the last were unrelated to him by blood. By 2008 there were already more than three dozen family clusters across Asia and beyond, representing about a quarter of all confirmed cases, and in the overwhelming majority these involved blood relations like siblings, parent and child, children and grandfather, or niece and aunt. Rarely did both husband and wife test positive.

One of the largest clusters occurred in January 2006 in Cipedung, a destitute, hard-bitten village along Java's north coast. Unlike in the case of Rafei, there was little question where this family caught the bug. They had two dozen chickens, which regularly straggled into their flimsy bamboo shack, sleeping on the dirt floor beneath the platform beds. In the previous weeks, the virus had raced through this small flock. One by one, the chickens got drowsy and died. When the last six birds developed symptoms, the father, a meatball peddler, helped his brother slit their throats beside a large palm in their front yard. The chickens were plucked and cooked in coconut milk for a family feast. By the time I caught up with the family at a hospital in the provincial capital, Bandung, the father was huddled on a cot wrapped in a gray blanket, haggard and unshaven, under treatment for the virus. In an adjacent room, a teenage daughter with a fever lay sprawled on a bed under observation. Two other children had died before they ever got to the hospital.

The question was why the mother had been spared. Just beyond the doors, in a quarantined waiting room, she kept vigil. "I don't know why I'm healthy," the woman, Buenah, whispered to me. She was a short, frizzy-haired peasant with tired brown eyes, wearing only a surgical mask for protection. For my part, the hospital staff had outfitted me in a white hooded jumpsuit with goggles and a respirator and sent me in alone to speak with Buenah. The rubber gloves were making it difficult to take notes. "I don't have a fever, cough, or symptoms," she related. "I really don't know why not."

Outside the isolation ward in Hasan Sadikin Hospital, her relatives were camped on the lobby floor, spending nights on thin, woven mats, wondering how she had escaped the curse. "It's really incomprehensible to us," said Surip, her husband's cousin.

Could the rest of Buenah's immediate family have had more contact with sick chickens than she did? That's doubtful. Relatives and local agriculture officials explained to me that as a rural homemaker, she was in daily contact with livestock. Could the three children have caught the virus by playing with chickens and then passed it on to their father but not their mother? Relatives and fellow villagers reported that it was Buenah who usually looked after the children and that for days she had carried her ailing son in a sling across her chest. Could Buenah, who complained of high blood pressure, have skipped the repast of chicken and coconut milk? No, she and the rest of the extended family all took part.

During the coming months, as epidemiologists turned up this intriguing pattern of clusters over and over, flu specialists came to suspect some kind of coding in the genes that made some people susceptible to infection and others not. If understood, this could help design ways to slow or even stop an emerging epidemic. But the genetic mechanism has remained unclear. And some researchers even countered that statistical chance alone could account for what appears to be genetic susceptibility.


Markets still made Samaan uneasy. When she first started investigating flu cases in Indonesia, she was always fretting about catching the virus. "I counted the sneezes I'd make," she recalled. As she got better acquainted with the behavior of bird flu, she worried less about catching it at victims' homes or from their families. But traditional Asian poultry markets remained scary places where butchers, birds, and buyers all converged, swapping their microbes among splattering blood and flying feathers. It took courage for Samaan to brave a live market even when she wore a mask. But the trail of her victim's killer now led back to one of the capital's largest, a vast covered complex encompassing several city blocks in East Jakarta known as the Kramat Jati Market.

Samaan had done the math. The victim had started feeling sick on New Year's Day or perhaps a day earlier. The typical incubation period for the virus was believed to be three to five days. During that time, she had worked two shifts at the hospital, December 27 and 28. They were both overnighters. Samaan deduced that the woman would have been hungry when she came off duty at 7:00 A.M. and likely stopped at Kramat Jati, which was on the way home. In the hours just after dawn, the market would have been brimming with fresh produce and live poultry.

"She probably got a big dose of something and got sick," Samaan surmised. She was back in the rear seat of the maroon minivan, and the van was again crawling through Jakarta streets, fighting Monday morning traffic. Samaan sensed she was getting close to her prey. The van passed pickup trucks stacked with plastic chicken cages, loose feathers stuck to the cruddy exterior, then pulled into a parking spot beside titanic sacks of chili peppers. Samaan got out, stepping carefully around the mud puddles, and plunged into the dim aisles of the market. Escorted by two colleagues from the Indonesian health ministry, she marched past the cassette stalls blaring the Indo-Arabic strains of working-class *dangdut* music, past jewelry shops heavy with gold bangles and necklaces, past stalls overflowing with dry goods, and then up the stairs to the second floor, where the *thud! thud!* of a meat cleaver welcomed her.

Samaan sloshed along tile floors slick with water, mud, and rivulets of blood. The footing was treacherous. On the chipped tile counters lay butchered chickens in a row. Their claws extended upward into the twilight of several naked lightbulbs. Toward the back, a few survivors clucked in dissent, their legs bound to makeshift wooden cages. Samaan kept her arms folded tightly in front of her, avoiding any contact.

"If she came here, could she buy the chicken here?" Samaan asked.

One of the health officials nodded.

"Her friends at the hospital said she really liked chicken feet," Samaan continued. "Like that?" She pointed toward a small pile on the counter amid other odd bits and pieces of chicken. Several men were hacking plucked birds into pieces.

Samaan and her colleagues approached another butcher, a husky man in a skullcap. His feet were bare and his pants legs were rolled up to the knees. He was busy grasping birds with his bare hands and slitting their throats. The investigators asked whether the market had been checked for flu.

*Swoosh* went the knife. Blood freckled the man's forearms.

He glanced up and assured them that local veterinary authorities tested the market once or twice a week.

"What did they find?" Samaan asked.

*Swoosh* went the knife.

"It's disease-free," he said curtly.

Samaan wasn't convinced. Local food and health inspectors were notoriously lax when they weren't outrightly corrupt. But she wasn't going to linger any longer. She was reasonably confident she had stalked the infection to its source. She retreated through the maze and emerged into the sunlight.

There, she spied something that instantly made her amend her conclusions. Four peasant women were seated on the blacktop hawking chicken off wooden crates. Stacked behind them were round bamboo chicken cages, all empty. The birds had obviously been butchered right on the crowded sidewalk. Even if the market inside was free of infection, this informal commerce was less likely to be. She paused to look.

"She didn't have to go all the way in there," Samaan speculated, retracing the victim's steps in her mind. "She could have bought the chicken right here. That might be more risky."

Samaan reflected on the fateful morning, reasoning that the woman would have been weary after coming off her shift. She would have been eager to get home quickly. "That's it," Samaan thought. "She probably didn't go inside. She would have bought it right here. There was plenty of potential for exposure."

Samaan climbed back into the van.

"That's my hypothesis," she reported. "Can I prove it? It's impossible."

# CHAPTER ELEVEN

# The Lights Go Out at Seven

When I met the man who might save the world, he was making thirty-eight dollars a month. Ly Sovann was a physician in the Cambodian capital, Phnom Penh. He was full-faced with dark, playful eyes and sloping shoulders. He had a tendency to lecture, and when he did, he would stretch out his arms and gesture with open hands. But he was also quick to laugh, often at his own straits.

The first time I encountered him, Ly Sovann was planted behind an aging metal desk in a tiny room that passed for the headquarters of Cambodia's disease surveillance bureau.

He was the director, responsible for spotting the stirrings of an epidemic in a country where the public health and veterinary systems were so impoverished that experts acknowledged at the time they were probably failing to detect most of the human cases and had no idea how rampant the virus was among poultry. He shared the twelve-by-ten-foot office with the rest of his ten-member team. The room was crammed with four other metal desks and tables, filing cabinets, shelves heavy with bound reports, and five boxes stuffed with the health ministry's stockpile of protective gear, including gloves, goggles, masks, and aprons. There was only enough space for three people at a time, so his staff rotated through. They all shared one Internet line, which was just about the sole way they could follow the inexorable progress of the virus in neighboring countries, and Ly Sovann had secured that connection only after prevailing on the health minister to seek help from the prime minister's office. Even at times of crisis, they could work only until 7:00 P.M. each night. That was when the power in the health ministry was shut off and Ly Sovann had to find his way down the stairs from the third floor and out of the darkened building by the faint glow of his mobile phone.

"We've had over thirty years of war," Ly Sovann said as a small air conditioner sputtered and whined in the window behind him. "We need time to build up our system of public health. We try our best to build up the system for detecting avian flu in Cambodia. Five years ago, it was nothing. Now I have computers, paper, and stationery. It's better."

Still struggling to recover from decades of conflict and political instability, Cambodia's government had only three dollars per person to spend on health care each year despite high rates of HIV/AIDS, tuberculosis, and infant and maternal mortality. The country lacked trained doctors, clinicians, laboratory facilities, referral wards, epidemiologists, and an overall health system tying them together. For a time, the government couldn't even afford to produce radio spots warning about the risks of bird flu.

Out in the provinces, the health system was even more primitive than Ly Sovann's operation. Local clinics couldn't recognize cases of bird flu when the sick came in for treatment, raising the prospect that the virus would evolve into a more pernicious form and spread before anything could be done to stop it. This posed a danger not just for Cambodians but for those well beyond the country's borders. "The chain is as strong as the weakest link," warned Klaus Stohr when I first asked him in 2005 about his concerns over Cambodia and its destitute neighbor Laos. As the virus raced westward during the following year, WHO began raising a similar alarm over the threat posed by sub-Saharan Africa.

The situation would be less dire in some countries confronting the virus, but only by degree. Along the breadth of the battlefront, from Vietnam and Indonesia to Bangladesh, Egypt, and Nigeria, public-health and veterinary services have remained precariously short of the money needed to corner the disease in birds, detect and treat those people who contract the virus, and stem its onward spread.

For the wealthy of the world, geographic distance affords little protection from an emerging flu epidemic. There is no strategic depth, as war planners say. But the danger posed by limited resources goes beyond the shared vulnerability of all countries.

Inequality itself has a corrosive effect on efforts to confront this disease. WHO and its wealthier member states have urged developing countries to battle the novel strain on everyone's behalf. Yet these countries have been told they must do so without any solid assurance they'll get a fair share of antiviral drugs, vaccines, or other medical aid if an epidemic erupts. This despite some projections that a pandemic would take a disproportionate toll on developing countries.

Some Asian countries have done what they've been asked, even as they appeal for more money to do it. Others, at times, have resentfully rebuffed instructions from abroad, vowing to pursue their own national interest even if that puts the wider world at risk. Vietnam, for instance, was so sure it would be neglected in the event of a pandemic that local scientists pursued a homegrown vaccine using unorthodox techniques, though WHO warned that this effort could lead to tragic consequences. In Indonesia an aggrieved government went even further, turning the tables on the developed world. Indonesian health officials discovered that they controlled some of the most precious resources of all—actual virus samples urgently required by WHO's labs to monitor mutations in the strain—and stopped supplying these specimens. Indonesia demanded that its claim to these virus samples be recognized and any benefits, for instance vaccines produced from them, be more equitably shared.

At the very bottom of the heap, Cambodia has been in no position to insist on anything. Beaten down by history, it was already heavily dependent on foreign assistance just to keep from closing down. A full half of the central government's budget was financed by aid.

Ly Sovann was born in Phnom Penh in 1969, the year that the United States began its secret bombing of eastern Cambodia during the Vietnam War. This withering aerial campaign was aimed at eliminating the base camps of the Vietnamese Communists. But the upheaval caused by the four-year bombardment fueled the insurgency of Cambodia's own Communists, the Khmer Rouge, who were fighting to topple the Phnom Penh government allied with the United States. When the Khmer Rouge captured Phnom Penh in 1975 and established their genocidal rule there, the capital was emptied. Ly Sovann's family, like most others, was banished to the countryside.

After the Vietnamese army ousted the Khmer Rouge from Phnom Penh four years later, he returned to the capital, where he went on to study medicine at a local college. This was a break with tradition. Like many Cambodians of Chinese ancestry, his was a family of merchants and traders. So was that of his future wife, and her relatives would later help support him as he pursued his medical passion. With few options for advanced study in Cambodia, he left for Bangkok, where he received a master's degree in clinical tropical medicine. Once he returned, he joined the health ministry. He was promoted to director of disease surveillance after distinguishing himself during the SARS outbreak by crafting an aggressive national response.

That was when he began devising an epidemic alert system tailored for austerity. Ly Sovann told me he realized the one thing Cambodia had going for it was cell phones. They were in wide use because landlines were so rare, and cellular coverage had already reached two-thirds of the country. He'd taken advantage of this rare asset, he told me. Reaching backward to a bulletin board, he pulled down the roster of names and phone numbers he'd been compiling since SARS. The stapled sheets, worn and smudged with fingerprints, listed contacts for scores of health-care workers in Cambodia's cities and all twenty-four provinces. He had cobbled this network together with little more than charisma and extensive personal contacts. "He just knows everybody," a doctor in the local WHO office said to me. But calls cost money. He didn't have enough even to buy gas for his investigators' motorbikes, much less pay their salaries on time. So Cambodia applied for ten thousand dollars from foreign donors to purchase prepaid phone cards to allow local health workers to report suspicious respiratory cases that could be flu.

The effort stumbled at the start. Local doctors missed what would be Cambodia's first confirmed case of avian flu. It would have been overlooked altogether if the victim's family had not brought the twenty-four-year-old woman across the border for treatment in Vietnam, where the health system was more advanced. Though Vietnamese doctors could not save her, they did identify the virus. This first reported case in January 2005 drew intense international concern and several weeks later brought me to Cambodia's southern Kampot province.

When I arrived, I discovered that the woman was not the only one in her family who'd been stricken. I tracked down her father squatting in a sandy lot by the side of the road. With a homemade sledgehammer, he was pounding into place the wooden foundation of a new house. He was barefoot, and his narrow eyes squinted in the sun. He wasn't sure what was cursing his home in the parched rice fields across the road, but cursed it was. He had also lost a teenage son, he told me, and two others in his family had fallen ill.

The man, Uy Ngoy, related that his fourteen-year-old son was the first to get sick, complaining of a fever, diarrhea, and trouble breathing. The boy was brought to a storefront clinic with peeling paint and muddy tile floors in the local town. The clinician took the boy's temperature and blood pressure. His condition continued to deteriorate. Two days later, suspecting that the disease was somehow caused by an affront to the spirits, the clinician sent the boy home so his family could pray to their ancestors. The boy died soon after.

At the funeral, the boy's older sister had embraced his body. Soon she came down with the same symptoms. The family took her to a slightly better clinic, where an ultrasound scan revealed lung damage, and then across the nearby border for medical care in Vietnam. It was too late. By the time the next two fell sick, Uy Ngoy had lost faith in modern medicine. "I decided to try another way," he recounted. He dispatched them first to a Buddhist priest and then to a witch doctor in the mountains. These family members later recovered.

There was no public education about the virus, and local public health was hardly any better. The health workers at the two local clinics told me in separate interviews that they had believed the siblings had routine pneumonia, common among villagers. The clinicians never thought to report the cases to Ly Sovann's bureau or any other official. After Ly Sovann learned about the woman's death from media reports, he rushed to Kampot with his team and stayed for a week.

Blood samples were taken from family members, villagers were canvassed, and health warnings were broadcast from loudspeakers mounted on motorbikes. Ly Sovann's mobile phone rang relentlessly.

Several weeks later, he returned to the province, setting out from Phnom Penh before dawn on the three-hour drive. He had put on a white dress shirt with sleeves buttoned to the wrists and a dark striped tie fastened with a clip. His black hair was slicked to the side. He wanted to look as authoritative as possible. He had to convince the villagers to start reporting suspicious illnesses. Too many lives could be at stake. When he arrived at a community hall in Kampot town, he set up his laptop computer for a slide presentation and fished his PDA from the breast pocket. Then, over the hum of the ceiling fans, he made his pitch. The farmers, provincial officials, and community activists in the audience were skeptical. Some approached the microphone to question whether bird flu was even real.

Foreign governments were far more unnerved than these locals. Flu specialists worried that cases were being missed and warned of a stealth outbreak in an utterly unprepared land. Over the following two years, foreign donors provided $13 million in emergency assistance. It wasn't a lot, but in a small country it went a long way. Cambodia established rapid-response squads in every province. Composed mostly of doctors, these two-member teams were trained to investigate any suspected human case. Twenty-nine thousand village volunteers were mobilized to report rumors of unusual illnesses. Antiviral drugs and protective gear for health-care workers were purchased and pre-positioned in each province. Building on Ly Sovann's cell-phone network, Cambodia pioneered a program in which local health officials could report a variety of diseases, including unusual respiratory outbreaks, by text message directly to a central computer.

Cambodia soon had more foreign aid for the flu fight than its tiny cadre of medical professionals could handle. Yet the funding was short-term, good only for about three or four years. Cambodian health officials and their WHO colleagues remained anxious. After the emergency funding for bird flu was exhausted, who would pay to keep the rapid-response teams trained and send them to the field? Who would pay to replenish the antiviral drugs as they expired and conduct lab tests on samples? Who would pay to print the flyers and brochures distributed to villagers warning of the virus? Who would pay for the cell-phone calls?

Nearly a year after I wrote about Ly Sovann for the *Washington Post,* I happened across a surprising item in *Vanity Fair* magazine. It was a long piece about bird flu that mentioned this Cambodian hero. Turns out, the magazine reported, the *Post* article had done Ly Sovann some good. The publicity had gotten him a raise. Two dollars more a month.


David Nabarro was an effusive speaker. Yet when it came to money for fighting flu, the world's wealthiest countries got him to bite his tongue. Nabarro was an Oxford-trained physician, a veteran of development efforts in Asia and Africa, and, as a senior United Nations official, he had worked some of the world's most horrifying disasters, including the Indian Ocean tsunami and the genocidal conflict in Sudan's Darfur region. He had been in the UN's Baghdad headquarters in 2003 when it was devastated by a car bombing that killed twenty-two of his colleagues. After all that, he had become the global flu czar. As United Nations System Senior Coordinator for Avian and Human Influenza, he was responsible for coordinating the alphabet soup of UN agencies involved with flu.

He was also the money guy. By 2007, donor countries and international institutions like the World Bank had pledged $2.3 billion toward fighting bird flu and preparing for pandemic. That was a significant sum, equivalent to what I later learned was being spent in the Washington area to replace an aging interstate highway bridge over the Potomac River. About $1 billion of the promised funding had already been delivered by the end of 2007, mostly to poorer countries. But as Nabarro reviewed the figures in advance of an avian-flu conference in New Delhi in December of that year, he realized that even if all the pledges were met, the needs would still far outstrip the available funds. He calculated the deficit would run into billions of dollars.

Some donor countries didn't want to hear it. They had already ponied up at a pair of similar conferences the previous year. They didn't want to want to get tapped for more. "The donors convinced me not to come out with a strong statement about funding gaps," Nabarro told me soon after.

So in a briefing for reporters in New York before flying to New Delhi, he pulled his punches. He told the press that the world's response to the pandemic threat was on the upswing but that "much more remains to be done." He gave no dollar figure. That was the same equivocal message offered by a ninety-one-page progress report jointly released at the time by Nabarro's office and the World Bank.

Less than two weeks after the conference was over, Nabarro sent out a new version of the report. It was practically identical except for two extra pages inserted toward the beginning. They said what Nabarro had not. "There are signs of declining donor interest," the updated edition warned. It disclosed that pledges were tumbling while the shortfall in emergency funding was on the rise. (The number of donors and amount of new money pledged would decline even further in 2008 as a senior U.S. diplomat warned of growing "flu fatigue.") The updated document warned that the immediate fight against bird flu was now short about $1.3 billion. Most of that was money required by countries on the front line. In late 2006, the World Bank had helped estimate the funding these countries would need through the end of 2008. The document reported that nearly half of this was still unmet. Most notably, countries in sub-Saharan Africa still needed $462 million and those in East Asia were short $341 million. Those figures did not even address the longer-term needs that Ly Sovann in Cambodia and his counterparts across the developing world would soon face.

To careful readers of the original ninety-one pages, this staggering deficit should have been no surprise. The report detailed sector after sector where frontline countries were outgunned by the virus. More than one-third of Asian and African countries surveyed said they had no lab capacity to confirm human flu infections. About half of these countries said they didn't have enough antiviral drugs to cover even 1 percent of their population. The situation was equally grim on the livestock side, where many countries said they lacked basic veterinary services and lab facilities. On average, these countries required more than a week to identify poultry outbreaks and notify international authorities, a potentially catastrophic time lag. About a quarter of African and Asian countries reported they had no lab capacity to detect bird flu viruses in poultry. Only a small percentage paid farmers enough compensation to get them to report infected flocks rather than cover them up.

Looking five to ten years ahead, the report concluded, "Adequate financial support for long term technical assistance and for integrated country programs is essential." How much would it all run? Nabarro declined to name a figure. He said an honest number would scare donor governments.

With outside assistance uncertain, some in the developing world have sought their own solutions, their initiatives driven by pride and pragmatism. Professor Nguyen Thu Van was one. Behind the high gray double doors of her Hanoi laboratory, Van was on a quest for the Holy Grail: a vaccine against the novel flu strain. Her team of young Vietnamese researchers scurried around the small, second-floor lab, some nights breaking only long enough to steal a few hours of sleep in some corner of the elegant French colonial building that houses her institute. They were attempting a new, unorthodox approach to accelerate their effort. It involved developing a vaccine strain with fast-growing cancer cells despite the risk of deadly contamination. Two months before I met her in April 2005, the team had successfully tested the prototype on monkeys. The researchers were now preparing to try it on themselves.

At age fifty, Van was a veteran vaccine scientist with a proven record that had won her the top post at Vabiotech, a pharmaceutical company affiliated with Vietnam's National Institute of Hygiene and Epidemiology. There was little about her appearance to suggest Van's august place in her country's emerging drug industry. She dressed simply in a plaid sweater over black slacks, with her plain, straight hair held back by a barrette. She had a quiet confidence to match. But her brown eyes gleamed when she predicted that Vietnam would soon outpace more advanced countries by developing the world's first effective vaccine against the virus. She offered a warm smile that tempered the audacity of her boast.

Flu specialists outside Vietnam did not share her enthusiasm. To outsiders, her unconventional methods looked like reckless endangerment. Yet to Vietnam, already facing recurring bird flu outbreaks, it was necessity, a clear case of national security. "We cannot wait," Van told me.

Few pharmaceutical endeavors could save as many lives as the development of a pandemic flu vaccine. Yet the obstacles are many. For years, international research into any kind of flu vaccine languished, in large part because government investment was directed toward AIDS and other diseases considered more pressing. The drug industry had no incentive to put its own money into influenza at a time when anemic public demand even for seasonal flu shots often fell short of existing supply. As a result, the technique for making flu vaccines, an unwieldy process that uses fertilized chicken eggs, remained largely unchanged since the 1950s. Since bird flu began proliferating in Asia, researchers have been chipping away at the challenge. But without significant scientific advances, vaccines against a pandemic strain will only become available long after it has circled the globe because of delays inherent in the technology. An analysis in 2007 found it would take more than half a year after the pandemic strain emerged for the first doses to be delivered. This reflects a timeline that includes about a month for WHO to isolate and distribute the seed strain, a month and a half to ramp up manufacturing, and another four months to produce the vaccine and release it for use. And even after the initial doses debut, most people will still have a long wait because of the world's modest production capacity.

The H5N1 strain has been particularly nettlesome. The virus has continued to rapidly evolve, spinning off various subtypes that confound efforts to develop a single vaccine in anticipation of an epidemic. The strain has also proven unusually resistant to experimental vaccines. This means higher doses are required to produce immunity, and the higher the dose, the less vaccine there is to go around.

Scientists have been experimenting with new ways to get over these hurdles. One has centered on using cells, rather than chicken eggs, to incubate vaccines. This could cut the production time for a pandemic vaccine in half while increasing available doses. Initial clinical trials of a bird flu vaccine made by Baxter Bioscience using this technology showed promising results, according to a 2008 report.

Another approach has centered on the use of chemicals called adjuvants, which enhance the effectiveness of a given dose by stimulating a person's immune response. In 2007 the Belgian drug company GlaxoSmithKline reported the results of a study showing that an experimental vaccine against bird flu containing an adjuvant worked at lower doses than even seasonal flu shots. Later that year, WHO announced that these boosters could radically increase the global supply of pandemic vaccines. The agency predicted that by 2010, the world might be able to produce enough to immunize 4.5 billion people per year. But as WHO acknowledged, this would still fall short of what would be required to protect everyone on Earth.

Even with these scientific advances, Vietnam has long suspected it would be at the back of the line if it waited for someone else to come up with a vaccine. Health officials in Hanoi—as well as those in Bangkok, Jakarta, and other Asian capitals—often told me they were sure industrialized countries would look after their own people first. And even if there were enough to go around, who could afford it? The drug industry "can't provide vaccines to the world free of charge," stressed Wayne Pisano, chief executive of French vaccine-maker Sanofi Pasteur.

Van had no illusions. "If another country develops this vaccine, the cost will be very, very high," she explained. "Vietnam is still very poor and could not afford a vaccine with a high cost. We need to provide this essential vaccine through local production at an affordable cost."

She had been down this road before. In 1997 she had been involved in developing a local vaccine against hepatitis B, which at the time infected at least 15 percent of all Vietnamese. The availability of this low-cost alternative to imported vaccines allowed Vietnam to immunize millions of children against the potentially fatal disease. Five years later, using genetic engineering, Van's company produced a new generation of vaccines against both hepatitis A and B. She crowed at the time that Vietnam was one of only three countries able to make them. The potential savings were tremendous. At sixty cents per dose, the locally produced hepatitis A vaccine cost less than one-fifteenth the price of those on the international market.

In the 1970s, Van had studied biochemistry in the Soviet Union before returning to get her doctorate in Vietnam. She had done subsequent training in Japan, Russia, and twice at the CDC in Atlanta. For most of her three decades at the institute, hepatitis had been her specialty. But after bird flu erupted in 2003, one of her mentors, a senior Vietnamese virologist who had earlier developed the vaccine that helped eradicate polio in their country, suggested that Van tackle influenza. He told her to hurry. "It was difficult at the beginning, because we did not have the experience," she recounted, smiling and laughing softly. "But it isn't really new for me. Any vaccine has similar steps and similar techniques."

With Japanese assistance, Van's researchers obtained samples of the flu virus and then engineered a prototype vaccine strain by reverse genetics. The team began growing vaccine in monkey kidney cells, the same method the institute had used in making hepatitis A and polio vaccines. But before clinical trials could begin, WHO asked Vietnam in February 2005 to apply the brakes.

A special WHO delegation, which included officials from agency headquarters in Geneva and the U.S. Department of Health and Human Services in Washington, was sympathetic to Vietnam's predicament. "No licensed H5N1 vaccine for human use is available from vaccine manufacturers and future availability to Vietnam is doubtful," the mission wrote.

But the visiting officials objected that Van's approach was too hazardous. Her team had flouted international guidelines by using monkey kidney cells, which were unapproved for making flu vaccines and might allow the virus to mutate into epidemic form. The use of cancer cells to accelerate the growth of the vaccine strain could introduce another fatal ingredient. Moreover, her lab lacked strong enough safety measures to ensure that the new, genetically engineered strain would not escape. The WHO mission also raised "serious ethical reservations" about the institute's plan to ask its own scientists to volunteer as guinea pigs. The issue was coercion. "There are concerns that the volunteer 'spirit' may not be universally shared and some volunteers may feel uncomfortable and unable to state that for various reasons," WHO wrote.

Members of the delegation later told me they had received official guarantees from Vietnam that it would abandon the program. But Van had been in those two days of meetings and came away with a different impression. "I believe in our procedures and all the laboratory testing," she said. "I'm sure our vaccine is safe. So I'm not concerned."

Clinical trials would go ahead in five months, Van told me. The first phase would involve about twenty volunteers. She would be one of the first. If production stayed on schedule, her company could deliver a half-million doses by the end of the year.

When senior health officials in Geneva and Washington read Van's comments on the front page of the *Washington Post* several weeks later, they were taken aback. The United States dispatched the health attaché at its Hanoi embassy to privately confront NIHE's director and insist that Vietnam make good on its pledge to suspend the program. WHO officials made the same demand in public. Under duress, senior Vietnamese health officials sidetracked the vaccine program, and for a time Van's drive for national self-sufficiency ran aground on the conflicting anxieties of rich and poor.

Finally, in early 2008, clinical trials began. It was the same vaccine, developed in monkey kidney cells. And it was initially tested, as long planned, on researchers at the institute, ten in all. Then the vaccine was tested on what Vietnamese officials described as thirty student volunteers at the country's Military Medical Institute, with larger trials planned. "Good results," Van reported. Her institute planned to start mass production by late 2009. Each dose would cost 30,000 Vietnamese dong, or a mere $1.80.

If Van confronted global inequities with determination and forbearance, Indonesia's Siti Fadilah Supari was far less patient. And on a crisp Geneva morning on a climactic day in 2007, Dr. Supari was already running late. Her scheduled flight on a British airliner had been grounded. But she had been lucky enough to find a later flight to Geneva, this one on Lufthansa. She had flown through the night and, once on the ground, had to wait again, this time for nearly everyone else on board to file out. There had only been room for her at the very back of the plane, even though she was a cabinet member from the world's fourth-largest country.

Supari, Indonesia's health minister, was scheduled to give her speech in half an hour, and she was still on the plane. "Dear God, help me!" she thought. As she descended a stairway from the jetliner and headed finally for the VIP lounge, she was welcomed by a delegation of fellow Indonesians, including her country's ambassador at the UN mission in Geneva. They exchanged pleasantries. Then her colleagues handed the minister a draft of the address she would give at the meeting. When Supari read what they had written, she felt wounded. "I was thoroughly shocked," she recalled. "They had no idea of the core of the problem."

By that time, in November 2007, Indonesia had long since become the epicenter of the bird flu outbreak. More people were dying of the virus there than anywhere else on Earth, and many flu specialists feared Indonesia would ultimately be the source of a pandemic. Yet for more than a year, Indonesia had refused to provide virus samples from the overwhelming majority of its cases to WHO-affiliated labs, leaving the world blind to the evolution of the virus. The danger was grave. If the pathogen was mutating into a pandemic strain, no one might know. This could preclude the world from launching an effort at rapid containment before the virus spread or making emergency preparations in case it did. Without current samples, scientists would be hampered in developing vaccines against the pandemic strain, monitoring the effectiveness of antiviral drugs, or producing updated diagnostic tests to identify the virus elsewhere.

WHO had convened a special meeting of representatives from more than a hundred countries to address complaints raised specifically by Supari over how the developing world shared virus samples and what health benefits it received in return. The speech penned by her colleagues explained that Indonesia's objective was to safeguard the lives of those in poor countries by winning them better access to pandemic vaccines. But Supari's fight had gone beyond that by now. It was no longer a technical dispute over the distribution of vaccines. It was a wider struggle over fundamental inequities in the global health system embodied by WHO. In short, she told her colleagues, she was taking aim at the "oppression between nations." Supari chided them, informing them she had her own draft and that was the one she would deliver.

Supari stopped at her hotel only long enough to change clothes. Then she dashed for the Palais des Nations. Initially erected as the headquarters of the League of Nations in 1919, the monumental building now served as the European seat of the United Nations. The meeting itself was running late, and Supari was allowed to proceed with her speech.

For ten minutes she skewered the long-standing system under which countries like Indonesia freely provided virus specimens to WHO labs and they in turn supplied them to commercial drug makers. The countries of origin were rarely told what purpose their samples were put to. "We do not really know whether they are used for research and publication or they are shared with vaccine manufacturers for vaccine production. Or maybe they are utilized for the development of biological weapons," Supari told the audience. Her hyperbole was startling for those accustomed to the diplomatic palaver of Geneva. She pressed on, lambasting the arrangement in which developing countries were forced to pay market prices for vaccines, even if these countries were the original source of the vaccine strain. "If these oppression practices continue, poor countries will become poorer and rich countries will become richer," she warned. "This is more dangerous than an avian influenza pandemic itself—and even a nuclear explosion."

The gap between rich and poor that had provoked resentment elsewhere was here begetting full rebellion. Nor was this resistance at the margins of global efforts to stem a pandemic. Now it struck at the heart.

Barely three years earlier, few even in Indonesian public health circles had heard of Supari. She had been an obscure cardiologist and medical researcher at a Jakarta hospital. Then, one evening in October 2004, her cell phone rang. She never did find out who the caller was. But he told her that Indonesia's incoming president, retired general Susilo Bambang Yudhoyono, urgently had to meet her. He wanted her to be his health minister. She would be sworn in the following day. "Why me?" she asked. "I am just a woman. The president needs someone who is tough. Am I that?"

What she had were the right credentials. One day before his inauguration, Yudhoyono was still struggling to assemble his cabinet. He'd won a decisive electoral victory, but his own political party was tiny. His government would be stillborn unless he could line up backing from larger political forces. Supari's family had long been active in Muhammadiyah, a Muslim civic organization that claimed about 30 million members nationwide, and its support would be a boon to any politician. So when the group's chairman suggested Supari, Yudhoyono agreed. Besides, she was indeed a woman. The cabinet was short on those.

Supari had graduated with a medical degree from Indonesia's elite Gajah Mada University but, unlike many in the cabinet, had never studied abroad. She was a stout woman with large, round, rimless eyeglasses. She had jet-black hair and, though a devout Muslim, often appeared in public without concealing her bouffant beneath a traditional headscarf. She favored batik dresses and suit jackets, accessorizing generously with gold and pearls. At first she had difficulty being taken seriously as a minister. Jakarta's chattering classes dismissed her as giggly and prone to public gaffes. But she proved shrewd. She honed her public relations, even launching her own Sunday evening television talk show. She also tapped into Indonesia's profound sense of national grievance.

Every way Indonesians looked, life seemed to be getting harder. That was particularly true for their health. The public health system Supari inherited as minister was sorely underfunded and had eroded sharply since the 1997 Asian financial crisis. In one typical clinic in southeastern Sumatra, the director explained to me that he could no longer offer routine immunization against childhood diseases. "For us, it's hard to answer the parents when they ask why the vaccines have run out," he complained. Apologizing for the rat droppings that littered the clinic floor, he shuffled into his tiny, tiled office and opened the rusty clasps on the fifteen-year-old freezer once stocked with vaccines. It was almost empty. He told me that health workers were forced to scavenge for unused syringes in other medical offices or scrape together money to buy their own. The refrigerator used for making ice to transport vaccines into the field was broken. "Money is our unending problem," he said. Already unable to provide basic care across much of the archipelago, Indonesia's health system then suffered a series of staggering calamities within months of Supari taking office: the tsunami that killed at least 150,000 in Sumatra, earthquakes, and a resurgence of polio. Finally, bird flu struck.

Indonesia was fortunate that it started with an isolated outbreak. At the premier hospital in Jakarta for treating infectious diseases, there were enough doses of Tamiflu on hand to treat no more than eight people. "When we have an epidemic, we cry for help," said Santoso Soeroso, a physician at Sulianti Saroso Infectious Disease Hospital, as he gave me a tour of his facility's spartan isolation wing. He acknowledged that he worried there was not enough of the drugs for preventive use by doctors and nurses who would care for flu patients. The balance of the national supply had already been divvied up among thirty-three other hospitals, with each receiving enough for just two patients. Soeroso said he had no budget for any more.

WHO eventually shipped over more supplies of Tamiflu. But these were just limited emergency stockpiles. Indonesia had no money to follow the lead of wealthier nations that were already ordering sufficient quantities of the drug to treat as much as half their population should the disease spread beyond Asia. Supari learned that Indonesia would have to wait months to buy more Tamiflu if it wanted any, and even then would have to pay up to forty dollars per treatment. She was flabbergasted.

In November 2006, David Heymann and Keiji Fukuda came calling in Jakarta. Heymann, who had helped stand up WHO's global strike force, was the agency's new assistant director general for communicable diseases. Fukuda, who had helped quarterback the world's response to the Hong Kong bird flu outbreak in 1997 and the Vietnam outbreak just over six years later, had been tapped to become WHO's new influenza chief. The agency was planning to spell out in an official resolution what had long been the informal process for sharing virus samples with WHO, and the pair wanted to run it by Supari.

What she told them in response brought them up short. Indonesia would no longer share, not at all.

Behind the scenes, Indonesia had already been locked in a running dispute with foreign scientists over access to flu virus samples. For a year, researchers from the U.S. Navy lab in Jakarta had been engaged in what an Indonesian health officer called a "cat-and-mouse game," trying to collect specimens in hospitals and stricken villages despite government efforts to stop this. "It was so difficult to get the damn virus out of Indonesia and analyzed," recalled the navy lab's influenza chief at the time. "You had to go around the system." Neither side ever publicly disclosed this subterranean contest, but the competing claims to local virus samples silently poisoned relations between Indonesian and foreign scientists.

The U.S. Naval Medical Research Unit 2, or NAMRU-2, was established in Jakarta in 1970 to help U.S. military forces research diseases they might encounter in the tropics. Along the way, the lab worked on maladies that also afflicted the local population, including malaria and dengue fever, and eventually set up a system to monitor for seasonal flu viruses in collaboration with six Indonesian hospitals. NAMRU was the most sophisticated infectious-disease lab in Indonesia, the only one with safeguards required to fully analyze pathogens like bird flu, and its staff, largely Indonesian, was the most technically proficient in the country.

In 2000 the agreement between the United States and Indonesia authorizing the lab to operate ended. But for several years, the lab continued its activities with the approval of Indonesian officials and even expanded its flu surveillance network to cover twenty hospitals on most of the country's main islands, stretching from Papua in the east to Sumatra in the west. In return for supplying virus samples from patients, Indonesian doctors were paid a monthly stipend by the lab. It also gave them money to come to training sessions, with daily allowances that often exceeded the actual expenses. The payments to each doctor could come to several hundred dollars a year, a generous sum in Indonesia. Plus, NAMRU helped the doctors pay for equipment, such as microscopes and refrigerators for storing specimens. As long as the navy lab was focused on seasonal flu, few Indonesian officials objected.

In July 2005 that relationship changed with the country's first recorded case of bird flu. NAMRU had dispatched Andrew Jeremijenko, the leader of its influenza surveillance project, to investigate the outbreak along with Gina Samaan of WHO. Jeremijenko already had a relationship with the pathologist at the local suburban hospital through the flu network and was able to secure samples of this new, mysterious virus taken from the victims. It was NAMRU scientists who initially identified H5N1 in Indonesia. But Indonesian officials barred Jeremijenko and his colleagues from disclosing these results. Moreover, the government insisted bird flu was different from seasonal flu and thus not covered by the protocol allowing the lab to conduct research. To the navy lab, the distinction was preposterous.

"We were doing human influenza surveillance in the hospital and that's all influenza. It doesn't matter what sort of influenza it is. The problem with H5N1 is that it is a political disease," recalled Jeremijenko, an Australian doctor who worked on contract at the lab between 2004 and 2006. "We were walking on eggshells. The whole time, I was afraid I was going to get thrown out of the country."

Indonesian officials were determined to keep control over both the samples and any findings about the disease's evolution, which were potentially staggering for the country's economy. The government tried to block NAMRU staff from investigating subsequent outbreaks. Then, in late October 2005, a senior health ministry official informed health agencies and more than two dozen hospitals around the country that the navy lab would have to cease all activities at the end of the year. No longer was any agency or hospital permitted to collaborate with the lab.

But the lab persevered. It continued to offer payments to doctors in the hospital network, and they continued to send samples. The lab also continued to dispatch staff to various hospitals, where they collected even more virus specimens and related material such as copies of X-rays, often beating government health officials to the site. "We had to push these things," Jeremijenko said. "We couldn't do anything in Indonesia if we didn't break the rules sometimes." On several occasions he and his colleagues even eluded government restrictions to visit the scene of outbreaks, arriving under the radar in a private van. It was always on a Sunday, when Indonesian officials were scarce. "We weren't officially anywhere," he put it to me. "We didn't have the right to go sample and investigate anything. We did it surreptitiously."

Indonesian health officials weren't completely in the dark. After one outbreak in Sumatra, a stricken man resisted giving mucus and blood samples to government disease investigators. "Some of you already took my blood," the patient told them, one official recounted. "Why would I give two times?" NAMRU had already been there. And when a girl on the eastern island of Sulawesi fell sick, her family refused to give samples to government health investigators, another official recalled. NAMRU staff, accompanied by a team from the provincial hospital, had already been through the village gathering specimens.

"They manipulated our agreement," a senior Indonesian health official later fumed. "They think they can just go to the field without proper procedure and violate the protocol."


Three months after their initial visit with Supari, Heymann and Fukuda were back at Indonesia's health ministry, a large, modern structure on a downtown boulevard lined with bank towers, embassies, and hotels. The meeting was off to a late start. That day in February 2007 was a Friday, the Muslim Sabbath, when everything in Jakarta runs even slower than usual. Supari and her staff had temporarily disappeared to make their midday prayers while the visitors kept their eyes on the clock. They were scheduled to fly out of Jakarta later that same day. When they were finally ushered into the minister's spacious second-floor office, there were far more people gathered around the large wooden coffee table than three months before. The dispute over virus sharing had escalated.

Supari had been stunned weeks earlier to learn that an Australian drug maker, CSL Limited, had used an Indonesian virus strain provided by WHO to manufacture an experimental human bird flu vaccine. "I never gave permission to any Australian company to produce a vaccine based on the Indonesia strain of the virus," she protested.

Even before that, she had opened talks with U.S. drug maker Baxter International over acquiring 2 million doses of bird flu vaccine for Indonesians. Baxter had offered to sell the vaccine, which was based on a separate H5N1 strain from Vietnam, but at the steep market price. That was when Supari gained her first glimpse of what she called the "neocolonialism" of WHO. "The situation is ironic," Supari later told me. "The virus obtained by vaccine manufacturers came from dead Vietnamese people who had been grieved by their brothers and sisters and parents, and then it is commercialized by other nations without compensating Vietnam." Supari had made Baxter a counteroffer. Indonesia would give permission to Baxter to develop a vaccine from the Indonesian subtype in return for 2 million vaccine doses and help in establishing a vaccine plant in Indonesia itself. Indonesia and Baxter signed a memorandum of agreement just days before Heymann and Fukuda returned to see the minister. The suggestion from Jakarta was that no one else would get access to the Indonesian strain, now the deadliest on Earth, unless they met Supari's demands.

For half a century, countries had freely provided samples of circulating flu viruses to WHO's collaborating labs under a system called the Global Influenza Surveillance Network. These routine viruses were analyzed by WHO each year to predict which would cause the next round of seasonal flu. Then the seed viruses were turned over to drug companies to make annual flu vaccines. The developed world has been home to both the labs and the drug makers. But since most of those getting flu shots were in Temperate Zone industrialized countries anyhow, poorer nations had paid the process little mind. Now, facing a possible pandemic, Supari was challenging not only this traditional system but the integrity of WHO itself. Her attack was unprecedented in the agency's history. Further, she was making the claim that viruses were biological resources owned by the countries where they circulate, not public health information that must be shared freely with the world. She insisted that WHO acknowledge Indonesia's sovereignty over its viruses and sign a material transfer agreement, which would limit what could be done with Indonesia's virus samples and potentially entitle it to compensation if they were used commercially.

But senior WHO officials feared that such conditions could wreck the global surveillance system. Flu specialists depended on unencumbered access to virus specimens to watch for menacing twists in viral mutation, keep the tests for the virus updated, and ensure that vaccine research remained current.

"WHO has become a target for Indonesia," Fukuda later lamented. "That's not us. We support Indonesia."

There were already signs the Indonesian rebellion might spread. Thailand, for one, had recently signaled its sympathy in remarks at a WHO executive board meeting. Heymann and Fukuda aimed to keep the disagreement from escalating any further. They wanted to frame the debate around proposals for reform, like making the virus-sharing system more transparent and enhancing the access of poorer countries to vaccines and vaccine production technology. Heymann balked at Indonesia's more radical demands. The meeting was deadlocked, and yet they kept talking and arguing for five hours, right until the last minute when the visitors had to leave for the airport.

As Supari escorted them from her office, they were unexpectedly confronted by a pack of reporters. She hadn't bothered to tell her guests she'd invited the media. It felt like an ambush, and two things immediately became clear: They would miss their flight. And there was now little hope of keeping the dispute from boiling over into a heavily publicized political row that could sabotage global cooperation between rich and poor.

Heymann struck a diplomatic note for reporters, telling them,

"Indonesia's leadership alerted the international community to the needs of developing countries to benefit from sharing virus samples, including access to quality pandemic vaccines at affordable prices." He said he hoped Indonesia would soon resume sending its specimens to WHO.

Supari turned and glared at him. "Without the new regulations," she said, "until our last drop of blood, the answer is still no, no, no!"

It was time for Margaret Chan as director general to weigh in. She could speak with unquestioned authority for WHO and, by virtue of her position, demonstrate to Supari that she was being accorded all due respect. Chan, moreover, could connect with Supari as an Asian woman. But if Supari wanted to rumble, Chan was an adroit inside fighter well seasoned by Hong Kong's contentious politics.

At the end of February 2007, the women spoke by telephone. The conversation seemed to end in agreement, with Supari pledging to begin sharing samples. But within hours, they were disputing what they'd agreed to. Supari told the press that Chan had acceded to demands that specimens be used only to monitor the evolution of the virus and not for making commercial vaccines. WHO officials adamantly denied the account. No samples were sent.

A month later, Chan made an unscheduled detour from Singapore to pursue the issue with President Yudhoyono himself at the Istana Negara palace in Jakarta. The Indonesian leader stressed that international drug companies had to make affordable, high-quality vaccines available to developing countries. Chan agreed, vowing to do what she could. Afterward, Supari, who had attended the talks, came out and announced the disagreement was essentially resolved. "We will resume the sending of virus samples for the sake of global interests," she told reporters. "The delivery will take place this year, within two months from now at the latest." Again, it didn't happen.

And so it continued over the months, with each side accusing the other of bad faith. One grueling negotiating session followed the next, often grinding far into the night and deep into the fine print of international law. In public Chan would at times express sympathy for Indonesia. "I believe the developing countries are right to ask us to address the issue of equitable access now," she told a meeting of drug industry executives and laboratory directors. "To date, developing countries have suffered the most from this virus." But in private, she was increasingly exasperated. Sometimes she let it show. "If you do not share the virus with us, I want to be absolutely honest with you, I will fail you," Chan warned ominously during a speech at the annual meeting of WHO member states in May 2007. She didn't name Indonesia, but she didn't have to. "I will fail you because you are tying my hands. You are muffling my ears. You are blinding my eyes."

The prospect of a flu pandemic had deepened the fissure between rich and poor, which in turn was subverting international efforts to head off the coming plague. Poverty had always hamstrung the fight against disease, leaving humanity vulnerable to epidemics that arise in the developing world. The emergence of AIDS from Africa was but one tragic example. But now the world was also at risk because Indonesia had realized that viruses were a resource potentially more prized than oil and could be used to press its claims against the wealthy. The challenge represented such a threat to global health that some veteran observers suggested the UN Security Council might ultimately be asked to break the impasse.

When the two women met next, in November 2007, the Indonesian minister was coming off her impassioned speech in Geneva's Palais des Nations. The jet lag and drama were taking their toll. "I was exhausted," she recounted. But at the moment, it was Chan she pitied. "I knew I made her feel uneasy," Supari later said. "My move had made her tremendously busy."

Supari steeled herself for their private encounter, telling herself to be strong. The two women hugged tightly. Then they started in, speaking in little more than whispers.

Chan urged the minister to trust her.

She did, Supari answered. "But I don't trust the WHO system."

Chan appealed for understanding, explaining that she and Heymann were trying to transform the antiquated system they'd inherited. She needed Supari's help to do it.

Supari was struck by Chan's apparent good will. But the minister refused to bow. "I tell you once again, Madame Chan, my trust in the

WHO will resume only if they materialize a new mechanism which is equitable and transparent. We, the Third World, have long been suffering from the inequity, Madame Chan. It is time to change." Supari felt compassion for Chan, an Asian sister who she believed had been forced to do the bidding of rich, powerful countries like the United States. "She was a brave warrior, like me, actually," Supari thought. "But she had to give up her principles, her conscience, to keep her position in the WHO."

The minister apologized, repeating that Indonesia would not compromise on its demands.

"I need a total and fundamental transformation, Madame Chan. And I will continue to speak about this all over the world," she continued. "I am your friend and I do not deliberately put you into this difficult position. However, the bigger concern of humanity made me do this. I am very sorry, Madame Chan."


The dispute metastasized much as senior WHO officials had feared. The distrust engendered by the clash over virus sample sharing and associated benefits infected other facets of the effort to contain bird flu. Moreover, during 2008 and on into 2009, Indonesia's grievance grew into a crippling distraction. Jakarta continued to press its campaign on behalf of the world's underprivileged while largely ignoring a virus that was assiduously putting down ever deeper roots across the Indonesian archipelago. Other governments were entangled in the diplomatic quarrel with Indonesia while frittering away years crucial for their own pandemic preparations.

As the negotiations dragged on, the United States became Indonesia's main interlocutor. In their respective capitals, the thinking was that if two governments that were so far apart on the issue could find a resolution, everyone else would agree. But other countries, such as Brazil, India, and Thailand, were now starting to agitate for sovereign control over their viruses, suggesting that a deal with Indonesia alone would no longer settle the matter. At the same time, the disagreement between Washington and Jakarta was becoming ever more shrill. Then Supari released her book.

In early 2008 she published *It's Time for the World to Change: In the Spirit of Dignity, Equity, and Transparency, Divine Hand Behind Avian Influenza.* Drawn from her diary, the book was a blistering critique of the industrialized world and WHO's global system for virus sharing. The most sensational attack was saved for the United States: her claim in the first chapter that the U.S. government could use bird flu samples to fashion weapons of mass destruction. Her evidence was that genetic data from some virus samples was stored in a database at Los Alamos National Laboratory, a U.S. government lab that conducts advanced research on such diverse subjects as national security, climate change, traffic management, and disease dynamics.

"It was the same laboratory that designed the atomic bomb to destroy Hiroshima in 1945," Supari wrote. "It is likely that they utilize the same facility to research and develop chemical weapons. What a terrifying fact! The DNA sequence data had been the privilege for the scientists in Los Alamos. Whether they used it to make vaccine or develop biological weapons would depend on the need and the interest of the U.S. government."

The notion that the United States would weaponize bird flu is ludicrous to most Americans. "I think it's the nuttiest idea I ever heard," quipped U.S. Defense Secretary Robert Gates. But in developing countries, some were inclined to believe the worst, and in some quarters of Jakarta, Supari's book got a rapturous reception.

The dispute over virus sharing also spilled over into negotiations over the status of the NAMRU lab. The U.S. and Indonesian governments, which were quietly discussing a new agreement to authorize its activities, had been hung up over the finite question of how many U.S. personnel at the lab would be accorded diplomatic immunity. Full of fury, Supari abruptly announced in 2008 that it was time to expel the lab altogether. "NAMRU-2's presence is of no use to us," she averred in June of that year. "In fact, its operations are an encroachment on our country's sovereignty." She urged the Indonesian parliament to close it down. One lawmaker called for a probe into reports circulating among Indonesians that the navy lab was a front for spying. The fate of the most advanced disease lab in Indonesia grew even more uncertain.

Next Supari disclosed that the government would no longer announce human cases of bird flu on a routine basis as before. Days after she made that declaration in June 2008, she assured me Indonesia would still report the cases to WHO. But no more would the government release details to the public with each new infection. "The families of victims of avian flu have very, very fresh wounds," she explained, and the practice of announcing cases had been "very insensitive to them."

Maybe so. But yet another source of vital intelligence had been choked off. Indonesia had severed the supply of virus samples to the world's labs. It had shut down the one lab inside the country capable of fully analyzing specimens. And now the government had deprived flu hunters, and Indonesians themselves, of information needed to confront a budding epidemic. Supari promised that all details would be released in time. But time is the most precious commodity when it comes to flu.

The fracture between haves and have-nots now yawned wide. Even as the novel strain increasingly marked its territory in Indonesia, the world became ever less able to chart its progress.

# CHAPTER TWELVE

# Peril on the Floodplain

The figures first appeared on the ridgeline. They emerged one after another from behind a bluff in the middle distance, more than a dozen of them. Then, like ants, they started down a dirt track hewn from the lush, sculpted mountains separating Vietnam from China. The descent was steep, the footing treacherous. The slopes above were densely forested. But the trail itself was broad and exposed, the deep brown earth well trampled. As the smugglers drew closer, their stooped forms became visible in the afternoon light. Their backs heaved under the weight of their freight. Soon the cargo came into view. They were hauling bamboo cages crammed with live, bootleg chickens.

On the paved road below, two young men waited, mounted on a pair of red dirt bikes. They were lookouts. My Vietnamese driver had pulled our car to the side so I could check out the foot traffic coming over the border. Now, as I stood beside the guardrail, staring past a cornfield up into the craggy cliffs, I was the one being checked out. The two sentinels revved their engines and brazenly approached, slowing briefly as they buzzed by. Several dozen yards away, they stopped their bikes. One man produced a two-way radio and barked into it. Though his words were inaudible to my translator and me, within moments the figures on the slopes above began to shift to the edge of the track and melt into the surrounding brush. Yet almost instantly, more traffickers appeared over the ridge from China. Even more were bounding down a second path about a hundred yards to the left. This trail was narrower, largely concealed by banana palms and other trees. These smugglers apparently figured I couldn't see them and continued their progress undeterred by the alarm.

I had come to the village of Dong Dang escorted by two provincial officials. Now, with some urgency, they were inviting me to get back in the car. The smugglers were a violent lot, known to set upon outsiders with stones, even guns. In recent weeks, the traffickers had battled soldiers dispatched to intercept them. In one case, five troops had been injured and their car destroyed. I quickly understood who had the upper hand. "You can put helicopters up there, really mobilize the army and put all kinds of resources in, and it would still go on," Jeffrey Gilbert of the UN Food and Agriculture Organization told me after I returned to Hanoi. "It's like the Ho Chi Minh trail."

Every day, despite an import ban, the smugglers were hauling more than a thousand contraband chickens into Lang Son, one of six Vietnamese provinces along the untamed Chinese frontier. In Lang Son alone, the jagged border runs for 150 miles through angular, misty mountains that seem drawn from a stylized Oriental painting. The highest peak, Mau Son, rises nearly 4,500 feet. It also lends its name to the local rice wine, widely sold in unmarked, five-hundred-milliliter plastic bottles for about sixty cents each. My government escorts had knocked back enough shots over lunch that it had taken some of the edge off their anxiety.

For centuries, extended tribal families straddling the border have navigated highland footpaths to run goods from one side to the other. In recent years these had come to include electronics, DVDs, exotic wildlife, and all sorts of clothes and shoes. The illicit poultry business had turned lucrative in 2004 after Vietnam began slaughtering about 50 million chickens to contain its bird flu epidemic. The resulting shortage of chicken meat, a favorite protein source for the Vietnamese, sent prices soaring on their side of the border. But with the increase in the illegal poultry trade, the traffickers had also unknowingly—and repeatedly—smuggled the virus from its source in southern China into Vietnam, at times introducing altered strains that bedeviled efforts to contain the outbreaks.

Two months before my visit to Lang Son in July 2006, Vietnamese veterinary officials had disclosed they'd identified the virus in a sample taken there from smuggled chickens during a bust on the border. Two years later, provincial health authorities reported they were discovering H5N1 in nearly a quarter of all the illegally trafficked chickens they were confiscating. Researchers had already uncovered lab evidence implicating cross-border commerce in spreading the disease. In mid-2005 they had isolated a strain of the H5N1 virus in Vietnam that was entirely new to the country—different from the subtype that had burned through farms starting in 2003 and killed dozens of people—but similar to one found months earlier in China's Guangxi region, just over the mountains from Lang Son. Another study published in 2008 found genetic evidence that the virus may have been introduced from Guangxi to northern Vietnam on "multiple occasions," most likely by poultry trade.

An average of 1,500 birds came over the mountains into Long San each day, provincial officials reported. Along the entire Vietnam-China border, the total could run well into the thousands. The syndicates running the smuggling rings were paying local villagers about thirty cents a bird to haul the contraband along mountain trails that could snake for more than ten miles. Some smugglers, especially women and children, could carry only a few birds. But hardy highland men lugged as many as twenty at a time. Their earnings could far outstrip the salaries of animal-health officers, inspectors, and others charged with stemming the commerce.

Once the smugglers came down from the slopes, they often transferred their haul to motorbikes, which ferried it to local farms serving as transit depots. From there, the chickens were loaded onto trucks for transport, in many cases to the markets of Hanoi, five hours away, and points even farther south. The smugglers were repeatedly seeding new outbreaks, and each outbreak was affording the virus a new chance to ensnare human victims and, even more ominously, mutate. This was how the novel strain continued to press its offensive.

Do Van Duoc was the director of animal health in Lang Son, a friendly man with full cheeks and silver hair that sat atop his head like a mushroom cap. He explained it would be nearly impossible to stem the smuggling as long as prices on either side of the border were so different. On average, chicken that sold for thirty cents a pound in China was fetching a dollar or more in his country. But that wasn't the whole explanation. He accused Chinese farmers of unloading chickens from areas struck by bird flu at bargain-basement prices.

China's agriculture ministry confirmed for me that poultry was being illegally transported into Vietnam. An investigation by Guangxi animal-health investigators had discovered three clandestine routes originating in different areas adjacent to Lang Son. But Chinese officials, true to form, denied that any birds coming from their side of the border were infected. Duoc wasn't buying that. "We have evidence," he told me. "We've tested and we can prove there's H5N1."

The farther influenza goes, the closer it comes to hitting the microbial jackpot. Extent means opportunity, more chances to mutate or swap genes. By 2009 the virus had stricken birds in at least sixty countries and spread to people in fifteen of them.

The strain made its debut in Europe in October 2004 when customs officers at Brussels airport discovered two infected eagles in a passenger's hand luggage. A Thai traveler had smuggled them from Bangkok, wrapping the creatures in cotton cloth and shoving them headfirst into a pair of two-foot-long woven bamboo tubes. Then he had tucked these into an athletic bag, left unzipped slightly so the birds could breathe. His delivery had been destined for a Belgian falconer who had paid nearly $1,900 for the pair. They would have arrived unnoticed but for the passenger's bad luck. He was stopped and searched as part of a random drug check. "We were very, very lucky," Rene Snacken, the flu chief at Belgium's Scientific Institute of Public Health, said at the time. "It could have been a bomb for Europe."

But before long the virus indeed exploded out of East Asia. The startling outbreak among migratory birds at China's Qinghai Lake in April 2005 had left Hong Kong's Yi Guan and other scientists wondering where the virus might next wing. Over the following year, the disease struck birds in more than forty countries in Europe, Africa, South Asia, and the Middle East, and each time researchers checked, they found the distinctive genetic signature of the Qinghai subtype. By that winter, a dozen people had fallen sick in Turkey, four fatally, bringing human casualties to Europe's doorstep for the first time. Wild birds and domestic poultry were succumbing in more than two dozen European countries. Panic spread. In Paris, the famed bird market on the Île de la Cité removed all live fowl, and France's annual livestock festival took the unprecedented step of banning poultry. In Britain, where legend claims that the monarchy will survive only as long as the ravens at the Tower of London, these celebrated birds were locked up for their own protection for the first time in history.

The virus struck in the Middle East amid the conflicts in Iraq and the Palestinian territories, where health officials were hard-pressed to respond. The disease also turned its wrath on Egypt, where denizens of Cairo, terrified to learn that chicken carcasses were being dumped in the Nile River, initially stopped drinking tap water. Soon Egypt would record more human cases and deaths than anywhere but Indonesia and Vietnam.

Then the disease crossed the desert to sub-Saharan Africa, infecting commercial farms in Nigeria and stoking fears that this indigent, ill-governed continent could become an entirely new breeding ground for a pandemic strain. Nigeria's information minister told reporters, "While it was originally suspected that migratory birds may have been the purveyor of this disease, preliminary reports recently obtained from relevant security agencies indicate that there is a strong basis to believe that avian flu may have been introduced into Nigeria through illegally imported day-old chicks." Scientists who later decoded the genetics of Nigerian samples concluded the virus had been introduced to the country three separate times along routes coinciding with the flight paths of wild birds. But the study also said imports could not be ruled out as the cause.

Once the virus was established in Nigeria, the epidemic spread, most likely along internal trade routes, and put neighboring countries in jeopardy because of widespread cross-border commerce. By the close of 2007, eleven African governments had reported outbreaks in birds, some in countries no better able than the poorest in Asia to confront the disease.

A rancorous debate has erupted over exactly how avian flu spreads.

Are wild birds the culprit, conveying the virus on their seasonal migrations? Or is it trade in poultry, legal or not, that inadvertently extends the global reach? Some wildlife conservation groups have said that migratory patterns aren't a good match for the distribution of outbreaks, suggesting instead that the virus spread from Asia to Europe by commerce along the route of the Trans-Siberian Railroad. But other researchers have noted outbreaks near wintering sites for migratory birds and far from any farms or markets that could account for contamination. After more than a decade, there is now enough evidence to conclude that the novel strain has taken ample advantage of both these opportunities to advance across the Eastern Hemisphere.

Some countries have been able at times to roll back the tide of infection, notably those in Europe and more developed Asian states like Japan and South Korea. But elsewhere, the disease refused to surrender its foothold.

No one is better positioned to evaluate the viral storm gathering in the animal kingdom than Joseph Domenech, chief veterinarian of the UN Food and Agriculture Organization. In a series of public warnings starting in the fall of 2007 and running into 2008, Domenech offered a disturbing, though not universally bleak prognosis. "Surveillance and early detection and immediate response have improved and many newly infected countries have managed to eliminate the virus from poultry," he reported. "But," he continued, "the H5N1 avian influenza crisis is far from over."

He singled out Indonesia as the place where bird flu had become most stubbornly entrenched. "I am deeply concerned that the high level of virus circulation in birds in the country could create conditions for the virus to mutate and to finally cause a human influenza pandemic," Domenech said in another assessment in March 2008. "The avian influenza situation in Indonesia is grave—all international partners and national authorities need to step up their efforts for halting the spread of the disease in animals and making the fight against the virus a top priority." He faulted a lack of money, poor coordination among different levels of government in Indonesia's decentralized political system, and insufficient commitment. With the strain in Indonesia actively undergoing genetic changes, Domenech warned that the virus was spinning off new subtypes that could elude the poultry vaccines meant to contain it.

Even when outbreaks ebb, it doesn't mean flu has been beaten. It goes to ground, smolders, waits for an opening. Vietnam adopted a raft of eradication measures in late 2005, including a drive to vaccinate tens of millions of chickens and ducks. Officials also imposed a ban on live markets and poultry farming in cities, tightened regulations on transporting birds, and restricted the raising of ducks and quail, which were thought to be spreading the disease without getting sick themselves. The virus went silent for a time. Vietnam boasted it had cornered the virus. But then it resurfaced, establishing a new beachhead in ducks that had not been properly vaccinated. In 2008 and then again in 2009, poultry outbreaks were reported from the north of the country to the south. After more than two dozen provinces were struck in 2008, Vietnam's agriculture minister acknowledged that only a few localities were completely capable of controlling the disease and blamed their slow response for the recurring epidemic. His deputy said the country's poultry vaccination program was flagging. While both poultry outbreaks and human cases were still few relative to 2004 and 2005, FAO officials cautioned that the Vietnamese government would be unable to keep paying for the vaccination drive. Researchers, meantime, noted the separate strains circulating in northern and southern Vietnam were both becoming more lethal.

In China, human infections spiked in early 2009, striking provinces across the breadth of the country. As in earlier years, these cases were occurring without corresponding accounts of poultry outbreaks. WHO officials again concluded that the epidemic among birds was worse than China was reporting and promised, along with FAO, to press their concerns with the Beijing government. "We still have a very serious situation in the agriculture sector," said Hans Troedsson, who had become WHO's senior representative in Beijing after leaving Hanoi. "The virus is well-entrenched and circulating in the environment."

Thailand appeared to fare better. After repeatedly claiming it had banished the bug only to see it return, the Thai government launched an ambitious campaign to crush it in 2005. About seven hundred thousand village health volunteers were mobilized to watch for any hint of emerging infection. Twice a year, the government conducted nationwide door-to-door inspections, dubbed X-ray surveys, searching for infected poultry. Stricken flocks were culled, their owners compensated. Duck grazing was barred. Reported outbreaks tailed off. But researchers continued to detect the virus in bird samples, and suspicions lingered. Were Thai farmers secretly doping their chickens to look healthy? "Some commercial producers in Thailand are apparently using unauthorized vaccines to protect their flocks and vaccinating without proper oversight from health authorities," FAO's infectious disease chief wrote in August 2006.

Elsewhere, the virus kept coming back. The disease reappeared in 2008 and 2009 among the birds of Hong Kong, India, Pakistan, and the West African countries of Benin and Togo. In mid-2008, Nigeria reported its first new H5N1 outbreaks in almost a year, in a pair of poultry markets. When scientists tested the virus, they were dismayed to find it was different from those previously circulating in sub-Saharan Africa, a baffling transplant from either Europe or the Middle East. And after nearly a year, a new case surfaced in ducks in Germany. "Somewhat surprising," German authorities admitted.

The virus was already entrenched in a few places outside East Asia. Egypt officially acknowledged in July 2008 that H5N1 was endemic in its flocks, primarily in the Nile Delta. Bangladesh was equally unable to root out the disease. Domenech called the crises in those two countries "particularly worrying."

He also offered a chilling admonition for Europe. Tens of millions of apparently healthy ducks and geese that graze along the Danube River and in the wetlands surrounding the Black Sea could be spreading the infection. "It seems," he said, "a new chapter in the evolution of avian influenza may be unfolding silently in the heart of Europe."

Even more chilling was the prospect raised by the debut of H1N1 swine flu in early 2009. For many disease specialists, there was a new nightmare scenario: the combination of this recent arrival with bird flu. "We must never forget that the H5N1 avian influenza is now firmly established in poultry in several countries," Margaret Chan told an assembly of the world's top health officials in May 2009. "No one can say how this avian virus will behave when pressured by large numbers of people infected with the new H1N1 virus." What new virus might emerge from the encounter of these two novel strains? The swine flu virus had already demonstrated an uncanny ability to swap genes and shift shape. Might this highly contagious strain now acquire the attributes that make avian flu so deadly? Or might swine flu finally provide its avian cousin with the keys required to break loose among people? "Do not drop the ball on monitoring H5N1," Chan urged Asian health ministers at a separate session in Bangkok called to address swine flu. "We have no idea how H5N1 will behave under the pressure of a pandemic."

WHO and other health agencies watched closely as swine flu spread from its apparent source in North America to countries where bird flu was deeply rooted. In particular, WHO monitored Vietnam and Egypt, two places where both flu strains were circulating widely. China and Indonesia were also of great concern as the swine flu epidemic washed across their vast hinterlands. A year into the swine flu pandemic, scientists had yet to find proof that the pair of strains had reassorted with each other. But even as the swine flu outbreak appeared to be losing steam, global health experts offered warnings anew. As long as the two strains remained active, they could potentially breed a hybrid that was both highly contagious and devastatingly lethal. "We are certainly aware of the risk," said Dr. Shin Young-soo, WHO's regional director for East Asia. "We are on alert for this development." In China, Dr. Nanshan Zhong, the prominent medical researcher who'd confronted his government over its cover-up of SARS and helped Yi Guan acquire SARS samples for testing, added his voice in late 2009 to those urging vigilance. "Inside China, H5N1 has been existing for some time, so if there is really a reassortment between H1N1 and H5N1, it will be a disaster," Zhong said. "This is something we need to monitor, the change, the mutation of the virus."

---

In the five years after the virus reemerged in mid-2003, it tallied nearly 390 confirmed human cases and 245 confirmed deaths. For each one of those fatalities, almost precisely one million birds either succumbed to the disease or were slaughtered to contain it. That tremendous disparity between human and avian deaths underscores where the action has been. The bulk of the battle has consisted of culling, vaccinating, testing, and protecting poultry with the soldiers drawn from the ranks of the agricultural and veterinary services. The spread of disease among animals has long been beyond the mandate of public health, receiving little of the attention that a menace of such magnitude deserves.

Yet if there is any possibility of postponing future human flu pandemics, whether born of H5N1 or another strain, it means reducing the circulation of these viruses in animals. The less the microbe spreads and replicates, the fewer times the dice are rolled. And the less often the dice are rolled, the longer it will likely take for the virus to mutate or reassort into an epidemic strain.

By contrast, once a pandemic strain emerges and begins passing easily from one person to another, the only hope will be to retard its ineluctable progress by using antiviral drugs, quarantines, and other measures to keep people from mixing. A delay of even a few days would be no mean achievement. This could provide more time to start producing and distributing vaccines. Though they wouldn't come soon enough for those infected at the beginning of the pandemic, millions of lives could be spared if vaccines arrive in time for those sickened toward the tail end. Yet ultimately, this is a rear-guard action. "You accept a victory as slowing the virus down," offered Michael Ryan, WHO's head of epidemic alert and response.

For much of their history, the fields of human and animal health have been divorced and introspective. When bird flu outbreaks were reported across Southeast Asia in early 2004, relations were strained between WHO and its sister agency, the FAO. Though officials on the ground tried at times to craft ad hoc alliances, tensions simmered over turf and money. Since then, these two UN agencies and the independent World Organization for Animal Health have taken steps to enhance cooperation, convening joint strategy sessions and setting up a common early-warning system to share information about bird flu and other diseases. But for many in public health, the fight in the animal kingdom has remained an afterthought.

"I don't think the medical community has paid enough attention to the veterinary community in terms of the risks for humans," said veteran scientist Dr. Michael Perdue. "It is the responsibility for human health, for the medical community, to reach out." Perdue told me that means, for instance, pressing to get money for cash-strapped veterinary counterparts and helping set up more labs for testing animal diseases.

In 2006 Perdue joined the U.S. Department of Health and Human Services, where he took a leading role in preparing the United States for the coming epidemic. By the end of 2008, the department had spent about $1 billion toward creating a stockpile of antiviral drugs sufficient to treat about a quarter of the U.S. population. Another $1.5 billion went toward the development of advanced technologies for making pandemic vaccines, while nearly $1 billion more was invested in setting up a stockpile of prepandemic vaccines, which are based on current H5N1 subtypes and might afford limited protection to medical staff and other vital personnel in the early days of an epidemic. These measures, too, could save millions of lives, though they are powerless to preclude the inevitable.

Perdue, a genial Mississippi native, has a rare perspective on flu. He has hunted it on both sides of the species divide. As a microbiologist, he long probed the mysteries of avian influenza at the U.S. Department of Agriculture's elite poultry research lab in Athens, Georgia. Once the disease launched its unprecedented attack on Southeast Asia, WHO called for reinforcements, and Perdue crossed to the human health side, signing on to the agency's global influenza program in Geneva. From there, he was repeatedly dispatched to investigate human outbreaks. WHO is in a bind, he told me. As long as the virus remains primarily an animal disease, the agency must often defer to agriculture officials, who have priorities beyond fighting infectious disease—for instance, promoting livestock development. "It's sort of animal health versus human health. It's challenging," Perdue said. "The problem is, once it becomes a human virus and WHO is clearly engaged, then maybe it is too late, because then it's off and running."


The trail of the fatal strain led me from one end of Southeast Asia to the other. I tracked the virus across nine countries, through hospitals and laboratories, into chicken coops, rice paddies, wet markets, and cockfighting rings. Yet few places were as remote and none as awesome as the floodplain of the Tonle Sap.

For much of the year, this lake in northwestern Cambodia is modest, covering slightly more than a thousand square miles at depths of about a yard. But each June, monsoon rains swell the Mekong River and its tributaries, forcing the water to back up into the lake and transforming it into the largest freshwater body in all of Southeast Asia. The lake's waters overflow the banks and inundate another five thousand square miles, covering nearly a tenth of Cambodia's area. Fields and forests are submerged. Then, with the coming of the dry season each October, the waters drain away.

Over the generations, Cambodians have adapted to the furious but predictable mood swings of their habitat. Many have built their homes on stilts to stay clear of the rising waters. Some inhabitants migrate with the shoreline, dismantling their dwellings as the water advances and reassembling them farther out. Others live on houseboats or floating homes of thatch, palm fronds, and clapboard mounted on bamboo rafts. This ecosystem has also made the shores of the lake and the surrounding floodplain a unique wintering ground for Asia's wild birds.

It was the birds that brought me to the Tonle Sap. The monsoon was only weeks away, and soon they would migrate to summer breeding grounds in China, Japan, Siberian Russia, and even across the Bering Sea to North America. What if the birds were infected? Would the virus follow the multitude of flyways that radiate out from the Tonle Sap?

Few outside of Cambodia had heard of this wilderness. Yet here, in utter silence but for the occasional ruffle of wings, with no other sign of human life on this vast, flat expanse, I wondered whether a plague could be taking shape. If so, no place on Earth would be left untouched.

I came to the Tonle Sap floodplain looking for the virus. I found an insight. I realized we're all living on a floodplain.

Several times each century, a novel flu strain emerges, often from the fountainhead of Southeast Asia. Death and disorder wash across the face of the planet. Sometimes they just skim the surface; other times they deluge all that lies before them. Then they recede. This cruel cycle isn't as predictable as that of the Tonle Sap waters. But it is equally inevitable.

Those who live around the Tonle Sap have adapted with forethought and creativity to nature's challenge. They have learned to ride out the flood. We have yet to do so.

For much of the last generation, the silence has been deceiving. Not since 1968 had a new flu virus menaced humanity and circled the world. But the outbreaks of avian flu that multiplied over the early years of this new century made it hard to continue mistaking luck for a change in the laws of nature. "The present situation is unique," Margaret Chan has admonished from her bully pulpit in Geneva. "In the past, pandemics have always announced themselves with a sudden explosion of cases, and taken the world by surprise. For the first time in history, we have been given an advance warning." She calls this an "unprecedented opportunity" for countries and communities to get ready.

Then swine flu broke out. Researchers initially concluded it was spreading as vigorously as each of the three last pandemic strains. But soon it seemed the attack rate was no greater than that for ordinary flu bugs. And while swine flu caused swift and horrible death in a small percentage of cases, for the majority of people this bout of influenza turned out to be unremarkable. Once more, public health officials were making fateful decisions, whether, for instance, to embark on a crash vaccine program or to close schools, amid scientific uncertainty. Like all flu viruses, especially those so new to us, swine flu remained unpredictable. Its message, however, was unmistakable. Again, get ready.

Though governments have recently taken steps to gird the world for a pandemic, too little has been done. As I write, epidemic planning for hospitals and public health systems remains wanting. Preparations in other essential sectors, in particular those to ensure supplies of food, water, electricity, and fuel and to maintain public order, are even more deficient.

It's surprising how few people know the horrors of 1918. Perhaps that cataclysm is overshadowed in memory by the final months of World War I, a conflict that recast the geopolitics of the world and defined a generation. Perhaps epidemics exist outside of history, ideology, and meaning, and their imprint vanishes like footsteps on the beach, especially when people have no one to blame for their tragedy, no grievance to harbor. It could be that people simply want to forget suffering. Or maybe we believe, as an advanced civilization, we've moved beyond plagues. By that reasoning, 1918 belongs to another, remote era of little relevance to today.

After I started this book, I told friends and colleagues what I was writing. The response wasn't what I expected. Time after time, they would mention relatives who had died during the Spanish flu, a grandfather's brother, a distant cousin. My architect told me his uncle and two other kin are buried in Jamaica after perishing in 1918. Then, several months before I finished the first draft, I discovered evidence of another visitation.

Growing up, I'd heard the story about how my grandmother's mother, Yetta, had died as a young woman back in Poland. No one seemed to know precisely when, and no one ever mentioned the word *influenza*. We never connected it to anything else going on in the world at the time. All we knew was that she had been killed by some respiratory disease, maybe tuberculosis, maybe what they called grippe back then.

My grandmother, we heard, was eight at the time. It turned her life upside down. Soon after, Grandma's father abandoned the family, leaving her to care for her younger sister and brother. They drifted from place to place, sleeping one night here, another there. Grandma found odd jobs to feed them, even working for a time picking fruit in orchards around Warsaw. Ultimately, her two maternal uncles sent for the children. The pair had earlier immigrated to New York, where one became a political pamphleteer and the other, according to family lore, hosted the longest continuously running pinochle game in the Bronx. Grandma left her homeland and crossed the ocean.

We'd never done the math to determine what year great-grandmother Yetta died. We couldn't, because we didn't know when Grandma was born. Grandma had lied on her American immigration papers to pass for a few years older so she could line up work. Over the rest of her life, the birth date on her government papers was fiction. When she finally passed away in 2006, she herself wasn't sure how old she was. But my mother later came across Grandma's marriage contract in an old file. On the back of this crumbling religious document, the date of the wedding was noted in pencil. With that clue, we could calculate Grandma's birth date. She had been born in 1910. It meant Yetta had died, and our history pivoted, eight years later in 1918.

The timing wasn't absolute proof that flu had been the killer. But the odds were overwhelming. The last great epidemic suddenly felt much closer. So did the next.

# Acknowledgments

One of the secrets of foreign correspondence is how indebted we are to our local assistants. They translate not only language, when necessary, but culture. They help us conceive our stories, arrange logistics, navigate politics, identify the sources and subjects to interview, and, after all that, do old-fashioned reporting, sometimes at great personal risk, with rarely a byline to show for it. So the initial thanks have to go to Yayu Yuniar, a dogged, delightful journalist in Jakarta. Long before Indonesia confirmed its first human case of bird flu, Yayu dove into the basics of microbiology and went out chasing chickens with me. I'm also tremendously grateful to Lilian, Ira, and Sindi Pramudita, who did so much more than manage our office and local affairs in Indonesia. They became our second family. Noor Huda Ismail and Natasha Tampubolon rounded out our team in Jakarta. I also benefited from the hard work, generosity, and insights of local reporters and fixers elsewhere in Asia, in particular Somporn Panyastianpong in Bangkok, Phann Ana in Phnom Penh, K. C. Ng in Hong Kong, and Ling Jin in Beijing. Back in Washington, Jean Hwang was a tireless assistant, working into the wee hours transcribing often inaudible tapes.

*The Washington Post* has always been a special place. Yet with each passing year, as other newspapers succumb to the financial pressures of our troubled industry, the *Post*'s commitment to great journalism becomes ever more exceptional. That dedication has flowed from the top, from Donald E. Graham, Boisfeuillet Jones Jr., and now

Katharine Weymouth. For the opportunity to report from Southeast Asia, I'm grateful to former executive editor Leonard Downie Jr. and former managing editor Steve Coll, as well as to Philip Bennett, who was assistant managing editor for foreign news when I shipped out to Asia and managing editor when I returned. On the foreign desk, a string of talented editors had a hand in my writing about flu, including David Hoffman, Peter Eisner, John Burgess, Kathryn Tolbert, Tiffany Harness, and Jason Ukman. Thanks also to Nils Bruzelius, the deputy national editor for science and health, and news researchers Bob Lyford and Rob Thomason for their help.

The fraternity of flu writers is surprisingly small, given the drama and stakes of the subject. Yet in recent years I've been fortunate to count myself in the company of some terrific journalists. I've benefited from the work in Asia of Margie Mason of the Associated Press and Nicholas Zamiska of the *Wall Street Journal*. Closer to home, the standard has been set by Helen Branswell of the Canadian Press, Maggie Fox of Reuters, and Maryn McKenna, formerly of the *Atlanta Journal-Constitution*. In Jakarta, I was blessed with colleagues who were both good friends and reliable road companions at the toughest of times, in particular Rick Paddock of the *Los Angeles Times*, Shawn Donnan of the *Financial Times*, and John Aglionby of the *Guardian* and later the *Financial Times*. A special note of gratitude goes to Dean Yates of Reuters and to Mary Binks.

This would have been a very different book if not for the cooperation of countless individuals at the World Health Organization. I cannot name them all, and a fair number wouldn't want me to. But I would like to acknowledge several members of the agency's public affairs staff, past and present, including Maria Cheng, Peter Cordingley, Bob Dietz, Sari Setiogi, and Roy Wadia. Both in reporting for the *Post* and in researching this book, I found them consistently helpful. Mary Kay Kindhauser's efforts in facilitating my reporting in Geneva were invaluable. And I'm particularly grateful to Dick Thompson and Kristin Thompson for their professional assistance and personal kindness. There were also many at other agencies who took time to share their expertise and give me crash tutorials, including at the UN Food and Agriculture Organization, the World Animal Health Organization, and the Centers for Disease Control and Prevention. Nor can I say enough about the helpful staff of the U.S. National Library of Medicine or the team at the University of Minnesota that maintains the awesomely comprehensive Center for Infectious Disease Research and Policy (CIDRAP) Web site, www.cidrap.umn.edu.

It's not possible to credit the scores of scientists, public health experts, and medical specialists who agreed to be interviewed or offered me guidance. But I would be remiss if I didn't single out the following for thanks. Michael Perdue, formerly of WHO in Geneva and now at the U.S. Biomedical Advanced Research and Development Authority in Washington, has been a great resource for me on the science of the virus. Michael Osterholm, director of the Center for Infectious Disease Research and Policy in Minneapolis, has steered me right when I've been stumped by questions about the evolution of the threat. I'm also smarter for the repeated insights of Malik Peiris in Hong Kong, one of the world's premier researchers into respiratory viruses.

This book never would have been written at all without the intervention of three *Post* colleagues. David Maraniss encouraged me to embark on this project when I was still unsure whether to do so. Dana Milbank offered sage advice across a kitchen counter on how to frame the book, finally unlocking its potential. And Sandy Sugawara gave me the time to pursue the project even if the timing was inopportune. I'm beholden to all three.

I owe a major debt of gratitude to my very wise agent, Raphael Sagalyn. He immediately spotted the potential of this project, but for months kept pushing me to rethink and refine the conception until I got it right. Then he ran with it. At Viking Penguin, my editor Alessandra Lusardi has been a very sharp reader and elevated the writing by helping me bring the main themes to the fore while pruning back overgrowth in the storytelling.

In everything, I'm indebted to my parents for their love and gift of intellectual curiosity. Most of all, I'm grateful to Ellen, my partner in all things, for her understanding and encouragement. There's no one else I'd rather travel to the ends of the world with.

# Notes

## Prologue

**2**    **All flu viruses in fact emanate:** See, for instance, R. J. Webby and R. G. Webster, "Emergence of Influenza A Viruses," *Philosophical Transactions of the Royal Society of London B: Biological Sciences,* 356, no. 1416 (Dec. 29, 2001): 1817–28.

**6**    **a repeat of the Great Influenza:** For a definitive account of the 1918 pandemic, see John M. Barry, *The Great Influenza: The Epic Story of the Deadliest Plague in History* (Viking Penguin: New York, 2004). See also Alfred W. Crosby, *America's Forgotten Pandemic: The Influenza of 1918,* 2nd ed. (New York: Cambridge University Press, 2003).

**6**    **at least 50 million lives:** N. P. Johnson and J. Mueller, "Updating the Accounts: Global Mortality of the 1918-1920 'Spanish' Influenza Pandemic," *Bulletin of the History of Medicine* 76, no. 1 (Spring 2002): 105–15.

**7**    **"accelerated number of near-misses":** Anita Manning, "New, Deadly Flu Pandemic 'Inevitable,' Experts Warn," *USA Today,* Mar. 2, 2004.

**7**    **theater of conflict that is Asia:** Dr. Michael T. Osterholm of the University of Minnesota has called Asia "the genetic roulette table for H5N1 mutations."

**7**    **first documented global outbreak:** C. W. Potter, "A History of Influenza," *Journal of Applied Microbiology* 91 (2001): 572–79.

**7**    **since the twelfth century:** August Hirsch, *Handbook of Geographical and Historical Pathology* (London: New Sydenham Society, 1883).

## Chapter One: The Revenge of Begu Ganjang

This chapter draws on interviews with disease specialists from the World Health Organization and the U.S. Centers for Disease Control and Prevention, officials from the governments of Indonesia, North Sumatra province, and Karo district, and medical professionals and residents in North Sumatra province, as well as internal documents from WHO and the governments of Indonesia, North Sumatra, and Karo.

**13**    **cruise ships in Alaska:** An account of the investigation can be found in Timothy M. Uyeki et al., "Large Summertime Influenza A Outbreak Among Tourists in Alaska and the Yukon Territory," *Clinical Infectious Diseases* 36 (2003): 1095–1102.

13   **island nation of Madagascar:** Accounts of the investigation into the outbreak can be found in "Influenza Outbreak—July–August 2002," *Morbidity and Mortality Weekly Report,* 51, no. 45 (Nov. 15, 2002): 1016–18; and *Weekly Epidemiological Record* 2002, no. 46 (Nov. 15, 2002): 77, 381–88.

18   **At Imperial College London:** Neil M. Ferguson et al., "Strategies for Containing an Emerging Influenza Epidemic in Southeast Asia," *Nature* 437 (Sept. 8, 2005): 209–14.

18   **A separate team:** Ira M. Longini Jr. et al., "Containing Pandemic Influenza at the Source," *Science* 309, no. 5737 (Aug. 12, 2005): 1083–87.

19   **"No attempt has ever been made":** "WHO Activities in Avian Influenza and Pandemic Influenza Activities, WHO, Jan.–Dec. 2006, 16. The approach was quickly embodied in WHO planning. See "WHO Pandemic Influenza Draft Protocol for Rapid Response and Containment," updated draft, WHO, May 30, 2006.

19   **WHO's emergency containment plan:** The strategy is described in the finalized WHO protocol. See WHO, "WHO Interim Protocol: Rapid Operations to Contain the Initial Emergence of Pandemic Influenza," updated Oct. 2007.

20   **"It will require excellent surveillance":** WHO, "WHO Activities in Avian Influenza and Pandemic Influenza Activities: January–December 2006," p. 16.

25   **by the time Puji was buried:** A local account of the cases and grassroots response is "Report Regarding Bird Flu Disease in Karo Regency, from the Karo Regent Daulat Daniel Sinulingga to the Governor of North Sumatra in Medan, Kabanjahe," May 31, 2006.

28   **According to Batak lore:** Interview with Juara Ginting, an anthropologist who grew up in Karo district and studied Batak belief and superstition at North Sumatra University before pursuing a master's degree at Leiden University in the Netherlands.

29   **beheaded a chicken:** Margie Mason, "Officials Backtrack Bird Flu Cluster," Associated Press, May 26, 2005.

29   **scores of poultry traders:** Jason Gale and Karima Anjani, "Indonesian Bird-Flu Victim Sought Witchdoctor, Shunned Hospital," Bloomberg, May 26, 2005.

29   **"significantly delayed":** WHO, "Avian Influenza Cluster, Karo, North Sumatra, May 2006, WHO Interim Report."

30   **"universally refused":** Ibid.

38   **senior Indonesian officials:** See, for instance, Health Minister Siti Fadilah Supari, quoted in Tubagus Arie Rukmantara, "Awareness and Prevention Key in Bird Flu Fight," *Jakarta Post,* July 28, 2006.

38   **"From the lessening of the tension":** Siti Fadilah Supari, *It's Time for the World to Change: In the Spirit of Dignity, Equity, and Transparency, Divine Hand Behind Avian Influenza* (Jakarta: Sulaksana Watinsa Indonesia, 2008), 23.

38   **Supari would continue:** See, for example, "Minister Denies Bird Flu in RI Spreading by Human-to-Human Transmission," Antara news agency, Sept. 3, 2007.

38   **were quickly convinced:** Georg Petersen, "Investigative Report of a Cluster of Human Avian Influenza Cases, North Sumatra, May 2006," WHO.

39   **"If he turns out to be positive":** Internal WHO communication from Jakarta, May 21, 2006.

39   **"In response to the possibility":** Ibid.

40   **tested positive for the virus:** The specimens collected by Uyeki and his colleagues later showed that the virus had been aggressively mutating as it moved

from son to father. See, for example, Santoso Soroeso, "Epidemiology and Clinical Features of Avian Influenza in Indonesia, Questions and Lessons Learnt," presented at Australia-Indonesia Symposium in Science and Technology 2006, Sept. 13–14, 2006, Jakarta; and also Declan Butler, "Family Tragedy Spotlights Flu Mutations," *Nature* 442 (July 13, 2006): 114–15. Initial analysis of the specimens at Hong Kong University also showed that Dowes's sample had the same signature mutation as the sample from his son. This seemed to be further evidence that Dowes caught the virus from his son, marking the third generation of transmission. This finding was reported at the time by several media and cited by U.S. Secretary of Health and Human Services Michael O. Leavitt. See, for example, Department of Health and Human Services, Pandemic Planning Update II, a report from Secretary Michael O. Leavitt, June 29, 2006. But researchers familiar with the results later said the initial findings regarding a signature mutation were not as conclusive as at first thought. The case for third-generation spread still rests on the case histories of the Ginting family members and the timing of their illnesses.

42   **also onto a third:** Politics long prevented researchers from publishing their analysis of human-to-human-to-human spread in the Sumatra cluster. But half a year later, another instance of third-generation transmission was confirmed in Pakistan and later described in a publication. See: "Human Cases of Avian influenza A (H5N1) in North-West Frontier Province, Pakistan, October–November 2007," *Weekly Epidemiological Record*, no. 40 (October 3, 2008): 359–64.


## Chapter Two: A Visitation from Outer Space

This chapter draws on interviews with current and former public health, infectious-disease, and laboratory specialists at the Hong Kong Department of Health and the CDC, as well as animal health researchers in Hong Kong and the United States.

45   **a three-year-old boy:** The case is described in J. C. de Jong et al., "A Pandemic Warning?" *Nature* 389, no. 6651 (Oct. 9, 1997): 554; and in Kanta Subbarao et al., "Characterization of an Avian Influenza A (H5N1) Virus Isolated from a Child with a Fatal Respiratory Disease," *Science* 279, no. 5349 (Jan. 16, 1998): 393–96.

48   **far more than a runny nose and chills:** A thorough overview of the clinical spectrum is provided in J. S. Malik Peiris, Menno D. de Jong, and Yi Guan, "Avian Influenza Virus (H5N1): A Threat to Human Health," *Clinical Microbiology Review* 20, no. 2 (Apr. 2007): 243–67; and in K. Y. Yuen and S. S. Y. Wong, "Human Infection by Avian Influenza A H5N1," *Hong Kong Medical Journal* 11, no. 3 (June 2005): 189–99. WHO has described the symptoms and clinical course of the disease in reports by the agency's writing committee. See Writing Committee of the Second World Health Organization Consultation on Clinical Aspects of Human Infection with Avian Influenza A (H5N1) Virus, "Update on Avian Influenza A (H5N1) Virus Infection in Humans," *NEJM* 358 no. 3 (Jan. 17, 2008): 261–73. The cases in individual countries have also been surveyed and described. See, for example, Hongjie Yu et al., "Clinical Characteristics of 26 Human Cases of Highly Pathogenic Avian Influenza A (H5N1) Virus Infection in China," *PLoS One* 3, no. 8 (Aug. 21, 2008): e2985; and Sardikin Giriputro et al., "Clinical and Epidemiological Features of Patients with Confirmed Avian Influenza Presenting to Sulianti Saroso Infectious Diseases Hospital, 2005–2007," *Annals of the Academy of Medicine* (Singapore) 37 (2008): 454–57.

49   **a counterattack so furious:** This aggressive response has been much discussed.

The following is a sampling of the research: M. C. W. Chan et al., "Proinflamma-
tory Cytokine Responses Induced by Influenza A (H5N1) Viruses in Primary
Human Alveolar and Bronchial Epithelial Cells," *Respiratory Research* 6 (Nov. 11,
2005): 135; C. Y. Cheung et al., "Induction of Proinflammatory Cytokines in
Human Macrophages by Influenza A (H5N1) Viruses: A Mechanism for the
Unusual Severity of Human Disease?" *Lancet* 360, no. 9348 (Dec. 7, 2002):
1831–37; Menno D. de Jong et al., "Fatal Outcome of Human Influenza A (H5N1)
Is Associated with High Viral Load and Hypercytokinemia," *Nature Medicine* 12,
no. 10 (Oct. 2006): 1203–07; J. S. Malik Peiris et al., "Re-emergence of Fatal
Human Influenza A Subtype H5N1 Disease," *Lancet* 363, no. 9409 (Feb. 21,
2004): 617–19; Ka-Fai To et al., "Pathology of Fatal Human Infection Associated
with Avian Influenza A H5N1 Virus," *Journal of Medical Virology* 63 (2001):
242–46; and Jianfang Zhou et al., "Differential Expression of Chemokines and
Their Receptors in Adult and Neonatal Macrophages Infected with Human or
Avian Influenza Viruses," *Journal of Infectious Diseases* 194 (2006): 61–70.

**50　inviting a suicidal counterattack:** There has been debate about whether the
immune response or the virus itself is more directly responsible for death. See,
for instance, Kristy J. Szretter et al., "Role of Host Cytokine Responses in the
Pathogenesis of Avian H5N1 Influenza Viruses in Mice," *Journal of Virology* 81,
no. 6 (Mar. 2007): 2736–44; and Rachelle Salomon, Erich Hoffman, and Robert
G. Webster, "Inhibition of the Cytokine Response Does Not Protect Against
Lethal H5N1 Influenza Infection," *PNAS* 104, no. 30 (July 24, 2007):
12479–81.

**50　enters the human body:** For an overview of how the microbe operates, see
J. S. Malik Peiris, Menno D. de Jong, and Yi Guan, "Avian Influenza Virus
(H5N1): A Threat to Human Health," *Clinical Microbiology Review* 20, no. 2
(Apr. 2007): 243–67; and R. G. Webster and D. J. Hulse, "Microbial Adaption
and Change: Avian Influenza," *Revue Scientifique et Technique, Office Interna-
tional des Épizooties* 23, no. 2 (2004): 453–65.

**52　receptors in the human respiratory tract:** There has also been extensive
discussion about the preferences that different strains have for human and avian
receptors and the crucial role these play in transmission. The following is a sam-
pling of the research: Susan J. Baigent and John W. McCauley, "Influenza Type
A in Humans, Mammals and Birds: Determinants of Virus Virulence, Host-
Range and Interspecies Transmission," *BioEssays* 25, no. 7 (2003): 657–71; Aarthi
Chandrasekaran et al., "Glycan Topology Determines Human Adaptation of
Avian H5N1 Virus Hemagglutinin," *Nature Biotechnology* 26, no. 1 (Jan. 2008):
107–13; A. Gambaryan et al., "Evolution of the Receptor Binding Phenotype of
Influenza A (H5) Viruses," *Virology* 344, no. 2 (Jan. 20, 2006): 432–38; Thijs
Kuiken et al., "Host Species Barriers to Influenza Virus Infections," *Science*
312, no. 5772 (Apr. 21, 2006): 394–97; Masato Hatta et al., "Growth of H5N1
Influenza A Viruses in the Upper Respiratory Tracts of Mice," *PLoS Pathogens* 3,
no. 10 (Oct. 2007): 1374–79; John M. Nicholls et al., "Sialic Acid Receptor Detection in the Human Respiratory Tract: Evidence for Widespread Distribution
of Potential Binding Sites for Human and Avian Influenza Viruses," *Respiratory
Research* 8 (2007): 73; J. M. Nicholls et al., "Tropism of Avian Influenza A (H5N1)
in the Upper and Lower Respiratory Tract," *Nature Medicine* 13 (2007): 147–49;
Kyoko Shinya et al., "Influenza Virus Receptors in the Human Airway," *Nature*
440 (Mar. 23, 2006): 435–36; Debby van Riel et al., "H5N1 Virus Attachment to
Lower Respiratory Tract," *Science* 312, no. 5772 (Apr. 23, 2006): 399; Terrence
M. Tumpey et al., "A Two-Amino Acid Change in the Hemagglutinin of the 1918
Influenza Virus Abolishes Transmission," *Science* 315, no. 5812 (Feb. 2, 2007):

655–59; Shinya Yamada et al., "Haemagglutinin Mutations Responsible for the Binding of H5N1 Influenza A Viruses to Human-type Receptors," *Nature* 444 (Nov. 16, 2006): 378–82 and *Influenza Research at the Human and Animal Interface: Report of a WHO Working Group*, WHO, Geneva, Sept. 21–22, 2006.

52    **a few other genetic tweaks:** For discussion of possible changes in viral proteins that can lead to an avian virus attacking humans and becoming more lethal, see Christopher F. Basler and Patricia V. Aguilar, "Progress in Identifying Virulence Determinants of the 1918 H1N1 and the Southeast Asian H5N1 Influenza A Viruses," *Antiviral Research* 79 (2008): 166–78; Andrea Gambotto et al., "Human Infection with Highly Pathogenic H5N1 Influenza Virus," *Lancet* 371, no. 9622 (Apr. 26, 2008): 1464–75; and Neal Van Hoeven et al., "Human HA and Polymerase Subunit PB2 Proteins Confer Transmission of an Avian Influenza Virus Through Air," *PNAS,* published online before print February 11, 2009, doi: 10.1073/pnas.0813172106.

55    **Growing up in Hong Kong:** Miriam Shuchman, "Improving Global Health—Margaret Chan at the WHO," *NEJM* 356, no. 7 (Feb. 15, 2007): 653–56; and Lawrence K. Altman, "Her Job: Helping Save the World from Bird Flu," *New York Times*, Aug. 9, 2005.

56    **a baffling plague:** On the connection between Hoi-ka's case with the earlier poultry outbreak, see Eric C. J. Claas et al., "Human Influenza A H5N1 Virus Related to a Highly Pathogenic Avian Influenza Virus," *Lancet* 351, no. 9101 (Feb. 14, 1998): 472–77; and David L. Suarez et al., "Comparisons of Highly Virulent H5N1 Influenza A Viruses Isolated from Humans and Chickens from Hong Kong," *Journal of Virology* 72, no. 8 (Aug. 1998): 6678–88.

58    **The Spanish flu:** For a scientific investigation of the 1918 pandemic, see Jeffrey K. Taubenberger and David M. Morens, "1918 Influenza: The Mother of All Pandemics," *Emerging Infectious Diseases* 12, no. 1 (Jan. 2006): 15–22.

58    **two subsequent pandemics:** On flu pandemics of the last century, see Edwin D. Kilbourne, "Influenza Pandemics of the 20th Century," *Emerging Infectious Diseases* 12, no. 1 (Jan. 2006): 9–14. WHO estimates that the 1957 pandemic killed two million and the 1968 pandemic one million.

60    **If two different flu strains:** For a discussion of the compatibility of genes from H5N1 and human viruses, see Li-Mei Chen et al., "Genetic compatibility and Virulence of Reassortants Derived from Contemporary Avian H5N1 and Human H3N2 Influenza A Viruses," *PLoS Pathogens* 4, no. 5: e1000072.

60    **the recent, seemingly improbable encounter:** For early discussions of the H1N1 swine flu virus, see Rebecca J. Garten et al., "Antigenic and Genetic Characteristics of Swine-Origin 2009 A (H1N1) Influenza Viruses Circulating in Humans," *Science,* published online before print May 22, 2009, doi: 10.1126/science.1176225; Novel Swine-Origin Influenza A (H1N1) Virus Investigation Team, "Emergence of a Novel Swine-Origin Influenza A (H1N1) Virus in Humans," *NEJM,* published online before print May 7, 2009, doi: 10.1056/NEJMoa0903810; and Robert B. Belshe, "Implications of the Emergence of a Novel H1 Influenza Virus," *NEJM,* published online before print May 7, 2009, doi: 10.1056/NEJMe0903995. On the triple reassortant virus, see Vivek Shinde, et al., "Triple-reassortant swine influenza A (H1) in Humans in the United States, 2005–2009," *NEJM,* published online before print May 7, 2009, doi: 10.1056/NEJMoa0903812.

60    **even infecting mammals:** See, for example, Juthatip Keawcharoen et al., "Avian Influenza H5N1 in Tigers and Leopards," *Emerging Infectious Diseases* 10, no. 12 (Dec. 2004): 2189–91; and Guus F. Rimmelzwaan et al., "Influenza A Virus (H5N1) Infection in Cats Causes Systemic Disease with Potential Novel

Routes of Virus Spread Within and Between Hosts," *American Journal of Pathology* 168, no. 1 (Jan. 2006): 176–83.

60 **The dice were being rolled:** Alice Croisier et al., "Highly Pathogenic Avian Influenza A (H5N1) and Risks to Human Health," Background Paper at the Technical Meeting on Highly Pathogenic Avian Influenza and Human H5N1 Infection, June 27–29, 2007, Rome.

60 **"appear out of control":** I. Capua and S. Marangon, "Control and Prevention of Avian Influenza in an Evolving Scenario," *Vaccine* 25, no. 30 (July 26, 2007): 5645–52.

60 **it returns:** Antonio Petrini, "Global Situation: HPAI Outbreaks in Poultry—A Synthesis of Country Reports to the OIE," Background Paper at the Technical Meeting on Highly Pathogenic Avian Influenza and Human H5N1 Infection, June 27–29, 2007, Rome.

60 **"a distant and unlikely prospect":** Joseph Domenech et al., "Trends of Dynamics of HPAI—Epidemiological and Animal Health Risks," Background Paper at the Technical Meeting on Highly Pathogenic Avian Influenza and Human H5N1 Infection, June 27–29, 2007, Rome.

60 **not the only avian virus menacing humanity:** J. S. Malik Peiris, Menno D. de Jong, and Yi Guan, "Avian Influenza Virus (H5N1): A Threat to Human Health," *Clinical Microbiology Review* 20, no. 2 (April 2007): 243–67.

60 **avian strain called H9N2:** K. M. Xu et al., "Evolution and Molecular Epidemiology of H9N2 Influenza A Viruses from Quail in Southern China, 2000 to 2005," *Journal of Virology* 81, no. 6 (Mar. 2007): 2635–45; and K. M. Xu et al., "The Genesis and Evolution of H9N2 Influenza Viruses in Poultry from Southern China, 2000 to 2005," *Journal of Virology* 81 no. 19 (Oct. 2007): 10389–10401.

61 **"The establishment and prevalence":** Hongquan Wan et al., "Replication and Transmission of H9N2 Influenza Viruses in Ferrets: Evaluation of Pandemic Potential," *PLoS One* 3, no. 8 (Aug. 2008): e2923.

61 **"continued surveillance and study":** Jessica A. Belser et al., "Contemporary North American Influenza H7 Viruses Possess Human Receptor Specificity: Implications for Virus Transmissibility," *PNAS* 105 no. 21 (May 27, 2008): 7558–63.

61 **Some medical scholars dissent:** Dennis Normile, "Avian Influenza: Pandemic Skeptics Warn Against Crying Wolf," *Science* 310, no. 5751 (Nov. 18, 2005): 1112–13; and Declan Butler, "Yes, But Will It Jump?" *Nature* 439, no. 12 (Jan. 2006): 124–25.

62 **"Such complacency":** Robert G. Webster et al., "H5N1 Outbreaks and Enzootic Influenza," *Emerging Infectious Diseases* 12, no. 1 (Jan. 2006): 3–8.

62 **"The virus has evolved":** Remarks in a speech tape for Business Preparedness for Pandemic Influenza, Second Annual Summit, sponsored by the University of Minnesota Center for Infectious Disease Research and Policy, Feb. 5, 2007.

62 **"If you put a burglar":** Margaret Chan, "Pandemics: Working Together for an Effective and Equitable Response," address to the Pacific Health Summit, Seattle, June 13, 2007.

66 **"There's a possibility":** Cindy Sui, "Hospital Staff Ill After Treating Bird Flu Victims," *Hong Kong Standard,* Dec. 8, 1997.

66 **reached double digits:** For a clinical discussion of the Hong Kong cases, see K. Y. Yuen et al., "Clinical Features and Rapid Viral Diagnosis of Human Disease Associated with Avian Influenza A H5N1 Virus," *Lancet* 351, no. 9101 (Feb. 14, 1998): 467–71; and Paul K. S. Chan, "Outbreak of Avian Influenza A (H5N1) Virus Infection in Hong Kong in 1997," *Clinical Infectious Diseases* 34 (2002): S58–S64.

68    **The parallels were eerie:** David M. Morens and Anthony S. Fauci, "The 1918 Influenza Pandemic: Insights for the 21st Century," *Journal of Infectious Diseases* 195 (2007): 1018–28; Jeffrey K. Taubenberger, "The Origin and Virulence of the 1918 'Spanish' Influenza Virus," *Proceedings of the American Philosophical Society* 150, no. 1 (Mar. 2006); Jeffrey K. Taubenberger and David M. Morens, "1918 Influenza: The Mother of All Pandemics," *Emerging Infectious Diseases* 12, no. 1 (Jan. 2006): 15–22; and L. Simonsen et al., "Pandemic Versus Epidemic Influenza Mortality: A Pattern of Changing Age Distribution," *Journal of Infectious Diseases* 178, no. 1 (July 1998): 53–60.

68    **this disquieting pattern:** "Epidemiology of WHO-Confirmed Human Cases of Avian Influenza A (H5N1) Infection," *Weekly Epidemiological Record* 81, no. 26 (June 30, 2006): 249–57; and "Update: WHO-Confirmed Human Cases of Avian Influenza A (H5N1) Infection, 25 November 2003–24 November 2006," *Weekly Epidemiological Record*, 82, no. 6 (Feb. 9, 2007): 41–47.

68    **"most important unsolved mystery":** David M. Morens and Anthony S. Fauci, "The 1918 Influenza Pandemic: Insights for the 21st Century," *Journal of Infectious Diseases* 195 (2007): 1018–28.

69    **tremendous cytokine storms:** See, for example, John C. Kash, et al., "Genomic Analysis of Increased Host Immune and Cell Death Responses Induced by 1918 Influenza Virus," *Nature* 443 (Oct. 5, 2006): 578–81; Darwyn Kobasa et al., "Enhanced Virulence of Influenza A Viruses with the Hemagglutinin of the 1918 Pandemic Virus," *Nature* 431, no. 7009 (Oct. 7, 2004): 703–7; and Darwyn Kobasa et al., "Aberrant Innate Immune Response in Lethal Infection of Macaques with the 1918 Influenza Virus," *Nature* 445 (Jan. 18, 2007): 319–23. A study in 2008 comparing the immune response to H5N1 and the 1918 virus in mice showed "considerable similarities" but found that the H5N1 strain actually elicited significantly higher levels of cytokines and macrophages. See Lucy A. Perrone et al., "H5N1 and 1918 Pandemic Influenza Virus Infection Results in Early and Excessive Infiltration of Macrophages and Neutrophils in the Lungs of Mice," *PLoS Pathogens* 4, no. 8 (2008): e1000115.

69    **"kissing cousin":** Remarks at Business Preparedness for Pandemic Influenza, Second Annual Summit, University of Minnesota Center for Infectious Disease Research and Policy, Feb. 5, 2007.

69    **a wholly avian virus:** Jeffrey K. Taubenberger et al., "Characterization of the 1918 Influenza Virus Polymerase Genes," *Nature* 437 (Oct. 6, 2005): 889–93; and Terrence M. Tumpey et al., "Characterization of the Reconstructed 1918 Spanish Influenza Pandemic Virus," *Science* 310, no. 5745 (Oct. 7, 2005): 77–80.

69    **"a number of the same changes":** Jeffrey K. Taubenberger et al., "Characterization of the 1918 Influenza Virus Polymerase Genes," *Nature* 437 (Oct. 6, 2005): 889–93.

69    **more like the Spanish flu strain:** For example, see James Stevens et al., "Structure and Receptor Specificity of the Hemagglutinin from an H5N1 Influenza Virus," *Science* 312, no. 5772 (Apr. 21, 2006): 404–10.

69    **A series of studies:** H. Chen et al., "The Evolution of H5N1 Influenza Viruses in Ducks in Southern China," *PNAS* 101, no. 28 (July 13, 2004): 10452–57; Taronna R. Maines et al., "Avian Influenza (H5N1) Viruses Isolated from Humans in Asia in 2004 Exhibit Increased Virulence in Mammals," *Journal of Virology* 79, no. 18 (Sept. 2005): 11788–11800; Hui-Ling Yen et al., "Virulence May Determine the Necessary Duration and Dosage of Oseltamivir Treatment for Highly Pathogenic A/Vietnam/1203/04 Influenza Virus in Mice," *Journal of Infectious Diseases* 192 (2005): 665–72; and Adrianus C. M. Boon et al., "Role of Terrestrial

Wild Birds in Ecology of Influenza A Virus (H5N1)," *Emerging Infectious Diseases* 13, no. 11 (Nov. 2007): 1720–24.

69 **"a process of rapid evolution":** "Mouse Studies of Oseltamivir Show Promise Against H5N1 Influenza Virus," *NIH News*, July 18, 2005.

69 **already become more ferocious:** Carole R. Baskin et al., "Early and Sustained Innate Immune Response Defines Pathology and Death in Nonhuman Primates Infected by Highly Pathogenic Influenza Virus," *PNAS*, Published online before print February 13, 2009, doi 10.1073/pnas.0813234106.

70 **If the virus continued to develop:** J. S. Malik Peiris, "H5N1 Pathogenesis in Humans: An Update," Power Point presentation to the WHO working group, Sept. 21–22, 2006.

70 **though later reported :** WHO, "Influenza Research at the Human and Animal Interface: Report of a WHO Working Group," Geneva, September 21–22, 2006.

70 **62 million:** Christopher J. L. Murray et al., "Estimation of Potential Global Pandemic Influenza Mortality on the Basis of Vital Registry Data from the 1918–1920 Pandemic: A Quantitative Analysis," *Lancet* 368, no. 9554 (Dec. 23, 2006): 2211–18.

70 **$3.13 trillion during the first year:** The figures for severe, mild, and moderate pandemics are based on numbers included in Andrew Burns, Dominique van der Mensbrugghe, and Hans Timmer, "Evaluating the Economic Consequences of Avian Influenza," updated in September 2008. An earlier version of this report, which had calculated the costs using a lower figure for global GDP, put the toll of a severe pandemic at $2.38 trillion. The study was originally published in a slightly different form in the World Bank's June 2006 edition of *Global Development Finance*. For further discussion, see Milan Brahmbatt, "Economic Impacts of Avian Influenza Propagation," speech at the First International Conference on Avian Influenza in Humans, June 29, 2006.

71 **"It's a possibility in this case":** Jane Moir, "Cousins of Child Victim in Flu Alert: Human Transmission Suspected," *South China Morning Post*, Dec. 17, 1997.

71 **"They live together at Grandma's":** Edward A. Gargan, "Chicken-Borne Flu Virus Puts Hong Kong on Alert," *New York Times*, Dec. 17, 1997.

71 **"working at breakneck pace":** Jane Moir, "Cousins of Child Victim in Flu Alert: Human Transmission Suspected," *South China Morning Post*, Dec. 17, 1997.

71 **barely three hundred square feet:** Rhonda Lam Wan, "Bird Flu Cousins' Flat Behind Pile of Rubbish," *South China Morning Post*, Dec. 19, 1997.

71 **they were rebuffed:** Ibid.

72 **a city under siege:** See, for example, the following accounts, all from the *South China Morning Post*: Rhonda Lam Wan and Billy Wong Wai-Yuk, "Doctors Scramble for Special Drug," Dec. 13, 1997; Andrea Li and Alex Lo, "Fears Force Changes to Menus," Dec. 16, 1997; Stella Lee, "18 Private Doctors to Join Bird Flu Probe," Dec. 18, 1997; Rhonda Lam Wan, "Flood of Requests Prompts Promise of A-Strain Testing," Dec. 20, 1997; and Ng Kang-Chung, "Rush for Bird Flu Tests As Seven More Suspected Victims Found," Dec. 26, 1997.

73 **Fukuda had never before missed:** Patricia Guthrie, "Focus on Hong Kong Flu," *Atlanta Journal and Constitution*, Dec. 25, 1997.

73 **"will stop or spread":** "Hong Kong Tests Show Human-to-Human Transmission of Bird Flu Difficult," Agence France Presse, Dec. 27, 1997.

73 **"measures are sufficient":** Keith B. Richburg, "Hong Kong Killing All Chickens in Fight Against 'Bird Flu' Virus," *Washington Post*, Dec. 29, 1997.

73 **more bad news:** For discussion of the poultry outbreaks in late December 1997, see L. D. Sims et al., "Avian Influenza in Hong Kong 1997–2002," *Avian Diseases* 47, no. s3 (2003): 832–38; and Kennedy F. Shortridge, "Poultry and the Influenza H5N1 Outbreak in Hong Kong, 1997: Abridged Chronology and Virus Isolation," *Vaccine* 17 (1999): s26–s29.

75 **would kill every last chicken:** For an overview of the 1997 poultry outbreaks and government response, see Kennedy F. Shortridge et al., "Interspecies Transmission of Influenza Viruses: H5N1 Virus and a Hong Kong SAR Perspective," *Veterinary Microbiology* 74 (2000): 141–47.

75 **The government pressed:** Robin Ajello and Catherine Shepherd, "The Flu Fighters," *Asiaweek,* Jan. 2008.

75 **bloody chaos:** Billy Wong Wai-Yuk, "Tears of Anger, Rivers of Blood," *South China Morning Post,* Dec. 30, 1997; Stella Lee, "Slaughter Held Up by Inexperience," *South China Morning Post,* Dec. 31, 1997; Keith B. Richburg, "Chicken Sightings Frighten Hong Kong, *Washington Post,* Jan. 3, 1998; and Keith B. Richburg, "Hong Kong Faulted on Handling of 'Bird Flu' Crisis, *Washington Post,* Jan. 4, 1998.

76 **On the third day of the slaughter:** "Tung Handling of Flu Crisis Attacked," *South China Morning Post,* Jan. 1, 1998.

76 **In a front-page editorial:** "Only Time Will Prove Wisdom of Dramatic Move," *South China Morning Post,* Dec. 29, 1997.

76 **the "botched" operation:** "Taking Charge," *South China Morning Post,* Jan. 3, 1998.

76 **nearly 350 chickens:** Kennedy F. Shortridge, "Poultry and the Influenza H5N1 Outbreak in Hong Kong, 1997: Abridged Chronology and Virus Isolation," *Vaccine* 17 (1999): s26–s29.

76 **more widespread than expected:** Kennedy F. Shortridge et al., "Characterization of Avian H5N1 Influenza Viruses from Poultry in Hong Kong," *Virology* 252, no. 2 (Dec. 20, 1998): 331–42.

77 **a pandemic had been averted:** Kennedy F. Shortridge, J. S. Malik Peiris, and Yi Guan, "The Next Influenza Pandemic: Lessons from Hong Kong," *Journal of Applied Microbiology* 94 (2003): 70S–79S.

77 **its most successful:** Several researchers have held up Hong Kong as the model. See, for example, Robert Webster and Diane Hulse, "Controlling Avian Flu at the Source," *Nature* 435 (May 26, 2005): 415–16. Yet the model may be hard to apply elsewhere. See Les Sims, "Achievements, Issues and Options on Strategies for HPAI Control and Prevention," Background Paper at the Technical Meeting on Highly Pathogenic Avian Influenza and Human H5N1 Infection, June 27–29, 2007, Rome.

78 **Their main exposure:** Anthony W. Mounts et al., "Case-Control Study of Risk Factors for Avian Influenza A (H5N1) Disease, Hong Kong, 1997," *Journal of Infectious Diseases* 180 (1999): 505–8.

78 **would resurface in 2001:** On this outbreak and Hong Kong's response, see Yi Guan et al., "Emergence of Multiple Genotypes of H5N1 Avian Influenza Viruses in Hong Kong SAR," *PNAS* 99, no. 13 (June 25, 2002): 8950–55; and N. Y. Kung et al., "The Impact of a Monthly Rest Day on Avian Influenza Virus Isolation Rates in Retail Live Poultry Markets in Hong Kong," *Avian Diseases* 47 (2003): 1037–41.

78 **the virus struck yet again:** For an overview of the poultry outbreaks in 2001–2002, see L. D. Sims et al., "Avian Influenza in Hong Kong 1997–2002," *Avian Diseases* 47, no. s3: 832–38; L. D. Sims et al., "An Update on Avian Influenza in Hong Kong 2002," *Avian Diseases* 47 (2003): 1083–86; and Kennedy F.

Shortridge, J. S. Malik Peiris, and Yi Guan, "The Next Influenza Pandemic: Lessons from Hong Kong," *Journal of Applied Microbiology* 94 (2003): 70S–79S.

## Chapter Three: The Elephant and the Lotus Leaf

This chapter draws on interviews with WHO infectious disease specialists and other officials in both Geneva and Asia, public health officials and medical professionals in Thailand, and internal documents from WHO.

81  **"Influenza has been an epidemic illness":** Prasert Thongcharoen, *Influenza* (Bangkok: Mahidol University, 1998).

82  **blame the spiraling death toll on the weather:** Newin Chidchob, deputy agriculture minister, is quoted in "Thailand Declared Free of Bird Flu," *Nation* (Thailand), Jan. 15, 2004. Yukol Limlamthong, director-general of the Livestock Department, is quoted in "Bird Flu: Govt to Sue over 'False Report,'" *Nation* (Thailand), Jan. 17, 2004.

85  **researchers had confirmed:** Arthit Khwankhom and Sirinart Sirisunthorn, "Govt Ignored Chula Warning," *Nation* (Thailand), Jan. 30, 2004.

86  **"We were fighting":** "What Happened When the H5N1 Virus Visited Thailand," lecture at the Asia Medical Forum, Lancet 2006, Singapore, May 4, 2006.

87  **"Irresponsible media":** Tini Tran, "WHO Says SARS Helped Asia Prepare for Bird Flu; Poultry Culls Continue," Associated Press, Jan. 16, 2004.

87  **"There's absolutely no evidence":** Alisa Tang, "Thai Cabinet Seeks to Boost Confidence of Chicken-Wary Public," Associated Press, Jan. 19, 2004.

88  **a confidential tip:** Internal WHO report, Jan. 20, 2004.

90  **broader resurgence of infectious disease:** For an excellent exploration of what was optimistically called the Health Transition and subsequent setbacks, see Laurie Garrett, *The Coming Plague* (Penguin: New York, 1995).

91  **Storm clouds were gathering:** See, for example, David L. Heymann and Guenael R. Rodier, "Hot Spots in a Wired World: WHO Surveillance of Emerging and Re-emerging Infectious Diseases."

91  **WHO's rapid response:** David L. Heymann and Guenael Rodier, "Global Surveillance, National Surveillance and SARS: Commentary," *Emerging Infectious Diseases* 10, No. 2 (Feb. 1, 2004); and David L. Heymann, Mary Kay Kindhauser, and Guenael Rodier, "Coordinating the Global Response," in *SARS: How a Global Epidemic Was Stopped* (Manila: WHO Western Pacific Regional Office, 2006).

93  **Subsequent study:** Kulkanya Chokephaibulkit et al., "A Child with Avian Influenza A (H5N1) Infection," *Pediatric Infectious Disease Journal* 24, no. 2 (Feb. 2005): 162–66; and Mongkol Uipprasertkul et al., "Influenza A H5N1 Replication Sites in Humans," *Emerging Infectious Diseases* 11, no. 7 (July 2005): 1036–41.

94  **followed right behind:** For a synopsis of Thailand's experience with avian flu in 2004, see Thanawat Tiensin, et al., "Highly Pathogenic Avian Influenza H5N1, Thailand, 2004," *Emerging Infectious Diseases* 11, no. 11 (Nov. 2005): 1664–72.

95  **"It's not a big deal":** Sutin Wannabovorn, "Thailand Confirms Two Human Cases of Bird Flu as Infection spreads Through Asia," Associated Press, Jan 23, 2004.

97  **teams were running short:** Tipawayan Kwankhauw, "Anger and Tears as Thailand's Farmers Cull Millions of Chickens," Agence France Presse, Jan. 25, 2004.

100  **mixing vessel:** See, for instance, S. Scholtissek et al., "The Nucleoprotein as a Possible Major Factor in Determining Host Specificity of Influenza H3N2 Viruses," *Virology* 147 (1985) 287–94; H. Kida et al., "Potential for Transmission of Avian Influenza Viruses to Pigs," *Journal of General Virology* 75, no. 9 (Sept. 1994): 2183–88; and Ian H. Brown, "The Epidemiology and Evolution of Influenza Viruses in Pigs," *Veterinary Microbiology* 74, nos. 1–2 (May 22, 2000): 29–46.

101  **"Are the doctor and the media":** "PM Derides Doctor over Pig Comments," *Nation* (Thailand), Jan. 28, 2004.

### Chapter Four: Into the Volcano

This chapter draws on interviews with current and former infectious-disease specialists, investigators, and other officials at WHO and CDC in the United States, Geneva, and Asia, with Vietnamese, Thai, and Hong Kong disease specialists, and on documents from WHO and CDC and personal notes kept by participants in the events described.

104  **When SARS broke out:** The results of the outbreak investigation in Vietnam are discussed in Hoang Thu Vu et al., "Clinical Description of a Completed Outbreak of SARS in Vietnam, February–May 2003," *Emerging Infectious Diseases* 10, no. 2 (Feb. 2004): 334–38; and Mary G. Reynolds et al., "Factors Associated with Nosocomial SARS-CoV Transmission Among Healthcare Workers in Vietnam, 2003," *BMC Public Health* 6 (2006): 207.

109  **The flu outbreak that began that fall:** For more discussion, see Niranjan Bhat et al., "Influenza-Associated Deaths Among Children in the United States, 2003–2004," *NEJM* 353, no. 24 (Dec. 15, 2005): 2559–67; and Laura Jean Podewils et al., "A National Survey of Severe Influenza-Associated Complications Among Children and Adults, 2003–2004," *Clinical Infectious Diseases* 40 (June 1, 2005):1693–96.

109  **flooded with the infirm:** See, for example, Rob Stein, "Shortage of Flu Shots Prompts Rationing," *Washington Post,* Dec. 9, 2003; Rob Stein, "24 States Hit Hard by Flu Outbreak," *Washington Post,* Dec. 12, 2003; and Anita Manning and Tom Kenworthy, "Flu and Fear Run Rampant," *USA Today,* Dec. 10, 2003.

109  **give up their beds:** "Influenza: Last Bad Flu Season Killed Nearly 65,000; Will This Season Be Worse?" *Drug Week,* Jan. 2, 2004.

109  **made its first recorded appearance:** Alfred W. Crosby, *America's Forgotten Pandemic: The Influenza of 1918,* 2nd ed. (New York: Cambridge University Press, 2003), 71.

109  **sailors transferred days earlier:** John M. Barry, *The Great Influenza: The Epic Story of the Deadliest Plague in History* (Viking Penguin: New York, 2004), 192.

110  **Fourth Annual Liberty Loan parade:** *Philadelphia Inquirer,* Sept. 29, 1918.

110  **an old photograph:** www.history.navy.mil/photos/images/h41000/h41730.jpg.

110  **every hospital bed:** Barry, *Great Influenza,* 220.

110  **"When they got there":** Selma Epp, transcript of unaired interview for "Influenza 1918," *American Experience,* Feb. 28, 1997, quoted in Barry.

110–11  **"historic records of the plague":** Ellen C. Potter, letter to Miss M. Carey Thomas, Oct. 3, 1918, M. Carey Thomas Papers, Special Collections Department, Bryn Mawr College.

111  **254 deaths in a single day:** Barry, *Great Influenza,* 221.

111  **daily toll was 759:** Ibid., 329.

111  **"none to replace them in the wards":** Francis Edward Tourscher, *Work of the Sisters During the Epidemic of Influenza, October, 1918* (Philadelphia: American

Catholic Historical Society, 1919), p. 18, accessed through Villanova University Digital Library Browser, reprinted from the *Records of the American Catholic Historical Society of Philadelphia* 30s, nos. 1–3 (Mar.–Sept. 1919).

111 **Almost half the doctors and nurses:** Barry, *Great Influenza*, 226.

111 **"had no attention for over 18 hours":** Tourscher, *Work of the Sisters*, 18.

111 **"After gasping for several hours":** Ira Starr, "Influenza in 1918: Recollections of the Epidemic in 1918," *Annals of Internal Medicine* 145, no. 2 (July 18, 2006).

111 **at the poorhouse:** Tourscher, *Work of the Sisters*, 50.

112 **the residence of a wealthy family:** Ibid., 62.

112 **cars bearing medical insignia:** Starr, "Influenza in 1918."

112 **so they could help fill prescriptions:** Eileen A. Lynch, "The Flu of 1918: It Started with a Cough in the Summer of 1918," *Pennsylvania Gazette*, Nov. 1998.

112 **Nearly 500 police officers:** *Philadelphia Inquirer*, Oct. 20, 1918.

112 **About 1,800 telephone employees:** Barry, *Great Influenza*, 328.

112 **"no other than absolutely necessary calls":** *Philadelphia Inquirer*, Oct. 18, 1918.

112 **one Fishtown home:** Tourscher, *Work of the Sisters*, 74.

112 **During the second week:** Great Britain Ministry of Health, *Report on the Pandemic of Influenza 1918–1919*, Reports on Public Health and Medical Subjects no. 4 (London: His Majesty's Stationery Office, 1920), 319–20, quoted in Crosby, *America's Forgotten Pandemic*.

112 **abandoned corpses were stacked:** "Emergency Service of the Pennsylvania Council of National Defense in the Influenza Crisis," 35, quoted in Crosby, *America's Forgotten Pandemic*.

113 **piling up on the porches:** Harriet Ferrell, transcript of unaired interview for "Influenza 1918," *American Experience*, Feb. 28, 1997, quoted in Barry, *Great Influenza*.

113 **"The smell would just knock you":** Interview by Charles Handy for WHYY-FM program "The Influenza Pandemic of 1918: Philadelphia, 1918."

113 **"They were taking people out":** Ibid.

113 **"They had so many died":** Ibid.

113 **dispatched a steam shovel:** "Emergency Service of the Pennsylvania Council of National Defense in the Influenza Crisis," 35, quoted in Crosby, *America's Forgotten Pandemic*; and the *Philadelphia Inquirer*, Oct. 12, 1918.

113 **people were stealing them:** Michael Donohue, transcript of unaired interview for "Influenza 1918," *American Experience*, Feb. 28, 1997, quoted in Barry, *Great Influenza*.

113 **under armed guard:** Barry, *Great Influenza*, 327.

113 **12,897 Philadelphians:** Great Britain Ministry of Health, *Report on the Pandemic*, 319–320, quoted in Crosby, *America's Forgotten Pandemic*.

113 **"It was the fear and dread":** Tourscher, *Work of the Sisters*, 105.

114 **tremendous financial pressure:** One-third of hospitals were reported to be operating at a deficit. See John G. Bartlett and Luciano Borio, "The Current Status of Planning for Pandemic Influenza and Implications for Health Care Planning in the United States," *Clinical Infectious Diseases* 46 (Mar. 15, 2008): 919–25.

114 **Hospitals have been closing:** Neil A. Halpern, Stephen M. Pastores, and Robert J. Greenstein, "Critical Care Medicine in the United States 1985–2000: An Analysis of Bed Numbers, Use, and Costs," *Critical Care Medicine* 32, no. 6 (June 2004): 1254–59. Between 1993 and 2003, the United States saw a net loss of 703 hospitals, or 11 percent, and a decline in inpatient beds of 198,000 or 17 percent. See American Hospital Association figures cited in *Institute of Medicine,*

*Hospital-Based Emergency Care: At the Breaking Point* (Washington: National Academies Press, 2007), 38. Sixty percent of U.S. hospitals reported in 2001 that they were operating at or over capacity. See the Lewin Group, *Emergency Department Overload: A Growing Crisis,* the results of the AHA Survey of Emergency Department (ED) and Hospital Capacity.

114    **vacant ICU beds were rare:** Lewis Rubinson et al., "Augmentation of Hospital Critical Care Capacity After Bioterrorist Attacks or Epidemics: Recommendations of the Working Group on Emergency Mass Critical Care," *Critical Care Medicine* 33, no. 10 (2005): 2392–2403. In a severe pandemic, the demand for these ICU beds could outstrip capacity by nearly five times. See Eric Toner et al., "Hospital Preparedness for Pandemic Influenza," *Biosecurity and Bioterrorism* 4, no. 2 (2006): 207–14. Even in a moderately severe outbreak, half the states would run out of hospital beds within two weeks. See Trust for America's Health, *Ready or Not? Protecting the Public's Health from Diseases, Disasters and Bioterrorism,* Dec. 2006.

114    **a severe nursing shortage:** See Elizabeth Daugherty, Richard Branson, and Lewis Rubinson, "Mass Casualty Respiratory Failure," *Current Opinion in Critical Care* 13, no. 1 (Feb. 2007): 51–56; Derek C. Angus et al., "Current and Projected Workforce Requirements for Care of the Critically Ill and Patients with Pulmonary Disease," *Journal of the American Medical Association* 284, no. 21 (Dec. 6, 2000): 2762–70; Mark A. Kelley et al., "The Critical Care Crisis in the United States: A Report from the Profession," *Chest* 125 (2004): 1514–17; Gary W. Ewart et al., "The Critical Care Medicine Crisis: A Call for Federal Action," white paper from the Critical Care Professional Societies, *Chest* 125 (2004): 1518–21; and J. K. Stechmiller, "The Nursing Shortage in Acute and Critical Settings," *AACN Clinical Issues* 13, no. 4 (Nov. 2002): 577–84. The nationwide shortage of nurses has been estimated at between 100,000 and 291,000. All but ten states had a shortage of registered nurses in 2006. See John G. Bartlett and Luciano Borio, "The Current Status of Planning for Pandemic Influenza and Implications for Health Care Planning in the United States," *Clinical Infectious Diseases* 46 (Mar. 15, 2008): 919–25; and Trust for America's Health, *Ready or Not? Protecting the Public's Health from Diseases, Disasters and Bioterrorism,* Dec. 2006.

114    **Emergency rooms are being shuttered:** Eric W. Nawar, Richard W. Niska, and Jianmin Xu, "National Hospital Ambulatory Medical Care Survey: 2005 Emergency Department Summary," advance data from *Vital and Health Statistics,* no. 386, June 29, 2007. For a comprehensive overview of the crisis facing U.S. emergency departments, see Institute of Medicine, *Hospital-Based Emergency Care: At the Breaking Point* (Washington: National Academies Press, 2007). According to figures from the American Hospital Association cited in the IOM report, the number of hospitals with emergency departments declined by 425 over the decade ending in 2003.

114    **departments were routinely overcrowded:** "State of Emergency Medicine: Emergency Physician Survey," American College of Emergency Physicians, October 2003. Sixty-two percent of U.S. hospitals surveyed in 2001 said their emergency departments were operating at or over capacity. For large hospitals and those offering the most advanced trauma care, the percentage increased to about 90 percent. See the Lewin Group, *Emergency Department Overload: A Growing Crisis,* results of the AHA Survey of Emergency Department (ED) and Hospital Capacity, Apr. 2002. As the IOM writes, "In many cities, hospitals and trauma centers have problems dealing with a multiple-car highway crash, much less the volume of patients likely to result from a large-scale disaster." Institute of Medicine, *Hospital-Based Emergency Care,* 265.

114   **once every single minute:** Catharine W. Burt, Linda F. McCaig, and Roberto H. Valverde, "Analysis of Ambulance Transports and Diversions Among U.S. Emergency Departments," *Annals of Emergency Medicine* 47, no. 4 (2006): 317–26. See also Sally Phillips, "Current Status of Surge Research," *Academic Emergency Medicine* 13 (2006): 1103–8.

114   **hospital executives were too preoccupied:** "Emergency Preparedness: States are Planning for Medical Surge, but Could Benefit from Shared Guidance for Allocating Scarce Medical Resources," U.S. Government Accountability Office, June 2008.

114   **decreased 18 percent:** Ibid.

114   **producers of medical oxygen:** Michael D. Christian et al., "Definitive Care for the Critically Ill During a Disaster: Current Capabilities and Limitations," *Chest* 133, no. 5 (May 2008): 8S–17S.

114   **tremendous shortage of ventilators:** Ibid.; and Isaac Weisfuse, "Summary Background on Hospital Pandemic Preparedness in NYC," in Beth Maldin-Morgenthau et al., "Roundtable Discussion: Corporate Pandemic Preparedness," *Biosecurity and Bioterrorism: Biodefense Strategy, Practice, and Science* 5, no. 2 (2007): 171.

115   **about 740,000 people would require ventilation:** "States are Planning for Medical Surge," U.S. Government Accountability Office, June 2008.

115   **between 53,000 and 105,000:** Michael T. Osterholm, "Preparing for the Next Pandemic," *NEJM* 352, no. 18 (May 5, 2005): 1839–42; and Elizabeth Daugherty, Richard Branson, and Lewis Rubinson, "Mass Casualty Respiratory Failure," *Current Opinion in Critical Care* 13, no. 1 (Feb. 2007): 51–56. A study by New York State found that even in a moderate pandemic, there would be a statewide shortfall of 1,256 ventilators. In a severe pandemic, the total demand for ventilators in peak weeks would run to 17,844, almost three times the existing capacity. See NYS Working Group on Ventilator Allocation in an Influenza Pandemic, NYS DOH/NYS Task Force on Life and the Law, "Allocation of Ventilators in an Influenza Pandemic: Planning Document," Mar. 15, 2007.

115   **the Spanish flu's victims:** David M. Morens, Jeffrey K. Taubenberger, and Anthony S. Fauci, "Predominant Role of Bacterial Pneumonia as a Cause of Death in Pandemic Influenza: Implications for Pandemic Influenza Preparedness," *Journal of Infectious Diseases* 198, no. 7 (Oct. 1, 2008): 962–70; Jonathan A. McCullers, "Planning for an Influenza Pandemic: Thinking Beyond the Virus," *Journal of Infectious Diseases* 198, no. 7 (Oct. 1, 2008): 945–47; and John F. Brundage and G. Dennis Shanks, "Deaths from Bacterial Pneumonia During 1918–1919 Influenza Pandemic," *Emerging Infectious Diseases* 14, no. 8 (Aug. 2008): 1193–99.

115   **80 percent of all prescription drugs:** Michael T. Osterholm, "Unprepared for a Pandemic," *Foreign Affairs*, Mar.–Apr. 2007.

115   **"interconnectedness of the global economy":** Ibid.

116   **would run short on everything:** Michael T. Osterholm, "Preparing for the Next Pandemic," *Foreign Affairs*, July–Aug. 2005.

116   **an unpublicized conference call:** Personal notes of conference call, Jan. 15, 2004.

116   **an article for *Science*:** Richard J. Webby and Robert G. Webster, "Are We Ready for Pandemic Influenza?" *Science* 302, no. 5650 (Nov. 28, 2003): 1519–22. Webster also raised concerns in 2003 in Robert G. Webster and Elizabeth Jane Walker, "Influenza," *American Scientist* 91, no. 2 (Mar.–Apr. 2003): 122.

116   **"Klaus was very excited":** Interview with Dick Thompson.

117   **"Hitoshi suddenly came alive again":** Interview with Peter Cordingley.

118 **The cases continued to come:** WHO: "Preliminary Clinical and Epidemiological Description of Influenza A (H5N1) in Vietnam," Feb. 12, 2004; Tran Tinh Hien et al., "Avian Influenza A (H5N1) in 10 Patients in Vietnam," *NEJM* 350, no. 12 (Mar. 18, 2004): 1179–88; and Pham Ngoc Dinh et al., "Risk Factors for Human Infection with Avian Influenza A H5N1, Vietnam 2004," *Emerging Infectious Diseases* 12, no. 12 (Dec. 2006): 1841–47.

118 **Thailand finally stopped:** For an overview of cases in both Thailand and Vietnam, see "Avian Influenza A (H5N1)," *Weekly Epidemiological Record* 79, no. 7 (Feb. 13, 2004): 65–76. On Thailand specifically, see "Cases of Influenza A (H5N1)—Thailand 2004," *Morbidity and Mortality Weekly Report* 53, no. 5 (Feb. 13, 2004): 100–103; Darin Areechokchia et al., "Investigation of Avian Influenza (H5N1) Outbreak in Humans—Thailand, 2004," *Morbidity and Mortality Weekly Report* 55, suppl. 1 (Apr. 28, 2006): 3–6; and Anucha Apisarnthanarak et al., "Atypical Avian Influenza (H5N1)," *Emerging Infectious Diseases* 10, no. 7 (July 2004): 1321–24.

122 **release the findings:** WHO, "Avian Influenza A (H5N1)—Update 14: Two Additional Human Cases of H5N1 Infection Laboratory Confirmed in Vietnam, Investigation of a Family Cluster," Feb. 1, 2004.

123 **"the ethics of researchers":** "Thaksin Challenges WHO Statement," *Nation* (Thailand), Feb. 3, 2004.

123 **"temperatures were running high":** Nguyen Tran Hien, Jeremy Farrar, and Peter Horby, "Person-to-Person Transmission of Influenza A (H5N1)," *Lancet* 371, no. 9622 (Apr. 26, 2008): 1392–94.

125 **"think of it like a war":** Notes of WHO teleconference, Feb. 7, 2004.

125 **widespread in ducks:** Y. Guan et al., "H5N1 Influenza: A Protean Pandemic Threat," *PNAS* 101, no. 21 (May 25, 2004): 8156–61.

125 **permanent foothold in Asian poultry:** K. S. Li et al., "Genesis of a Highly Pathogenic and Potentially Pandemic H5N1 Influenza Virus in Eastern Asia," *Nature* 430 (July 8, 2004): 209–13.

126 **"no link could be established":** Internal WHO report, undated.

126 **"almost certainly H2H transmission":** E-mail, Nov. 6, 2004.

127 **more of the story:** The cluster is also described in Kumnuan Ungchusak et al., "Probable Person-to-Person Transmission of Influenza A (H5N1)," *NEJM* 352, no. 4 (Jan. 27, 2005): 333–40; and "Excerpts of the Meeting of the Expert Panel on Avian Influenza," Bangkok, Sept. 27, 2004.

130 **Thailand's health ministry announced:** "Avian Influenza Infection of Patients in Kamphaeng Phet," press release, Ministry of Health, Thailand, Sept. 28, 2004.

130 **WHO released a statement:** WHO, "Avian Influenza—Situation in Thailand," Sept. 28, 2004.

130 **But even as they accepted:** Likely cases of human transmission have occurred in at least a half-dozen countries, also including Indonesia, Cambodia, Pakistan, and China. On the last, for example, see Hua Wang et al., "Probable Limited Person-to-Person Transmission of Highly Pathogenic Avian Influenza A (H5N1) Virus in China," *Lancet* 371, no. 9622 (April 26, 2008): 1427–34.

## Chapter Five: Livestock Revolution

This chapter draws on interviews with dozens of villagers in Suphan Buri Province.

133 **another continent in the 1950s:** Interview with local historian Samreong Reaungrit.

134    **chicken made its debut:** For the history of the chicken industry in Thailand, see Christopher L. Delgado, Clare A. Narrod, and Marites M. Tiongco, *Policy, Technical, and Environmental Determinants and Implications of the Scaling-Up of Livestock Production in Four Fast-Growing Developing Countries: A Synthesis,* Final Research Report of Phase II, International Food Policy Research Institute, June 23, 2003, ch. 2.2; Christopher L. Delgado and Clare A. Narrod, *Impact of Changing Market Forces and Policies on Structural Change in the Livestock Industries of Selected Fast-Growing Developing Countries,* Final Research Report of Phase I, International Food Policy Research Institute, June 28, 2002, chapter 4.5; and Nipon Poapongsakorn et al., "Annex IV: Livestock Industrialization Project: Phase II—Policy, Technical, and Environmental Determinants and Implications of the Scaling-Up of Swine, Broiler, Layer and Milk Production in Thailand," July 25, 2003, included in Delgado, Narrod, and Tiongco, *Policy, Technical, and Environmental Determinants,* 2003.

134    **doubled the average amount of chicken:** Nipon Poapongsakorn et al., "Annex VIII: Livestock Industrialization, Trade and Social-Health-Environment Issues for the Thai Poultry, Dairy, and Swine Sector," May 2002, included in Delgado and Narrod, *Impact of Changing Market Forces,* 2002.

134    **an even cheaper source of protein:** Ibid.

135    **soaring demand for eggs:** Thailand's egg consumption doubled in a decade. Delgado, Narrod, and Tiongco, *Policy, Technical, and Environmental Determinants,* ch. 2.2.

137    **the first to fall sick:** Jared Diamond, *Guns, Germs, and Steel* (New York: W. W. Norton, 1997), 92.

137    **afflicting their livestock:** Ibid., 196–97.

137    **evolved from animal pathogens:** Jared Diamond, "Evolution, Consequences and Future of Plant and Animal Domestication," *Nature* 418 (Aug. 8, 2002): 700–707.

137    **about 60 percent also cause disease in animals:** S. Cleaveland, M. K. Laurenson, and L. H. Taylor, "Diseases of Humans and Their Domestic Mammals: Pathogen Characteristics, Host Range and Risk of Emergence," *Philosophical Transactions of the Royal Society of London B: Biological Sciences* 356, no. 1411 (July 29, 2001): 991–99.

137    **These microbes can hopscotch:** Willam B. Karesh and Robert A. Cook, "The Human-Animal Link," *Foreign Affairs,* July–Aug. 2005.

137    **An even higher proportion:** L. H. Taylor, S. M. Latham, and M. E. Woolhouse, "Risk Factors for Human Disease Emergence," *Philosophical Transactions of the Royal Society of London B: Biological Sciences* 356, no. 1411 (July 29, 2001): 983–89.

137    **"Similar to the time":** "Animal Health at the Crossroads: Preventing, Detecting, and Diagnosing Animal Diseases" (Washington: National Academy of Sciences, 2005), 27.

137    **a mystery illness erupted:** For an account, see Keith B. Richburg, "Malaysia Slow to Act on Virus," *Washington Post,* Apr. 29, 1999.

138    **opening of trade routes:** See, for example, William H. McNeill, *Plagues and Peoples* (New York: Anchor Books, 1998); Wu Lien-Teh et al., *Plague: A Manual for Medical and Public Health Workers* (Shanghai: National Quarantine Service, 1936); and John Kelly, *The Great Mortality* (New York: HarperCollins, 2005).

138    **"Pharoah's rats":** Wu Lien-Teh et al., *Plague.* I learned of this reference in Kelly, *Great Mortality.*

138    **plague erupted in southern China:** For a good account of the Yunnan outbreak and the subsequent spread of the disease, see Carol Benedict, *Bubonic*

*Plague in Nineteenth-Century China* (Stanford, CA: Stanford University Press, 1996).

139   **it ravaged Hong Kong:** For accounts, see Benedict, *Bubonic Plague;* and Edward Marriott, *The Plague Race: A Tale of Fear, Science and Heroism* (New York: Picador, 2003).

139   **"Little wonder, then":** Marriott, *Plague Race*, 52.

139   **reported that patients suffered:** James Cantlie, "The First Recorded Appearance of the Modern Influenza Epidemic," *British Medical Journal* 2 (1891): 491.

140   **"epicenter" of all influenza viruses:** Kennedy F. Shortridge and C. H. Stuart-Harris, "An Influenza Epicentre?" *Lancet* 2 no. 8302 (Oct. 9, 1982): 812–13; and Kennedy F. Shortridge, "Is China an Influenza Epicenter?" *Chinese Medical Journal* 110 no. 8 (1997): 637–41. More recently, researchers who studied the global spread of seasonal H3N2 flu strains between 2002 and 2007 have also suggested that the region of East and Southeast Asia is the annual source of the world's seasonal flu viruses. See Colin A. Russell et al., "The Global Circulation of Seasonal Influenza A (H3N2) Viruses," *Science* 320, no. 5874 (Apr. 18, 2008): 340–46.

140   **aquatic birds:** Robert G. Webster et al., "Evolution and Ecology of Influenza A Viruses," *Microbiological Review* 56, no. 1 (Mar. 1992): 152–79.

140   **actually been isolated earlier:** A picture of the Chinese scientist who Shortridge says first isolated the 1957 Asian flu virus is shown in Kennedy F. Shortridge, "Influenza—a Continuing Detective Story," *Lancet* 354 (1999): suppl. SIV 29.

141   **matter of greater dispute:** For a broader examination of competing hypotheses, see Gina Kolata, *Flu: The Story of the Great Influenza Pandemic of 1918 and the Search for the Virus that Caused It* (New York: Simon & Schuster, 1999), ch. 10.

141   **Haskell County, Kansas:** John M. Barry, "The Site of the Origin of the 1918 Influenza Pandemic and Its Health Implications," *Journal of Translational Medicine* 2 (Jan. 20, 2004): 3. Among the evidence cited by Barry against an Asian or European provenance is Edwin O. Jordan, *Epidemic Influenza: A Survey* (Chicago: American Medical Association, 1927). Jordan concluded, "The primary origin of the 1918 pandemic cannot be traced with any degree of plausibility to any one of these localized outbreaks," referring to India, China, Japan, France, Germany, or the military camps of the United States and Britain. Jordan himself could not pinpoint the origin.

141   **British army camp:** J. S. Oxford, "The So-called Great Spanish Influenza Pandemic of 1918 May Have Originated in France in 1916," *Philosophical Transactions of the Royal Society of London B: Biological Sciences* 356 (2001): 1857–59.

141   **a Chinese pedigree:** Interviews and e-mail exchanges with Kennedy Shortridge. See also Kennedy F. Shortridge, "The 1918 'Spanish' Flu: Pearls from Swine?" *Nature Medicine* 5, no. 4 (Apr. 1999): 384–85.

141   **medical accounts of an American missionary:** W. W. Cadbury, "The 1918 Pandemic of Influenza in Canton," *China Medical Journal* 34 (1920): 1–17.

142   **the Pearl River delta:** Zhao Shidong et al., "Population, Consumption, and Land Use in the Pearl River Delta, Guangdong Province," in National Academy of Sciences, *Growing Populations, Changing Landscapes: Studies from India, China and the United States* (Washington: National Academies Press, 2001).

142   **"greatest mass urbanization":** This description comes in his tale of another emerging disease to explode out of East Asia: SARS. Karl Taro Greenfeld, *China Syndrome: The True Story of the 21st Century's First Great Epidemic* (New York: HarperCollins, 2006), 9.

143   **fastest growth on Earth:** World Bank, *World Development Report 1997* (Washington: International Bank for Reconstruction and Development, 1997), table 1.

143     **admiration for Bill Gates:** Gates was rated seven times more popular than any sitting member of the Vietnamese Politburo in a survey for *Tuoi Tre* newspaper. The issue, released in Jan. 2001, was pulled from the newsstands by authorities.

143     **"The demand-driven Livestock Revolution":** Christopher Delgado et al., *Livestock to 2020: The Next Food Revolution* (Washington: International Food Policy Research Institute, 1999), 4. For a discussion of the "livestock revolution," see also Christopher L. Delgado, Mark W. Rosegrant, and Siet Meijer, "Livestock to 2020: The Revolution Continues," Jan. 11, 2001, paper presented at the annual meetings of the International Agricultural Trade Research Consortium in Auckland, New Zealand, Jan. 2001.

144     **doubled the average amount of meat:** Henning Steinfeld and Pius Chilonda, "Old Players, New Players," in Food and Agriculture Organization, Livestock Report 2006.

144     **surpassed that in developed ones:** Ibid.

144     **China alone has accounted:** Ibid.

144     **A large majority:** Figures on China's livestock production come from the Food and Agriculture Organization's database FAOSTAT. For discussion of China's demand for livestock products, see William P. Roenigk, "Keynote Address: World Poultry Consumption," *Poultry Science* 78 (1999): 722–28; and Frank Fuller, Francis Tuan, and Eric Wailes, "Rising Demand for Meat: Who Will Feed China's Hogs," in Fred Gale, ed., *China's Food and Agriculture: Issues for the 21st Century,* Agricultural Information Bulletin no. AIB-775, Economic Research Service, U.S. Department of Agriculture, Apr. 2002.

144     **Southeast Asia's record:** Figures on Southeast Asia's livestock production come from the Food and Agriculture Organization's database FAOSTAT.

144     **the Indonesian egg:** D.K.S. Swastika et al., *The Status and Prospect of Feed Crops in Indonesia,* UN Centre for Alleviation of Poverty Through Secondary Crops' Development in Asia and the Pacific, working paper, no. 81, p. 23.

144     **as meat prices dropped:** "Managing the Livestock Revolution: Policy and Technology to Address the Negative Impacts of a Fast-Growing Sector," World Bank, June 2005, p. 12.

144     **the record is more mixed:** On possible negative effects on poverty, equality, food security, and the environment, see Cornelius de Haan et al., "Livestock Development: Implications for Rural Poverty, the Environment and Global Food Security," World Bank, Nov. 2001; Hartwig de Haen et al., "The World Food Economy in the Twenty-first Century: Challenges for International Cooperation," *Development Policy Review* 21, nos. 5–6 (Sept. 2003): 683–96; and Hans Wagner, "Protecting the Eenvironment from the Impact of the Growing Industrialization of Livestock Production in East Asia," special presentation to UN Animal Production and Health Commission for Asia and the Pacific, 26th session, Subang, Malaysia, Aug. 2002. On possible positive effects, see Christopher L. Delgado, Mark Rosegrant, and Nikolas Wada, "Meating and Milking Global Demand: Stakes for Small-Scale Farmers in Developing Countries," in A. G. Brown, ed., *The Livestock Revolution: A Pathway from Poverty?* (Canberra: ATSE Crawford Fund, 2003); and Christopher Delgado et al., *Livestock to 2020: The Next Food Revolution* (Washington: International Food Policy Research Institute, 1999). On equity benefits in Thailand, see Christopher, Narrod, and Tiongco, *Policy, Technical, and Environmental Determinants,* ch. 3.2.

145     **jutting into the fishpond:** On the potential pandemic hazards associated with the mixed development of aquaculture and livestock production, see Christoph Scholtissek and Ernest Naylor, "Fish Farming and Influenza Pandemics," *Nature* 331 (Jan. 21, 1988): 215.

146 **A single gram of bird feces:** Christine Power, "The Source and Means of Spread of the Avian Influenza Virus in the Lower Fraser Valley of British Columbia During an Outbreak in the Winter of 2004: An Interim Report," Canadian Food Inspection Agency, Animal Disease Surveillance Unit, Feb. 15, 1004.

146 **how to prevent epidemic contagion:** See, for example, V. Martin, A. Forman, and J. Lubroth, *Preparing for Highly Pathogenic Avian Influenza* (Rome: Food and Agriculture Organization of the United Nations, 2006), ch. 5.

148 **There was a time:** John Steele Gordon, "The Chicken Story," *American Heritage*, Sept. 1996.

148 **nearly every four days:** "Poultry Slaughter 2006 Annual Summary," Agricultural Statistics Board, U.S. Department of Agriculture, Feb. 2007.

149 **But the watershed:** Interview with Carol Cardona, Associate Veterinarian, University of California at Davis School of Veterinary Medicine.

149 **safety measures to prevent disease:** Interview with Cardona. In fairness, biosecurity remains imperfect. A study in Maryland found most poultry workers are given neither protective clothing nor facilities for on-site decontamination and hygiene. See Lance B. Price et al., "Neurologic Symptoms and Neuropathologic Antibodies in Poultry Workers Exposed to *Campylobacter jejuni*," *Journal of Occupational and Environmental Medicine* 49, no. 7 (July 2007): 748–55.

149 **safeguard their investments:** Interview with Goosen van den Bosch, head of technical services at Intervet.

149 **generous avenue to infection:** On the dangers posed by intensive poultry farming, see J. Otte et al., "Industrial Livestock Production and Global Health Risks," Pro-Poor Livestock Policy Initiative Research Report, UN Food and Agriculture Organization, June 2007.

149 **In the unnatural setting:** B. Schmit, "Disease Prevention Crucial in Intensive Livestock Production," *Zootecnica International,* July 1987, 49–51.

149 **Thai commercial farms:** J. Otte et al., "Evidence-Based Policy for Controlling HPAI in Poultry: Bio-security Revisited," Pro-Poor Livestock Policy Initiative Research Report, UN Food and Agriculture Organization, Dec. 20, 2006.

149 **lack of genetic diversity:** "Managing the Livestock Revolution: Policy and Technology to Address the Negative Impacts of a Fast-Growing Sector," World Bank, June 2005, p. 9.

149 **"Once an influenza virus invades":** R. G. Webster and D. J. Hulse, "Microbial Adaption and Change: Avian Influenza," Rev. sci. tech. Off. int. Epiz., 2004, 23 (2), 453–65.

150 **a country in transition:** Jan Slingenbergh et al., "Ecological Sources of Zoonotic Diseases," *Revue scientifique et technique de l'Office International des Epizooties* 23, no. 2 (2004): 467–84; Marius Gilbert et al., "Livestock Production Dynamics, Bird Migration Cycles, and the Emergence of Highly Pathogenic Avian Influenza in East and Southeast Asia," paper presented at a conference of the Food and Agriculture Organization, Rome, May 30–31, 2006.

150 **are concentrated around:** Pierre Gerber et al., "Geographical Determinants and Environmental Implications of Livestock Production Intensification in Asia," *Bioresource Technology* 96 (2005): 263–76; and Pierre Gerber et al., "Geographical Shifts of the Livestock Production: Land Use and Environmental Impact Implications," paper presented at the conference "Structural Change in the Livestock Sector—Social, Health, and Environmental Implications for Policy Making," Bangkok, Thailand, Jan. 27–29, 2004.

150 **"Agricultural practices have become":** Slingenbergh, "Ecological Sources of Zoonotic Diseases." On the role of ecological factors in the evolution of zoonotic pathogens, see also Stephanie J. Schrag and Pamela Wiener, "Emerging

Infectious Disease: What Are the Relative Roles of Ecology and Evolution?" *Trends in Ecology and Evolution* 10, no. 8 (Aug. 1995): 319–24.

150   **"virtual time bomb":** Les Sims and Claire Narrod, *Understanding Avian Influenza* (Rome: Food and Agriculture Organization, 2008), 2.

151   **showed no symptoms:** D. J. Hulse-Post et al., "Role of Domestic Ducks in the Propogation and Biological Evolution of Highly Pathogenic H5N1 Influenza Viruses in Asia," *PNAS* 102, no. 30 (July 26, 2005): 10682–87.

151   **tested flocks of free-range ducks:** Thaweesak Songserm et al., "Domestic Ducks and H5N1 Influenza Epidemic, Thailand," *Emerging Infectious Diseases* 12, no. 4 (Apr. 2006): 575–81.

151   **Mekong River delta:** On the role of ducks in Vietnam's outbreaks, see Dirk U. Pfeiffer et al., "An Analysis of the Spatial and Temporal Patterns of Highly Pathogenic Avian Influenza Occurrence in Vietnam Using National Surveillance Data," *Veterinary Journal* 174, no. 2 (Sept. 2007): 302–9.

151   **outbreaks in the chicken population:** Marius Gilbert et al., "Free-Grazing Ducks and Highly Pathogenic Avian Influenza, Thailand," *Emerging Infectious Diseases* 12, no. 2 (Feb. 2006): 227–34. Further research, broadened to include Vietnam, provided additional confirmation of the link between avian influenza outbreaks on one hand and ducks and intensive rice cultivation on the other. Rice paddies were identified as the best predictor of outbreak locations. See Marius Gilbert et al., "Mapping H5N1 Highly Pathogenic Avian Influenza Risk in Southeast Asia," *PNAS* 105, no. 12 (Mar. 25, 2008): 4769–74.

154   **Thai government would bar:** Thanawat Tiensin et al., "Geographic and Temporal Distribution of Highly Pathogenic Influenza A Virus (H5N1) in Thailand, 2004–2005: An Overview," *Avian Diseases* 51 (2007): 182–88.

154   **flu outbreaks unexpectedly erupted:** See remarks by Dr. Hoang Van Nam, Department of Animal Health, Ministry of Agriculture and Rural Development, Vietnam, at the Technical Meeting on Highly Pathogenic Avian Influenza and Human H5N1 Infection, June 27–29, 2007, Rome; and UN Food and Agriculture Organization, "Ducks May Be Behind Unexpected HPAI Outbreaks," press release, Avian influenza newsroom, June 7, 2007.

154   **the fields of Kanchanaburi province:** The episode is discussed in Thaweesak Songserm et al., "Domestic Ducks and H5N1 Influenza Epidemic, Thailand," *Emerging Infectious Diseases* 12, no. 4 (Apr. 2006): 575–81.

154   **a peasant named Bang-on Benphat:** Rungrawee C. Pinyorat, "Thailand Confirms 13th Human Death from Bird Flu," Associated Press, Oct. 20, 2005; and WHO, Situation in Thailand—Update 35, Oct. 20, 2005.

155   **"Even insects can't get in":** Even modern, all-enclosed poultry houses have been found to be vulnerable to disease. See, for example, J. Otte et al., "Industrial Livestock Production and Global Health Risks," Pro-Poor Livestock Policy Initiative Research Report, UN Food and Agriculture Organization, Agriculture and Consumer Protection Department, Animal Production and Health Division, June 2007. All-enclosed houses are even vulnerable to insects that spread infection. See Kyoko Sawabe et al., "Detection and Isolation of Highly Pathogenic Avian Influenza A Viruses from Blow Flies Collected in the Vicinity of an Infected Poultry Farm in Kyoto, Japan, 2004," *American Journal of Tropical Medicine and Hygiene* 75, no. 2 (2006): 327–32; and Birthe Hald et al., "Flies and *Campylobacter* Infection of Broiler Flocks," *Emerging Infectious Diseases* 10, no. 8 (Aug. 2004): 1490–92. In some cases, the fans used for ventilating enclosed houses expel contaminated particles into the outside air, where they can infect other poultry houses and farms. See Christine Power, "The Source and Means of Spread of the Avian Influenza Virus in the Lower Fraser Valley of British

Columbia During an Outbreak in the Winter of 2004: An Interim Report," Canadian Food Inspection Agency, Animal Disease Surveillance Unit, Feb. 15, 2004; and T. A. Jones, C. A. Donnelly, and M. Stamp Dawkins, "Environmental and Management Factors Affecting the Welfare of Chickens on Commercial Farms in the United Kingdom and Demark Stocked at Five Densities," *Poultry Science* 84 (2005): 1155–65.

## Chapter Six: From a Single Spark

This chapter draws on interviews with public health officials and other disease specialists in Hong Kong and Guangdong.

158   **"strange contagious disease":** SARS: *How a Global Epidemic Was Stopped* (Manila: WHO Western Pacific Regional Office, 2006), 5.
158   **detected in a Hong Kong family:** The case is described in J. S. Malik Peiris et al., "Re-emergence of Fatal Human Influenza A Subtype H5N1 Disease," *Lancet* 363, no. 9409 (Feb. 21, 2004): 617–19; and Bernice Wuethrich, "An Avian Flu Jumps to People," *Science* 299, no. 5612 (Mar. 7, 2003): 1504.
161   **a medical conference in Beijing:** Ceci Connolly, "Four Months of Clues to Diagnosis," *Washington Post*, June 23, 2003; and Donald J. McNeil Jr. with Lawrence K. Altman, "As SARS Outbreak Took Shape, Health Agency Took Fast Action," *New York Times*, May 4, 2003.
161   **"He talked about deaths":** *Disclosure*, Canadian Broadcasting Corp., Nov. 18, 2003, cited in "Documentary Says WHO Missed Chances to Contain SARS in China," Canadian Press, Nov. 18, 2003.
161   **"put two and two together":** Michael Specter, "Nature's Bioterrorist: Is There Any Way to Prevent a Deadly Avian-Flu Pandemic?" *New Yorker*, Feb. 28, 2005.
163   **precisely what it was:** For a description of the Guangdong outbreak and analysis of the samples collected in Guangdong, see N. S. Zhong et al., "Epidemiology and Cause of Severe Acute Respiratory Syndrome (SARS) in Guangdong, People's Republic of China, in February, 2003," *Lancet* 362, no. 9393 (Oct. 25, 2003): 1353–58.
163   **With the vials stashed in his satchel:** A riveting account of this episode can be found in Karl Taro Greenfeld, *China Syndrome* (New York: HarperCollins, 2006), ch. 20.
165   **"We tried to do our best":** Cheung Chi-fai, "Margaret Chan Breaks Down Twice at Hearing," *South China Morning Post*, Jan. 14, 2004.
165   **"Usually, with other infectious diseases":** Mary Ann Benitez, "Health Chief Told Outbreak a State Secret," *South China Morning Post*, Jan. 13, 2004.
165   **She was faulted:** Report of the Select Committee to Inquire into the Handling of the Severe Acute Respiratory Syndrome Outbreak by the Government and the Hospital Authority, Hong Kong Legislative Council, July 2004, ch. 3, www.legco.gov.hk/yr03-04/english/sc/sc_sars/reports/sars_rpt.htm (accessed Feb. 16, 2009).
165   **a forty-four-year-old seafood seller:** SARS: *How a Global Epidemic Was Stopped* (Manila: WHO Western Pacific Regional Office, 2006), ch. 13.
166   **ninth floor of the Metropole:** For an excellent account of the Metropole episode, see Ellen Nakashima, "SARS Signals Missed in Hong Kong," *Washington Post*, May 20, 2003. See also SARS: *How a Global Epidemic Was Stopped*, ch. 14.
168   **Air China flight 112:** SARS: *How a Global Epidemic Was Stopped*, ch. 15; Brad

Evenson, "'Viral Bullets': SARS 'Super Spreader' Seemed to Infect All Those Around Him on Air China Flight 112," *National Post,* Mar. 29, 2003; Joseph Kahn with Elisabeth Rosenthal, "Even in Remote China, SARS Arrives in Force," *New York Times,* Apr. 22, 2003; and Indira A. R. Lakshmanan, "Health Experts Express Alarm at Nature of SARS Spread on Air China Flight," *Boston Globe,* May 18, 2003.

168  **more than 4,000:** Ellen Nakashima, "SARS Signals Missed in Hong Kong," *Washington Post,* May 20, 2003.

169  **islanders of the Pacific:** Alfred W. Crosby, *America's Forgotten Pandemic: The Influenza of 1918,* 2nd ed. (New York: Cambridge University Press, 2003), ch. 12.

169  **Eskimo villages of Alaska:** John M. Barry, *The Great Influenza: The Epic Story of the Deadliest Plague in History* (New York: Viking Penguin, 2004), ch. 30; Gina Kolata, *Flu: The Story of the Great Influenza Pandemic of 1918 and the Search for the Virus That Caused It* (New York: Simon & Schuster, 1999), ch. 4; and Crosby, *America's Forgotten Pandemic,* ch. 12.

170  **threat of infectious disease:** For a discussion of the positive and negative implications of globalization for infectious disease, see Karen J. Monaghan, "SARS: Down But Still a Threat," National Intelligence Council, 2003, reprinted in Stacey Knobler et al., eds., *Learning from SARS: Preparing for the Next Disease Outbreak—Workshop Summary* (Washington: National Academies Press, 2004).

170  **speed of jet aircraft:** John T. Bowen Jr. and Christian Laroe, "Airline Networks and the International Diffusion of Severe Acute Respiratory Disease, SARS," *Geographical Journal* 172, no. 2 (June 2006): 130–44.

170  **"real potential for rapid dissemination":** Statement of Mark A. Gendreau before the Committee on House Transportation and Infrastructure Subcommittee on Aviation, Apr. 6, 2005, *Congressional Quarterly: Congressional Testimony,* Apr. 6, 2005.

170  **"a wake-up call":** WHO news release, International Health Regulations Enter into Force, June 14, 2006.

170  **The Black Death:** David Herlihy, *The Black Death and the Transformation of the West* (Cambridge, MA: Harvard University Press, 1997).

170  **The last of three cholera epidemics:** G. F. Pyle, "The Diffusion of Cholera in the United States in the Nineteenth Century," *Geographical Analysis* 1 (1969): 59–75.

170–71  **Using data on the volume of travelers:** Rebecca F. Grais, Hugh Ellis, and Gregory E. Glass, "Assessing the Impact of Airline Travel on the Geographic Spread of Pandemic Influenza," *European Journal of Epidemiology* 18 (2003): 1065–72.

171  **a different statistical approach:** Ben S. Cooper et al., "Delaying the International Spread of Pandemic Influenza," *PLoS Medicine* 3, no. 6 (June 2006): e212.

171  **isolated a pathogen:** J. S. Malik Peiris et al., "Coronavirus as a Possible Cause of Severe Acute Respiratory Syndrome," *Lancet* 361, no. 9366 (Apr. 19, 2003): 1319–25.

172  **an unprecedented coup:** For a fuller discussion of WHO's success, see J. S. Mackenzie et al., "The WHO Response to SARS and Preparations for the Future," in Stacey Knobler et al., eds., *Learning from SARS: Preparing for the Next Disease Outbreak—Workshop Summary* (Washington: National Academies Press, 2004); David L. Heymann and Guenael Rodier, "SARS: Lessons from a New

Disease," in Knobler, *Learning from SARS; and SARS: How a Global Epidemic Was Stopped* (Manila: WHO Western Pacific Regional Office, 2006), ch. 2.

172 **"The quality, speed and effectiveness":** Knobler, *Learning from SARS,* 2.

174 **approached the traders:** For a dramatic account, see Karl Taro Greenfeld, *China Syndrome* (New York: HarperCollins, 2006), ch. 69.

175 **found the evidence:** Yi Guan et al., "Isolation and Characterization of Viruses Related to the SARS *Coronavirus* from Animals in Southern China," *Science* 302, no. 5643 (Oct. 10, 2003): 276–78. For a discussion of Guan's investigation, see Dennis Normile and Martin Enserink, "Tracking the Roots of a Killer," *Science* 301, no. 5631 (July 18, 2003): 297–99.

176 **"first emerging disease":** *SARS: How a Global Epidemic Was Stopped,* overview.

177 **its reproductive number was lower:** Marc Lipsitch et al., "Transmission Dynamics and Control of Severe Acute Respiratory Syndrome," *Science* 300, no. 5627 (June 20, 2003): 1966–70.

177 **three times greater or more:** Christophe Fraser et al., "Factors That Make an Infectious Disease Outbreak Controllable," *PNAS* 101, no. 16 (Apr. 20, 2004): 6146–51.

177 **virus in their nose and throat:** J. S. Malik Peiris et al., "Clinical Progression and Viral Load in a Community Outbreak of *Coronavirus*-Associated SARS Pneumonia: A Progressive Study," *Lancet* 361, no. 9371 (May 24, 2003): 1767–72.

177 **rarely contagious in the first few days:** Roy M. Anderson et al., "Epidemiology, Transmission Dynamics and Control of SARS: The 2002–2003 Epidemic," *Philosophical Transactions of the Royal Society of London B: Biological Sciences* 359 (2004): 1091–1105; and Lipsitch, "Transmission Dynamics."

177 **"very lucky this time":** Anderson, "Epidemiology, Transmission Dynamics and Control of SARS."

177 **between 30 and 50 percent:** Fraser, "Factors That Make an Infectious Disease Outbreak Controllable."

177 **"Once adapted to human-to-human transmission":** J. S. Malik Peiris and Yi Guan, "Confronting SARS: A View from Hong Kong," *Philosophical Transactions of the Royal Society of London B: Biological Sciences* 359 (2004): 1075–79.

## Chapter Seven: Cockfighting and Karma

188 **about 7,500 years:** Barbara West and Ben-Xiong Zhou, "Did Chickens Go North? New Evidence for Domestication," *Journal of Archaeological Science* 15 (1988): 515–33.

188 **most likely in Thailand itself:** Akishinonomiya Fumihito et al., "One Subspecies of the Red Junglefowl (*Gallus gallus gallus*) Suffices as the Matriarchic Ancestor of All Domestic Breeds," *PNAS* 91 (Dec. 1994), 12505–9.

188 **red jungle fowl:** Ibid.

189 **"fight for kingdoms":** "Commemoration of King Naresuan: The Nation's Great King 400 Years Ago," Welcome to Chiangmai and Chiangrai, n.d., www.chiangmai-chiangrai.com/king_n.html (accessed Feb. 16, 2009).

190 **Yuenyong included the song:** Charles Piller, "Squawking at Bird Flu Warning," *Los Angeles Times,* Sept. 1, 2005.

196 **clear their throats:** For the various details of this case, see Vijay Joshi, "Thai Man Dies of Bird Flu, Asian Toll Rises to 28," Associated Press, Sept. 9, 2004; "Bird Flu Kills 18-Year-Old Man in Thailand," Kyodo News Service, Sept. 9,

2004; and "Thai Man Dies of Bird Flu: Health Ministry," Agence France Presse, Sept. 9, 2004.

197   **had died in previous weeks:** "Breeder Dies from Bird Flu," *Nation* (Thailand), Sept. 10, 2004.

197   **"The victim failed to report":** Anusak Konglang, "Thailand Reports First Bird Flu Death in Over Seven Months," Agence France Presse, July 26, 2006.

197   **villagers had declined to notify officials:** "Thailand Tries to Improve Bird Flu Monitoring," Reuters, June 27, 2006.

198   **The state's chief minister:** "Repent, Nik Aziz Tells Cockfighting Buffs," *Bernama*, Sept. 23, 2004.

198   **Thailand shipped nearly six thousand:** Kasikorn Research Center, Feb. 10, 2004, cited in "Raising Domestic Chicken Breeds: Interesting," Thai Press Reports, Feb. 11, 2004.

198   **illegal cockfighting tours:** "Cock-fighting Birds Likely Culprit in Mukdahan H5N1 Outbreak," *Nation* (Thailand), Mar. 21, 2007.

200   **30 million households:** "Govt Defends Its Bird Flu Measures," *Jakarta Post*, Aug. 10, 2006.

200   **"greatest single challenge":** Comments from the U.S. Agency for International Development in appendix 2 of *Influenza Pandemic: Efforts to Forestall Onset Are Underway; Identifying Countries at Greatest Risk Entails Challenges*, Government Accountability Office, June 2007, GAO-07-604.

201   **force poultry farming underground:** Juan Lubroth at an FAO press conference in Bangkok, quoted in FAO news release, "New Bird Flu Outbreaks Require Strong Vigilance," Jan. 23, 2007.

202   **continue to take risks:** Sowath Ly et al., "Interaction Between Humans and Poultry, Rural Cambodia," *Emerging Infectious Diseases* 13, no. 1 (Jan. 2007): 130–32; and H. M. Barennes et al., "Avian Influenza Risk Perceptions, Laos," *Emerging Infectious Diseases* 13, no. 7 (July 2007): 1126–28.

203   **thirteen thousand live poultry markets:** Joseph Domenech et al., "Trends and Dynamics of HPAI—Epidemiological and Animal Health Risks," Background Paper at the Technical Meeting on Highly Pathogenic Avian Influenza and Human H5N1 Infection, Rome, June 27–29, 2007.

203   **a perilous nexus:** Robert G. Webster, "Wet Markets—a Continuing Source of Severe Acute Respiratory Syndrome and Influenza?" *Lancet* 363, no. 9404 (Jan. 17, 2004): 234–36. See also Writing Committee of the Second World Health Organization Consultation on Clinical Aspects of Human Infection with Avian Influenza A (H5N1) Virus, "Update on Avian Influenza A (H5N1) Virus Infection in Humans," *NEJM* 358, no. 3 (Jan. 17, 2008): 261–73; and L. D. Sims, "Lessons Learned from Asian H5N1 Outbreak Control," *Avian Diseases* 51 (2007): 182–88 (2007).

203   **"missing link":** D. A. Senne, J. E. Pearson, and B. Panigrahy, "Live Poultry Markets: A Missing Link in the Epidemiology of Avian Influenza," in B. C. Easterday, ed., *Proceedings of the Third Annual Symposium on Avian Influenza, Madison, Wisconsin* (Richmond, VA: U.S. Animal Health Association, 1992).

203   **1997 human outbreak in Hong Kong:** J. C. de Jong et al., "A Pandemic Warning?" *Nature* 389, no. 6651 (Oct. 9, 1997): 554; Eric C. J. Claas et al., "Human Influenza A H5N1 Virus Related to a Highly Pathogenic Avian Influenza Virus," *Lancet* 351, no. 9101 (Feb. 14, 1998): 472–77; and Anthony W. Mounts et al., "Case-Control Study of Risk Factors for Avian Influenza A (H5N1) Disease, Hong Kong, 1997," *Journal of Infectious Diseases* 180 (1999): 505–8.

203   **returned to Hong Kong's markets:** L. D. Sims et al., "Avian Influenza in Hong Kong 1997–2002," *Avian Diseases* 47 (2003), no. s3: 832–38.

**203**  **On the mainland:** Ming Liu et al., "The Influenza Virus Gene Pool in a Poultry Market in South Central China," *Virology* 305, no. 2 (Jan. 20, 2003): 267–75.

**203**  **six city dwellers:** Hongjie Yu et al., "Human Influenza A (H5N1) Cases, Urban Areas of the People's Republic of China, 2005–2006," *Emerging Infectious Diseases* 13, no. 7 (July 2007): 1061–64.

**207**  **a rapid-response unit:** Interview with Nick Marx, WildAid.

**207**  **imported by the tens of thousands:** Hong Kong government press release, Jan. 6, 2007.

**207**  **principal threat of reinfection:** Mary Ann Benitez, "Ban Wild Bird Imports, Experts Say," *South China Morning Post,* Jan. 20, 2007.

## Chapter Eight: Sitting on Fire

**213**  **related to the Guangdong goose isolate:** Xiyan Xu et al., "Genetic Characterization of the Pathogenic Influenza A/Goose/Guangdong/1/96 (H5N1) Virus: Similarity of Its Hemagglutinin Gene to Those of H5N1 Viruses from the 1997 Outbreaks in Hong Kong," *Virology* 261, no. 1 (Aug. 15, 1999): 15–19.

**213**  **At least three other academic papers:** H. Chen K. Yu, and Z. Bu, "Molecular Analysis of Hemagglutinin Gene of Goose Origin Highly Pathogenic Avian Influenza Virus," *Agricultural Sciences in China* 32 (1999): 87–92; X. Tang et al., "Isolation and Characterization of Prevalent Strains of Avian Influenza Viruses in China," *Chinese Journal of Animal and Poultry Infectious Diseases* 20 (1998): 1–5; and Y. Guo, X. Xu, and X. Wen, "Genetic Characterization of an Avian Influenza A (H5N1) Virus Isolated from a Sick Goose in China," *Chinese Journal of Experimental and Clinical Virology* 12, no. 4 (Dec. 1998): 322–25.

**213**  **continued to deny publicly:** See, for example, "Authorities Deny Claim on Disease," Chinadaily.com, Mar. 7, 2007.

**213**  **spawned the wider epidemic:** On Guangdong as the continuing source of H5N1 virus strains that spread internationally, see, for example, Robert G. Wallace et al., "A Statistical Phylogeography of Influenza A H5N1," *PNAS* 104, no. 11 (Mar. 13, 2007): 4473–78. For a discussion of the precursor flu viruses that gave birth to H5N1, see, for example, L. Duan et al., "Characterization of Low-Pathogenic H5 Subtype Influenza Viruses from Eurasia: Implications for the Origin of Highly Pathogenic H5N1 Viruses," *Journal of Virology* 81, no. 14 (July 2007): 7529–39; and Zi-Ming Zhao et al., "Genotypic Diversity of H5N1 Highly Pathogenic Avian Influenza Viruses," *Journal of General Virology* 89 (2008): 2182–93.

**213**  **just months before the Hong Kong cases:** See, for example, Rone Tempest, "Hong Kong to Extend Poultry Ban to Ensure Avian Virus Is Eradicated," *Los Angeles Times,* Dec. 31, 1997. Chinese authorities may also have covered up poultry outbreaks in Guangdong in Oct. 2003. See Dennis Chong, "Guangdong Hid Deaths," *Standard* (Hong Kong), Feb. 4, 2004.

**213**  **continuing to circulate:** Angela N. Cauthen et al., "Continued Circulation in China of Highly Pathogenic Avian Influenza Viruses Encoding the Hemagglutinin Gene Associated with the 1997 H5N1 Outbreak in Poultry and Humans," *Journal of Virology* 74, no. 14 (July 2000): 6592–99; and Robert G. Webster et al., "Characterization of H5N1 Influenza Viruses That Continue to Circulate in Geese in Southeastern China," *Journal of Virology* 76, no. 1 (Jan. 2002): 118–26.

**213**  **geese and ducks exported from Guangdong:** Yi Guan et al., "H5N1 Influenza Viruses Isolated from Geese in Southeastern China: Evidence for Genetic Reassortment and Interspecies Transmission to Ducks," *Virology* 292, no. 1 (Jan. 5, 2002): 16–23.

**213** **duck meat exported from Shanghai:** Terrence M. Tumpey et al., "Characterization of Highly Pathogenic H5N1 Avian Influenza A Virus Isolated from Duck Meat," *Journal of Virology* 76, no. 12 (June 2002): 6344–55; and X. H. Lu et al., "Pathogenesis of and Immunity to a New Influenza A (H5N1) Virus Isolated from Duck Meat," *Avian Diseases* 47 (2003): 1135–40.

**213** **had repeatedly come back positive:** H. Chen et al., "The Evolution of H5N1 Influenza Viruses in Ducks in Southern China," *PNAS* 101, no. 28 (July 13, 2004): 10452–57. For more discussion of the virus in Chinese ducks, see L. D. Sims et al., "Origin and Evolution of Highly Pathogenic H5N1 Avian Influenza in Asia," *Veterinary Record* 157, no. 6 (Aug. 6, 2005): 159–64.

**213** **China as the wellspring:** H. Chen et al., "Establishment of Multiple Sublineages of H5N1 Influenza Virus in Asia: Implications for Pandemic Control," *PNAS* 103, no. 8 (Feb. 21, 2006): 2845–50. China's health ministry dismissed this study, saying it jumped to conclusions. See Mary Ann Benitez and Joseph Ma, "H5N1 Kills 8th Person on Mainland," *South China Morning Post,* Feb. 11, 2006.

**214** **vaccinating their flocks against it:** Debora MacKenzie, "Bird Flu Outbreak Started a Year Ago," *New Scientist,* Jan. 31, 2004.

**214** **an even riskier strategy:** The account of amantadine use in Chinese poultry flocks is based on extensive interviews with animal-health experts and executives of pharmaceutical companies.

**214** **had become resistant to the drug:** On amantadine resistance among some H5N1 variants, see K. S. Li et al., "Genesis of Highly Pathogenic and Potentially Pandemic H5N1 Influenza Virus in Eastern Asia," *Nature* 430 (July 8, 2004): 209–13; T. T. Hien et al., "Avian Influenza A (H5N1) in 10 Patients in Vietnam," *NEJM* 350 (2004): 1179–88; Natalia A. Ilyushina, Elena A. Govorkova, and Robert G. Webster, "Detection of Amantadine Resistant Variants Among Avian Influenza Viruses Isolated in North America and Asia," *Virology* 341, no. 1 (Oct. 10, 2005), 102–6; and Chung-Lam Cheung et al., "Distribution of Amantadine-Resistant H5N1 Avian Influenza Variants in Asia," *Journal of Infectious Diseases* 193 (June 15, 2006): 1626–29.

**215** **used the drug in poultry:** Fu Jing, "Misuse of Antiviral on Poultry Must Stop," *China Daily,* June 21, 2005.

**216** **in a pair of Hanoi markets:** Doan C. Nguyen et al., "Isolation and Characterization of Avian Influenza Viruses, Including Highly Pathogenic H5N1, from Poultry in Live Bird Markets in Hanoi, Vietnam, in 2001," *Journal of Virology* 79, no. 7 (Apr. 2005): 4201–14.

**216** **a new variant of the H5N1 virus:** J. Wang et al., "Identification of the Progenitors of Indonesian and Vietnamese Avian Influenza A (H5N1) Viruses from Southern China," *Journal of Virology* 82, no. 7 (Apr. 2008): 3405–14.

**216** **how it decimated poultry:** Interviews with animal health officials in Ha Tay and Vinh Phuc provinces.

**216** **Japfa's annual corporate filings:** PT Japfa Comfeed Indonesia Tbk, Annual Report 2003; PT Multibreeder Adirama Indonesia Tbk, Annual Report 2003; and PT Japfa Comfeed Indonesia Tbk, Annual Report 2004.

**216** **hushed up their findings:** Karl Taro Greenfeld, "On High Alert," *Time Asia,* Jan. 26, 2005.

**217** **"the first signs of an epidemic":** Ibid.

**217** **first outbreak eventually confirmed:** T. Delquigny et al., "Evolution and Impact of Avian Influenza Epidemic and Description of the Avian Production in Vietnam," final report for FAO's TCP/RAS/3010, Emergency Regional Support for Post Avian Influenza Rehabilitation (Rome: UN Food and Agriculture Organization, 2004).

218    **at least three separate occasions:** H. Chen et al., "Establishment of Multiple Sublineages of H5N1 Influenza Virus in Asia: Implications for Pandemic Control," *PNAS* 103, no. 8 (Feb. 21, 2006): 2845–50.

218    **the government had not responded:** Ben Rowse, "Hospitalized Woman in Vietnam Tests Positive for Bird Flu," Agence France Presse, Aug. 16, 2004.

218    **were going unanswered:** E-mail, Aug. 13, 2004.

218    **"So basically, bugger all":** E-mail, Aug. 17, 2004.

218    **"grave concerns":** Avian influenza update, WHO, Report from Hanoi office, Mar. 16, 2005.

218    **they were flying blind:** E-mail exchange between senior WHO officials in Geneva, Mar. 13, 2005.

220    **widely praised by UN agencies:** See, for example, FAO news release, "Once Hard Hit by Bird Flu, Vietnam Consolidates Progress," Dec. 6, 2006; and WHO, "Successful Strategies in Controlling Avian Influenza," INFOSAN Information Note no. 4/2006, Aug. 14, 2006.

220    **called its performance "remarkable":** Comments from the U.S. Agency for International Development in appendix 2 of *Influenza Pandemic: Efforts to Forestall Onset Are Underway; Identifying Countries at Greatest Risk Entails Challenges,* Government Accountability Office, June 2007, GAO-07-604. The term *remarkable* was applied to both Vietnam and Thailand.

220    **"The situation is alarming":** "Vietnam Battles Three Bird Flu Outbreaks," DPA (German Press Agency) article published in the *Bangkok Post,* Dec. 22, 2006.

222    **Their revelations hit the streets:** Duc Trung and Hoai Nam, "Chances of Bird Flu to Break Out, Quarantine Papers Are Sold like Vegetables," *Thanh Nieh,* Aug. 16, 2005.

222    **spread to nearly one-third of Indonesia's provinces:** Remarks of Agriculture Minister Bungaran Saragih in Material of Consultative Meeting Between Minister of Agriculture and Commission III DPR on Avian Influenza, Jan. 29, 2004.

222    **provenance of the Indonesian strain:** J. Wang et al., "Identification of the Progenitors of Indonesian and Vietnamese Avian Influenza A (H5N1) Viruses from Southern China," *Journal of Virology* 82, no. 7 (Apr. 2008): 3405–14.

223    **separate findings of a pathologist:** Walujo Budi Priyono of the Disease Investigation Centre in Yogyakarta, quoted in FAO news release, "Virus Detective Work in Indonesia: The Case of the Mysterious Livestock Disease," 2005.

223    **"As of now, there are no findings":** Sofyan Sudrajat in *Republika,* Jan. 25, 2004.

224    **10 million chickens:** "Death of 10 Million Laying Hens in Indonesia from Bird Flu," *Kompas,* Jan. 25, 2004.

224    **trade in poultry and poultry products:** G. J. D. Smith et al., "Evolution and Adaptation of H5N1 Influenza Virus in Avian and Human Hosts in Indonesia and Vietnam," *Virology* 350 (2006): 258–68.

224    **"Indonesia is a time-bomb":** "Indonesia Is a Bird-Flu Time-Bomb, Animal Health Chief," Agence France Presse, Apr. 14, 2006.

225    **She repeated her allegations:** Agnes Aristiarini, "Fighting Against Bird Flu," *Kompas,* Sept. 20, 2005.

225    **because of party politics:** "Avian Flu Expert Speaks Out," *Van Zorge Report,* Apr. 12, 2006.

226    **"the lack of a national strategy":** Katia Dolmadjian, "Animal Health Experts Discuss Merits of Vaccination Against Bird Flu," Agence France Presse, Mar. 22, 2007. On how the poultry epidemic in Indonesia continued to pose a threat to

human health, see Endang R. Sedyaningsih et al., "Epidemiology of Cases of H5N1 Virus Infection in Indonesia, July 2005–July 2006," *Journal of Infectious Diseases* 196 (Aug. 15, 2007): 522–27.

226     **the commission formally acknowledged:** The commission wrote, "Indonesia reported its first case of H5N1 infection in poultry in Pekalongan and [the Jakarta suburb of] Tangerang Regencies in August 2003." Media release, "Indonesia after 2 years, 99 cases," KOMNAS FBPI, June 6, 2007.

226     **"I will remember the support":** Josephine Ma and Mary Ann Benitez, "Beijing Agrees to Share Bird Flu Samples Sooner," *South China Morning Post*, Dec. 2, 2006.

227     **"my nationality on my sleeve":** Mary Ann Benitez, "A Giant Responsibility," *South China Morning Post*, Nov. 12, 2006.

228     **"We will have to look":** Ma and Benitez, "Beijing Agrees."

228     **"No nation has the right":** "Health Diplomacy in the 21st Century," address to Directorate for Health and Social Affairs, Norway, Oslo, Feb. 13, 2007.

228     **a cause for concern:** Doubts about China's openness in dealing with bird flu escalated in June 2006, when Chinese researchers disclosed that mainland China had had its first human case in November 2003, two years earlier than authorities had previously reported. See Qing-Yu Zhu et al., "Fatal Infection with Influenza A (H5N1) Virus in China," *NEJM* 354, no. 25 (June 22, 2006): 2731–32.

228     **"What on earth is going on?":** E-mail, Apr. 19, 2006.

228     **a related outbreak in poultry:** WHO's chief representative in China, Hank Bekedam, expressed public frustration, saying, "That is not a good record." Audra Ang, "WHO: Bird Flu Continues to Be Public Health Threat in China as New Case Reported in Military," Associated Press, May 28, 2007.

228     **without fully disarming the virus:** A general warning about the difficulty of monitoring for bird flu in areas where poultry immunization is widely but imperfectly practiced can be found in Influenza Team, European Centre for Disease Surveillance and Control, "World Avian Influenza Update," *Eurosurveillance* 11, no. 6 (2006): 060622.

228     **exposure to this second strain, H9N2:** For a fuller discussion of the dangers posed by cocirculation of H9N2 and H5N1, see Alexey Khalenkov et al., "Modulation of the Severity of Highly Pathogenic H5N1 Influenza in Chickens Previously Inoculated with Israeli H9N2 Influenza Viruses," *Virology* 383 (2009): 32–38.

229     **the politics of China's public health system:** Huang's exploration of infectious-disease policy in China can be found in his writings, including "China's Response to Avian Flu," paper delivered at SAIS China Forum, Mar. 18, 2006, Washington; "The Political Challenges of Health Crises in China," speech at the Conference on Asia and the Science and Politics of Pandemics, CNA Corp., Feb. 3, 2005; and "The Politics of China's SARS Crisis," *Harvard Asia Quarterly*, Fall 2003.

229     **transformed the country's health sector:** See also Nan-Shan Zhong and Guang-Qiao Zeng, "Pandemic Planning in China: Applying Lessons from Severe Acute Respiratory Syndrome," *Respirology* 13, suppl. 1 (2008): S33–S35.

229     **did not extend to the agriculture ministry:** See, for example, "Sanitising the Record; Infectious Diseases in China," *Economist*, July 1, 2006.

229     **the central government wasn't sure:** Notes of WHO meeting with Hui Liangyu in e-mail, Feb. 6, 2004. Two years later, the Chinese health ministry was still complaining that local authorities were failing to report possible cases of bird flu in a timely fashion. See, for example, Nicholas Zamiska, "China Bird-Flu Data in Doubt," *Wall Street Journal*, Apr. 27, 2006.

229     **that changed his life:** For good accounts of Qiao Songju's ordeal, see "Whistle-

Blower Awaits Blackmail Verdict," Chinadaily.com, Apr. 29, 2006; Xu Xiang, "China Plagued by Bird-Flu Cover-ups," Asia Times Online, June 8, 2006; and Jane Cai, "Bird Flu Whistle-Blower Gets Jail Term for Graft," *South China Morning Post,* July 10, 2006.

230 **"Qiao Songju is a sinner":** Xu Xiang, "China Plagued by Bird-Flu Cover-ups."

231 **"block information from us":** Notes of telephone briefing for WHO headquarters, July 20, 2005.

232 **"from poultry in southern China":** H. Chen et al., "H5N1 Virus Outbreak in Migratory Waterfowl," *Nature* 436 (July 14, 2005): 191–92. A separate study of the Qinghai Lake outbreak published at about the same time, is Jinhua Liu et al., "Highly Pathogenic H5N1 Influenza Virus Infection in Migratory Birds," *Science* 309, no. 5738 (Aug. 19, 2005): 1206.

232 **No bird flu has broken out:** "Chinese Official Questions Credibility of *Nature's* Article on Bird Flu," Xinhua, July 8, 2005.

232 **could have been contaminated:** Nicholas Zamiska and Matt Pottinger, "Two Experts in China Dispute Bird-Flu Risks," *Wall Street Journal,* July 19, 2005.

232 **shuttered immediately:** "Highly Pathogenic Microbe Labs Must Operate Under Government Supervision: Ministry," Xinhua, Dec. 15, 2005.

233 **a new wave of disease:** G. J. D. Smith et al., "Emergence and Predominance of an H5N1 Influenza Variant in China," *PNAS* 103, no. 45 (November 7, 2006): 16936–41.

233 **"not based on science":** Lindsay Beck, "China Shares Bird Flu Samples, Denies New Strain Report," Reuters, Nov. 10, 2006. This time, the research faced a wider barrage from Chinese officials, including the directors of the National Influenza Centre and the National Avian Influenza Reference Laboratory. See "Experts Refute New Bird Flu Strain Claim," Chinadaily.com, Nov. 6, 2006; and "New US Bird Flu Report Lacks Evidence Base," Chinadaily.com, Nov. 10, 2006.

## Chapter Nine: The Secret Call

This chapter draws on interviews with current and former infectious-disease and laboratory specialists and other public health officials at WHO and the agency's consultants in Asia, North America, Europe, and Australia. The chapter also draws on interviews with infectious-disease and laboratory specialists, public health officials, and doctors and nurses in Vietnam at the national level and at the provincial and local levels in Hanoi and Thai Binh, as well as with victims and their families. In addition, material for this chapter is drawn from internal documents from WHO and personal notes kept by several participants in the events described.

241 **genetic signature of the pathogen:** Q. Mai Le et al., "Isolation of Drug-Resistant H5N1 Virus," *Nature* 437 (Oct. 20, 2005): 1108.

245 **shifting patterns of infection:** WHO Inter-country Consultation: Influenza A/H5N1 in Humans in Asia, Manila, May 6–7, 2005. On the genetic changes detected, see also WHO Global Influenza Program Surveillance Network, "Evolution of H5N1 Avian Influenza Viruses in Asia," *Emerging Infectious Diseases* 11, no. 10 (Oct. 2005): 1515–21.

246 **drafted a confidential report:** "Reassessment of the Current Situation of Influenza A (H5N1) in Vietnam," internal WHO report, June 2005.

248 **"We'll never have perfect data":** Personal notes of meeting.

249 **"If the results are correct":** Vietnam Pandemic Assessment, internal WHO memo, June 9, 2005.

249 **the conference call:** The account of the conference call is based on interviews with ten of the participants, personal notes of the call kept by several participants, and WHO documents describing it.

253 **diagnosing the virus:** See, for example, Pui Hong Chung et. al., the Global Influenza Program, "Expert Consultation on Diagnosis of H5N1 Avian Influenza Infections in Humans," *Influenza and Other Respiratory Viruses* 1, no. 4 (July 2007): 131–38; Writing Committee of the Second World Health Organization Consultation on Clinical Aspects of Human Infection with Avian Influenza A (H5N1) Virus, "Update on Avian Influenza A (H5N1) Virus Infection in Humans," *NEJM* 358, no. 3 (Jan. 17, 2008): 261–73; and WHO, "Influenza Research at the Human and Animal Interface," Report of WHO Working Group, Sept. 21–22, 2006.

253 **mountains of eastern Turkey:** Ahmet Faik Oner et al., "Avian Influenza A (H5N1) Infection in Eastern Turkey in 2006," *NEJM* 355, no. 21 (Nov. 23, 2006): 2179–85.

254 **Researchers in Indonesia:** I. Nyoman Kandun et al., "Three Indonesian Clusters of H5N1 Virus Infection in 2005," *NEJM* 355, no. 21 (Nov. 23, 2006): 2186–94.

254 **doctors in Thailand:** "H5N1 Virus Now Harder to Detect in Humans," *Nation* (Thailand), Aug. 18, 2006; and Writing Committee of the World Health Organization Consultation on Human Influenza A/H5, "Avian Influenza A (H5N1) Virus Infection in Humans," *NEJM* 353, no. 13 (Sept. 29, 2005): 1374–85.

254 **WHO says they are not sensitive enough:** WHO, "WHO Recommendations on the Use of Rapid Testing for Influenza Diagnosis"; and WHO, "Clinical Management of Human Infection with Avian Influenza A (H5N1) Virus," updated Aug. 15, 2007.

255 **"does not encourage immediate openness":** Angus Nicoll, "Human H5N1 Infections: So Many Cases—Why So Little Knowledge?" *Eurosurveillance* 11, nos. 4–6 (Apr.–June 2006): 74–75.

255 **fewer than a dozen victims:** One of the few autopsies was performed on Captan Boonmanut, the Thai boy whose admission to Siriraj Hospital in Bangkok helped alert Dr. Prasert Thongcharoen to the spreading virus. See Mongkol Uiprasertkul et al., "Influenza A H5N1 Replication Sites in Humans," *Emerging Infectious Diseases* 11, no. 7 (July 2005): 1036–41.

256 **Rini Dina:** Her case is discussed in: I. Nyoman Kandun et al., "Three Indonesian Clusters of H5N1 Virus Infection in 2005," *NEJM* 355, no. 21 (Nov. 23, 2006): 2186–94.

258 **on the afternoon of Tuesday, June 14:** The account of this meeting is based primarily on an internal WHO account, "Notes of Avian Influenza Meeting, Tuesday, June 14, 2005."

260 **when Troedsson reported back:** Personal notes of call, June 14, 2005.

260 **Stohr and his colleagues in Geneva:** Personal notes of discussion, June 14, 2005.

260 **sent an e-mail:** E-mail from Yan Li, June 13, 2005.

261 **Kinsmen Place Lodge:** The details in this account are drawn from extensive press coverage of the episode in August and September 2003, including multiple articles written by Helen Branswell of Canadian Press; Pamela Fayerman, Kim Pemberton, and Nicholas Read of the *Vancouver Sun*; Mark Hume of the *Toronto Globe and Mail*; and Lawrence K. Altman of the *New York Times*. See also Wayne Kondro, "Canadian Officials Watch SARS-like Mystery Bug," *Lancet* 362, no. 9385 (Aug. 30, 2003): 714.

262   **without his permission:** Nicholas Zamiska, "Avian Flu Puts WHO in a Bind," *Wall Street Journal Asia,* Oct. 18, 2005.

263   **Khai signaled his government's good intentions:** For Ellen Nakashima's full interview with Khai, see "Transcript: Interview with Phan Van Khai," washingtonpost.com, June 16, 2005.

264   **"They should be from different backgrounds":** Personal notes of discussion.

264   **say as little as necessary:** Personal notes of call.

264   **without actually lying to them:** For WHO's vaguely worded, four-paragraph press release about the mission, see "International Team of Avian Influenza Experts Visits Viet Nam," June 24, 2005.

266   **rogue bits of genetic material:** Several members of the WHO mission, including Tashiro, told me that the Canadian primers were at fault. Plummer later told me that he did not dispute the findings that the tests conducted with the Canadian-supplied primers had yielded false positives. But he added, "Given that both our lab and the CDC have concluded that our primers are effective, we remain confident that they were not the cause of the false positives."

266   **French press agency reported:** "Top Scientists Downgrade Risk of Imminent Bird Flu Pandemic," Agence France Presse, June 29, 2005.

### Chapter Ten: Let's Go Save the World

269   **more people had died:** On the stream of Indonesian cases in late 2005 and early 2006, see Endang R. Sedyaningsih et al., "Epidemiology of Cases of H5N1 Virus Infection in Indonesia, July 2005–June 2006," *Journal of Infectious Diseases* 196, no. 4 (Aug. 15, 2007): 522–27.

269   **keeping her identity confidential:** Though I continue to honor this request, the victim's name was disclosed in Indonesian press reports.

273   **humming with questions:** On the key steps and objectives of an avian flu investigation, see "WHO Guidelines for Investigation of Human Cases of Avian Influenza A (H5N1)," WHO, Jan. 2007.

274   **A study of Indonesia's first 127 confirmed cases:** I. Nyoman Kandun et al., "Factors Associated with Case Fatality of Human H5N1 Infections in Indonesia: A Case Series," *Lancet* 372, no. 9640 (Aug. 30, 2008): 744–49.

277   **deadliest outbreak of Marburg hemorrhagic fever:** Accounts of the Angola outbreak can be found in John Donnelly, "Deadly Virus, Anger Take Hold in Angola," *Boston Globe,* Apr. 12, 2005; Sharon LaFraniere and Denise Grady, "Stalking a Deadly Virus, Battling a Town's Fears," *New York Times,* Apr. 17, 2005; and M. A. J. McKenna, "CDC Team Sees Small Advances Against Disease," *Atlanta Journal-Constitution,* May 25, 2005.

277   **a flare-up in southern Sudan:** A good account is David Brown, "U.N. Team Studies Sudan Outbreak," *Washington Post,* Oct. 21, 1998.

278   **the mountains of Afghanistan:** For details, see WHO press release, "'Influenza-like' Acute Respiratory Infection Behind Deadly Afghan Outbreak," Mar. 2, 1999.

280   **suddenly erupted in Azerbaijan:** Some of the details in this account come from internal WHO situation updates, field reports, and notes of conference calls in March 2006. The outbreak is described in "Human Avian Influenza in Azerbaijan, February–March 2006," *Weekly Epidemiological Record,* no. 18, May 5, 2006, 183–88; A. Gilsdorf et al., "Two Clusters of Human Infection with Influenza A/H5N1 Virus in the Republic of Azerbaijan, February–March 2006," *Eurosurveillance* 11, nos. 4–6, Apr.–June 2006, 122–26; and Caroline Brown, "First

H5N1 Outbreak in Humans Associated with Dead Wild Birds: Azerbaijan, February–April 2006," paper presented at FAO/OIE International Scientific Conference on Avian Influenza and Wild Birds, Rome, May 30–31, 2006.

282 **rumors and emerging reports:** See Gina Samaan et al., "Rumor Surveillance and Avian Influenza H5N1," *Emerging Infectious Diseases* 11, no. 3 (Mar. 2005): 463–66.

285 **Five South Koreans:** "Five More in S. Korea Infected by Bird Flu," Agence France Presse, Sept. 15, 2006.

285 **Hong Kong's mass slaughter:** C. B. Bridges et al., "Risk of Influenza A (H5N1) Infection Among Poultry Workers, Hong Kong, 1997–1998," *Journal of Infectious Diseases* 185, no. 8 (Apr. 15, 2002): 1005–10. This article also reports that 10 percent of 1,525 poultry workers tested in Hong Kong were positive for H5N1 antibodies.

285 **In one telling study:** Sirenda Vong et al., "Low Frequency of Poultry-to-Human H5N1 Virus Transmission, Southern Cambodia, 2005," *Emerging Infectious Diseases* 12, no. 10 (Oct. 2006): 1542–47. A pair of studies that looked at Nigerian poultry workers with widespread exposure to likely infected poultry and Chinese workers in live poultry markets at the time of a human infection possibly contracted in a market also showed minimal evidence of antibodies to H5N1. See J. R. Ortiz et al., "Lack of Evidence of Avian-to-Human Transmission of Avian Influenza A (H5N1) Virus Among Poultry Workers, Kano, Nigeria, 2006," *Journal of Infectious Diseases* 196, no. 11 (Dec. 1, 2007), 1685–91; and Ming Wang et al., "Food Markets with Live Birds at Source of Avian Influenza," *Emerging Infectious Diseases* 12, no. 11 (Nov. 2006): 1773–75.

285 **A subsequent study:** Sirenda Vong et al., "Risk Factors Associated with Subclinical Human Infection with Avian Influenza A (H5N1) Virus—Cambodia, 2006," *Journal of Infectious Diseases* 199, no. 12 (June 15, 2009): 1744–52.

285 **enigma of the cullers:** There's a similar puzzle for health-care workers. Studies of health-care staff in Vietnam who treated avian-flu patients also found no evidence of exposure to the virus. This is in marked contrast to the experience with SARS, which took a heavy toll on health-care workers. See Nguyen Thanh Liem, World Health Organization Avian Influenza Investigation Team Vietnam, and Wilina Lim, "Lack of H5N1 Avian Influenza Transmission to Hospital Employees, Hanoi, 2004," *Emerging Infectious Diseases* 11, no. 2 (Feb. 2005): 210–15; and Constance Schultsz et al., "Avian Influenza H5N1 and Healthcare Workers," *Emerging Infectious Diseases* 11, no. 7 (July 2005): 1158–59.

286 **unable to identify a possible source:** Endang R. Sedyaningsih et al., "Epidemiology of Cases of H5N1 Virus Infection in Indonesia, July 2005–June 2006," *Journal of Infectious Diseases* 196, no. 4 (Aug. 15, 2007): 522–27.

286 **were ruled inconclusive:** Bayu Krisnamurti, head of Indonesia's National Avian Influenza Committee, quoted in "Indonesia Investigating Suspicious Bird Flu cases: Official," Agence France Presse, Dec. 18, 2007.

286 **behavior of the virus in Indonesia "mysterious":** "Mysterious Bird Flu Baffles Indonesian Scientists," Agence France Presse, Feb. 6, 2008.

286 **the country's first human case:** It is described in I. Nyoman Kandun et al., "Three Indonesian Clusters of H5N1 Virus Infection in 2005," *NEJM* 355, no. 21 (Nov. 23, 2006): 2186–94.

287 **a quarter of all confirmed cases:** Writing Committee of the Second World Health Organization Consultation on Clinical Aspects of Human Infection with Avian Influenza A (H5N1) Virus, "Update on Avian Influenza A (H5N1) Virus Infection in Humans," *NEJM* 358, no. 3 (Jan. 17, 2008): 261–73. On case clusters, see also Sonja J. Olsen et al., "Family Clustering of Avian A (H5N1),"

*Emerging Infectious Diseases* 11, no. 11 (Nov. 2005): 1799–1801. On Indonesian clusters specifically, see discussion in Endang R. Sedyaningsih et al., "Epidemiology of Cases of H5N1 Virus Infection in Indonesia, July 2005–June 2006," *Journal of Infectious Diseases* 196, no. 4 (Aug. 15, 2007): 522–27.

289    **made some people susceptible:** See, for example, WHO, Report of WHO Working Group, "Influenza Research at the Human and Animal Interface," Sept. 21–22, 2006.

289    **statistical chance alone:** V. E. Pitzer et al., "Little Evidence for Genetic Susceptibility to Influenza A (H5N1) from Family Clustering Data," *Emerging Infectious Diseases* 13, no. 7 (July 2007): 1074–76.

## Chapter Eleven: The Lights Go Out at Seven

294    **despite some projections:** Hitoshi Oshitani, Taro Kamigaki, and Akira Suzuki, "Major Issues and Challenges of Influenza Pandemic Preparedness in Developing Countries," *Emerging Infectious Diseases* 14, no.6 (June 2008): 875–80.

295    **"He just knows everybody":** Interview with Dr. Megge Miller.

296    **The health workers at the two local clinics:** Interviews with Ly Lai and Dr. Ou Sary, Kampot province.

297    **it went a long way:** Interviews with current and former WHO officials in Cambodia, including Drs. Michael O'Leary, Isabel Bergeri, and Megge Miller. See Richard Stone, "Combating the Bird Flu Menace, Down on the Farm," *Science* 311, no. 5763 (Feb. 17, 2006): 944–46.

298    **a surprising item:** William Prochnau and Laura Parker, "The Waiting Plague," *Vanity Fair,* Nov. 2005.

298    **had pledged $2.3 billion:** UN System Influenza Coordinator (SIC) and World Bank, "Responses to Avian Influenza and State of Pandemic Readiness, Third Global Progress Report," Dec. 2007 (1st printing, released Nov. 29, 2007). The total figure increased modestly to $2.7 billion in 2008. See UN SIC and World Bank, "Responses to Avian Influenza and State of Pandemic Readiness, Fourth Global Progress Report," Oct. 2008.

299    **ninety-one-page progress report:** UN SIC and World Bank, "Responses . . . Third Global Progress Report," Dec. 2007 (1st printing).

299    **a new version of the report:** UN SIC and World Bank, "Responses . . . Third Global Progress Report," Dec. 2007 (2nd printing, released Dec. 18, 2007), 8–9.

299    **would decline even further in 2008:** UN SIC and World Bank, "Responses . . . Fourth Global Progress Report," Oct. 2008. The report warned, "There is a risk that this decline in resources pledged, especially for countries with the greatest remaining needs, could undermine the sustainability of the investments made to date."

299    **warned of growing "flu fatigue":** Paula Dobriansky, undersecretary of state for democracy and global affairs, "Remarks at the International Partnership on Avian and Pandemic Influenza Ministerial, Sharm El Sheikh, Egypt," Federal News Service, Oct. 25, 2008.

299    **the World Bank had helped estimate:** World Bank, "Avian and Human Influenza: Update on Financing Needs and Framework," Nov. 30, 2006. The original estimates were in World Bank, "Avian and Human Influenza: Financing Needs and Gaps," Jan. 12, 2006.

299    **sector after sector:** UN SIC and World Bank, "Responses . . . Third Global Progress Report," Dec. 2007 (1st printing).

300    **"Adequate financial support":** Ibid.

301    **Yet the obstacles are many:** An excellent examination of the challenges facing

the development of a pandemic vaccine is the seven-part series "The Pandemic Vaccine Puzzle," written by Maryn McKenna for the Center for Infectious Disease Research and Policy and posted online beginning October 25, 2007. The articles are available at www.cidrap.umn.edu. Another fine overview is Joost H. C. M. Kreijtz, Albert D. M. E. Osterhaus, and Guus F. Rimmelzwaan, "Vaccination Strategies and Vaccine Formulations for Epidemic and Pandemic Influenza Control," *Human Vaccines* 5 (Mar. 2009): 3. See also WHO, "Global Pandemic Influenza Action Plan to Increase Vaccine Supply," Oct. 23, 2006; and National Institute of Allergy and Infectious Diseases, "Report of the Blue Ribbon Panel on Influenza Research," Sept. 11–12, 2006.

301 **research into any kind of flu vaccine:** See, for example, a pair of studies by the Institute of Medicine. Kathleen R. Stratton, Jane S. Durch, and Robert S. Lawrence, eds., *Vaccines for the 21st Century: A Tool for Decisionmaking* (Washington; National Academies Press, 2000); and Institute of Medicine staff, *New Vaccine Development: Establishing Priorities* (Washington: National Academies Press, 1985).

301 **An analysis in 2007:** Aeby Thomas, Niels Guldager, and Klaus Hermansen, "Pandemic Flu Preparedness: A Manufacturing Perspective," *BioPharm International,* Aug. 2, 2007. For further discussion of the delays inherent in developing a pandemic vaccine, see Jesse L. Goodman, "How Fast Can a New Vaccine for an Emerging Respiratory Virus Be Developed and Available for Use?" Presentation at the International Conference on Emerging Infectious Diseases, Atlanta, GA, Mar. 22, 2006.

301 **confound efforts to develop a single vaccine:** See, for example, WHO, Report of WHO Working Group, "Influenza Research at the Human and Animal Interface," Sept. 21–22, 2006; Steven Riley, Joseph T. Wu, and Gabriel M. Leung, "Optimizing the Dose of Pre-Pandemic Influenza Vaccines to Reduce the Infection Rate," *PLoS Medicine* 4, no. 6 (June 2007): e218; and G. J. D. Smith et al., "Emergence and Predominance of an H5N1 Influenza Variant in China," *PNAS* 103, no. 45 (Nov. 7, 2006): 16936–41.

301 **unusually resistant to experimental vaccines:** See, for example, Karl G. Nicholson et al., "Safety and Antigenicity of Non-adjuvanted and MF59-adjuvanted Influenza A/Duck/Singapore/97 (H5N3) Vaccine: A Randomized Trial of Two Potential Vaccines Against H5N1 Influenza," *Lancet* 357, no. 9272 (June 16, 2001): 1937–43; Jean-Louis Bresson et al., "Safety and Immunogenicity of an Inactivated Split-Virion Influenza A/Vietnam/1194/2004 (H5N1) Vaccine: Phase I Randomised Trial," *Lancet* 367, no. 9253 (May 20, 2006): 1657–64; Isabel Leroux-Roels et al., "Antigen Sparing and Cross-Reactive Immunity with an Adjuvanted rH5N1 Prototype Pandemic Influenza Vaccine: A Randomised Controlled Trial," *Lancet* 370, no. 9587 (Aug. 18, 2007): 580–89; and Nega Ali Gogi et al., "Immune Responses of Healthy Subjects to a Single Dose of Intramuscular Inactivated Influenza A/Vietnam/1203/2004 (H5N1) Vaccine After Priming with an Antigenic Variant," paper presented at Third WHO Meeting on Evaluation of Pandemic Influenza Prototype Vaccines in Clinical Trials, Geneva, Feb. 15–16, 2007.

302 **This could cut the production time:** Peter F. Wright, "Vaccine Preparedness—Are We Ready for the Next Influenza Pandemic?" *NEJM* 358, no. 24 (June 12, 2008): 2540–43.

302 **Initial clinical trials:** Hartmut J. Ehrlich et al., "A Clinical Trial of a Whole-Virus H5N1 Vaccine Derived from Cell Culture," *NEJM* 358, no. 24 (June 12, 2008): 2573–84.

302 **than even seasonal flu shots:** David Brown, "Bird Flu Vaccine Shows

Promise," *Washington Post,* July 27, 2006; Leroux-Roels, "Antigen Sparing"; and
Suryaprakash Sambhara and Gregory A. Poland, "Breaking the Immunogenicity
Barrier of Bird Flu Vaccines," *Lancet* 370, no. 9587 (Aug. 18, 2007): 544.

302    **could radically increase:** WHO, "Projected Supply of Pandemic Influenza
Vaccine Increases Sharply," press release, Oct. 23, 2007.

302    **"can't provide vaccines to the world free":** Wayne Pisano, quoted in a Council on Foreign Relations letter from Laurie Garrett, senior fellow for global health,
June 20, 2007.

302    **a local vaccine against hepatitis B:** Tini Tran, "Vietnam Struggles to Rein In
Hepatitis B," Associated Press, July 17, 2000.

302    **a new generation of vaccines:** "Health: Vietnam Successfully Produces Second Generation Hepatitis Vaccines," *Vietnam News Briefs,* Apr. 11, 2002.

303    **"future availability to Vietnam is doubtful":** "Report of WHO Mission to
Support Influenza A/H5N1 Vaccine Development in Vietnam," 2005.

303    **"serious ethical reservations":** Ibid.

304    **"the volunteer 'spirit' may not be universally shared":** Internal WHO
document, 2005.

304    **monkey kidney cells:** "Vietnam-made Bird Flu Vaccine Proves Effective," Vietnamese News Agency, Aug. 22, 2008.

304    **on researchers at the institute:** "Volunteers for H5N1 Vaccines Get Second
Injection," Vietnamese News Agency, May 17, 2008.

304    **student volunteers:** Ibid.

304    **"Good results":** Tranh Dinh Lam, "Vietnam: Bird Flu Vaccine for Humans May
Be Available by 2009," Interpress Service, June 4, 2008.

304    **mass production by late 2009:** "Vietnam-made Bird Flu Vaccine Proves Effective," Vietnamese News Agency, Aug. 22, 2008.

304    **30,000 Vietnamese dong:** Ibid.

304    **Siti Fadilah Supari was far less patient:** The accounts of Supari's political
rise, her battle over the sharing of viruses, and the wider international dispute
over samples and benefits are drawn from interviews with Supari, current and
former officials of the Indonesian health ministry, and other Indonesian public
health officials and political figures. It also draws on interviews with WHO officials and public health and diplomatic officials from the United States, Australia, and other countries, as well as documents from the Indonesian health
ministry, WHO, and the U.S. Department of Health and Human Services.

305    **Her scheduled flight:** Details of Supari's trip to Geneva are drawn from Siti
Fadilah Supari, *It's Time for the World to Change: In the Spirit of Dignity, Equity,
and Transparency, Divine Hand Behind Avian Influenza,* 3rd ed. (Jakarta: Sulaksana Watinsa Indonesia, 2008), 112ff.

306    **"We do not really know":** Statement by the Minister of Health of the Republic
of Indonesia H. E. Dr. Siti Fadilah Supari at the Inter-Governmental Meeting for
Pandemic Influenza Preparedness, Geneva, Nov. 20, 2007.

307    **her cell phone rang:** Yanto Soegiarto, "Diving into the Deep End," *Globe Asia,*
Oct. 2007.

309    **a running dispute with foreign scientists:** The account of the NAMRU
dispute is based on interviews with Andrew Jeremijenko, other NAMRU staff,
and Indonesian health ministry officials.

309    **"difficult to get the damn virus":** Interview with Jeremijenko.

310    **cease all activities:** "Circular Regarding the Status of NAMRU-2," Oct. 25,
2005, signed by Secretary Titie Kabul Adimidjaja, acting head of Badan
Litbangkes.

312   **"I never gave permission":** Wahya Dhyatmika and Pramono, "WHO's Virus?" *Tempo*, Feb. 20, 2007.

312   **unless they met Supari's demands:** The Indonesian position is best detailed in Endang R. Sedyaningsih et al., "Toward Mutual Trust, Transparency and Equity in Virus Sharing Mechanism: The Avian Influenza Case of Indonesia," *Annals of the Academy of Medicine* (Singapore) 37, no. 6 (June 2008): 482–88. Supari was one of the authors on this paper written by Indonesian health ministry officials. The position prevailing in many developing countries, in particular the United States, is well articulated in Richard Holbrooke and Laurie Garrett, "'Sovereignty' That Risks Global Health," *Washington Post*, Aug. 10, 2008. An Indonesian response to the column by Holbrooke and Garrett is Makarim Wibisono, "The Responsible Virus and Sharing Benefits," *Jakarta Post*, Aug. 27, 2008. Wibisono was the Indonesian ambassador to the United Nations in New York and in Geneva.

313   **Her attack was unprecedented:** For a discussion of the global health issues raised by Indonesia's decision to withhold virus samples, see Chan Chee Khoon and Gilles de Wildt, "Developing Countries, Donor Leverage, and Access to Bird Flu Vaccines," DESA Working Paper no. 41, UN Department of Economic and Social Affairs, June 2007.

313   **viruses were biological resources:** On two rival interpretations of international law governing virus samples, see David P. Fidler, "Influenza Virus Samples, International Law, and Global Health Diplomacy," *Emerging Infectious Diseases* 14, no. 1 (Jan. 2008): 88–94.

313   **"WHO has become a target":** Personal notes of conversation.

314   **"Indonesia's leadership alerted":** "RI 'Will Not Share' Flu Samples," *Jakarta Post*, Feb. 7, 2007. The joint statement issued after the meetings between Supari and Heymann is "Sharing of Avian Influenza Viruses and Pandemic Vaccine Production," Joint Statement from the Ministry of Health of Indonesia and the World Health Organization, Feb. 16, 2007.

314   **"the answer is still no, no, no":** Wahya Dhyatmika and Pramono, "WHO's Virus?" *Tempo*, Feb. 20, 2007.

314   **The Indonesian leader stressed:** "RI Stresses Need for Production of Affordable Bird Flu Vaccines," Antara news agency, Apr. 4, 2007.

314   **"We will resume the sending of virus samples":** "Indonesia Confirms Readiness to Send Bird Flu Virus to WHO," Xinhua, April 4, 2007.

314   **"I believe the developing countries are right":** Margaret Chan, Opening remarks at the Meeting on Options for Increasing the Access of Developing Countries to H5N1 and other Potential Pandemic Vaccines, Geneva, April 25, 2007.

315   **"I will fail you":** Laura MacInnis, "WHO's Chan Pledges Fair Access to Bird Flu Vaccines," Reuters, May 17, 2007.

315   **suggested the UN Security Council:** Laurie Garrett and David P. Fidler, "Sharing H5N1 Viruses to Stop a Global Influenza Pandemic," *PLoS Medicine* 4, no. 11 (Nov. 2007): e330, 1712–14.

315   **"I was exhausted":** Siti Fadilah Supari, *It's Time for the World to Change: In the Spirit of Dignity, Equity, and Transparency, Divine Hand Behind Avian Influenza* (Jakarta: Sulaksana Watinsa Indonesia, 2008), 129.

315   **their private encounter:** Accounts of the meeting were provided by several Indonesian and WHO officials. Supari's comments are drawn from Supari, *It's Time for the World to Change*, 129–31.

317   **weapons of mass destruction:** Supari, *It's Time for the World to Change*, 19.

317   **"the nuttiest idea I ever heard":** "Remarks by Defense Secretary Robert Gates

to the Indonesian Council on World Affairs, Jakarta, Indonesia," Federal News Service, Feb. 25, 2008.

317 **a rapturous reception:** See, for example, "Alarm as Indonesia Thumbs Nose at West over Bird Flu," Agence France Presse, Sept. 7, 2008.

317 **"of no use to us":** "RI Seeking Equality in Cooperation with NAMRU-2: Minister," Antara news agency, June 26, 2008.

317 **One lawmaker called for a probe:** "News Focus: Call for Closure of NAMRU-2 in Indonesia Increasing," Antara news agency, June 27, 2008.

## Chapter Twelve: Peril on the Floodplain

321 **a quarter of all the illegally trafficked chickens:** "Illegal Chickens in Northern Area Test Positive for Bird Flu," *Viet Nam News,* Mar. 22, 2008.

321 **uncovered lab evidence:** H. Chen et al., "Establishment of Multiple Sublineages of H5N1 Influenza Virus in Asia: Implications for Pandemic Control," *PNAS* 103, no. 8 (Feb. 21, 2006): 2845–50.

321 **on "multiple occasions":** Tien Dung Nguyen et al., "Multiple Sublineages of Influenza A Virus (H5N1), Vietnam, 2005–2007," *Emerging Infectious Diseases* 14, no. 4 (Apr. 2008): 632–36.

322 **The strain made its debut:** For more discussion, see Carl Suetens et al., "Eagles Testing Positive for H5N1 Imported Illegally into Europe from Thailand," *Eurosurveillance* 8, no. 44 (Oct. 28, 2004); and Debora MacKenzie, "Europe Has Close Call with Deadly Bird Flu," *New Scientist,* Oct. 26, 2004.

322 **"very, very lucky":** MacKenzie, "Europe Has Close Call."

322 **exploded out of East Asia:** See overviews in "Epidemiology of WHO-Confirmed Human Cases of Avian Influenza A(H5N1) Infection," *Weekly Epidemiological Record* 81, No. 26 (June 30, 2006): 249–57; and "Update: WHO-Confirmed Human Cases of Avian Influenza A(H5N1) Infection," *Weekly Epidemiological Record* 82, No. 6 (Feb. 9, 2007): 41–47.

322 **each time researchers checked:** See, for example, B. Pattnaik et al., "Phylogenetic Analysis Revealed Genetic Similarity of the H5N1 Avian Influenza Viruses Isolated from HPAI Outbreaks in Chickens in Maharashtra, India, with Those Isolated from Swan in Italy and Iran in 2006," *Current Science* 91, no. 1 (July 10, 2006): 77–81; M. F. Ducatez et al., "Multiple Introductions of H5N1 in Nigeria," *Nature* 442 (July 6, 2006): 37; and Siegfried Weber et al., "Molecular Analysis of Highly Pathogenic Avian Influenza Virus of Subtype H5N1 Isolated from Wild Birds and Mammals in Northern Germany," *Journal of General Virology* 88 (2007): 554–58.

323 **fallen sick in Turkey:** For details, see Ahmet Faik Oner et al., "Avian Influenza A (H5N1) Infection in Eastern Turkey in 2006," *NEJM* 355, no. 21 (Nov. 23, 2006): 2179–85.

323 **the famed bird market:** Elaine Sciolino, "In the Land of Coq au Vin, Soul Searching over Bird Flu," *New York Times,* Feb. 24, 2006.

323 **the ravens at the Tower of London:** Mary Jordan, "Bird Flu Fears Coop Up London's Famous Ravens," *Washington Post,* Feb. 22, 2006.

323 **dumped in the Nile River:** Daniel Williams, "Spooked by Bird Flu, Egyptians Horde [*sic*] Water," *Washington Post,* Feb. 26, 2006.

323 **"While it was originally suspected":** "Nigeria; Bird Flu: FG Links Source to Illegal Importation of Chicks," *Africa News,* Mar. 3, 2006.

323 **introduced to the country three separate times:** M. F. Ducatez et al., "Multiple Introductions of H5N1 in Nigeria," *Nature* 442 (July 6, 2006): 37.

323     **most likely along internal trade routes:** "FAO Urges Nigeria to Increase Bird Flu Control Measures," FAO news release, Feb. 20, 2007. In most of the affected Nigerian states, the flu strains continued to evolve into new forms. See Isabella Monne et al., "Reassortant Avian Influenza Virus (H5N1) in Poultry, Nigeria, 2007," *Emerging Infectious Diseases* 14, No. 4 (Apr. 2008): 637–40.

324     **Are wild birds the culprit:** The evidence implicating wild birds has been mounting. The following is a sampling: H. Chen et al., "H5N1 Virus Outbreak in Migratory Waterfowl," *Nature* 436 (July 14, 2005): 191–92; Dennis Normile, "Are Wild Birds to Blame?" *Science* 310, no. 5747 (Oct. 21, 2005): 426–28; Robert G. Webster et al., "H5N1 Outbreaks and Enzootic Influenza," *Emerging Infectious Diseases* 12, no. 1 (Jan. 2006): 3–8; H. Chen et al., "Establishment of Multiple Sublineages of H5N1 Influenza Virus in Asia: Implications for Pandemic Control," *PNAS* 103, no. 8 (Feb. 21, 2006): 2845–50; Dennis Normile, "Evidence Points to Migratory Birds in H5N1 Spread," *Science* 311, no. 5765 (Mar. 3, 2006): 1225; Bjorn Olsen et al., "Global Patterns of Influenza A Virus in Wild Birds," *Science* 312, no. 5772 (Apr. 21, 2006): 384–88; Marius Gilbert et al., "Anatidae Migration in the Western Palearctic and the Spread of Highly Pathogenic Avian Influenza Virus H5N1 Virus," *Emerging Infectious Diseases* 12, no. 11 (Nov. 2006): 1650–56; Raja Sengupta et al., "Ecoregional Dominance in Spatial Distribution of Avian Influenza (H5N1) Outbreaks," *Emerging Infectious Diseases* 13, no. 8 (Aug. 2007): 1269–70; Juthatip Kwawcharoen et al., "Wild Ducks as Long-Distance Vectors of Highly Pathogenic Avian Influenza Virus (H5N1)," *Emerging Infectious Diseases* 14, no. 4 (Apr. 2008): 600–606; Donata Kalthoff et al., "Pathogenicity of Highly Pathogenic Avian Influenza Virus (H5N1) in Adult Mute Swans," *Emerging Infectious Diseases* 14, no. 8 (Aug. 2008): 1267–70; and A. Townsend Peterson et al., "Influenza A Virus Infections in Land Birds, People's Republic of China," *Emerging Infectious Diseases* 14, no. 10 (Oct. 2008): 1644–46.

324     **Or is it trade in poultry:** There is also a body of evidence indicating that the role of wild birds in spreading the virus is less significant than that of commerce. See, for example, D. S. Melville and Kennedy F. Shortridge, "Influenza: Time to Come to Grips with the Avian Dimension," *Lancet Infectious Diseases* 4, no. 5 (May 2004): 261–62; Chris J. Feare, "The Role of Wild Birds in the Spread of HPAI H5N1," *Avian Diseases* 51, no. S1 (2007): 440–47; M. Gauthier-Clerc, C. Lebarbenchon, and F. Thomas, "Recent Expansion of Highly Pathogenic Avian Influenza H5N1: A Critical Review," *Ibis* 149 (2007): 202–14; Thomas P. Weber and Nikolaos I. Stilianakis, "Ecological Immunology of Avian Influenza (H5N1) in Migratory Birds," *Emerging Infectious Diseases* 13, no. 8 (Aug. 2007): 1139–43; and "Don't Blame Wild Birds for H5N1 Spread—Expert," Reuters, Jan. 23, 2008.

324     **both these opportunities:** See, for example, Joseph Domenech et al., "Trends and Dynamics of HPAI—Epidemiological and Animal Health Risks," Background Paper at the Technical Meeting on Highly Pathogenic Avian Influenza and Human H5N1 Infection, Rome, June 27–29, 2007; and A. Marm Kilpatrick et al., "Predicting the Global Spread of H5N1 Avian Influenza," *PNAS* 103, no. 51 (Dec. 19, 2006): 19368–73. For a wide-ranging examination of the role wild birds play in the spread of the virus, see the presentations at the FAO-OIE International Scientific Conference on Avian Influenza and Wild Birds, Rome, May 30–31, 2006.

324     **"far from over":** "New Avian Influenza Flare-ups," FAO news release, Jan. 24, 2008.

324 **"finally cause a human influenza pandemic":** "Bird Flu Situation in Indonesia Critical," FAO news release, Mar. 18, 2008.

325 **actively undergoing genetic changes:** On the endemicity and continuing evolution of the virus in Indonesia, see Tommy Tsan-Yuk Lam et al., "Evolutionary and Transmission Dynamics of Reassortant H5N1 Influenza Virus in Indonesia," *PLoS Pathogens* 4, no. 8 (Aug. 2008): e1000130.

325 **only a few localities were completely capable:** Minister of Agriculture and Rural Development Cao Duc Phat, quoted in "Vietnam Preventive Measures Are Not Enough," *Thai Press Reports*, Mar. 24, 2008.

325 **poultry vaccination program was flagging:** Deputy Minister of Agriculture and Rural Development Bui Ba Bong, quoted in "Half-Done Vaccination Poses High Bird-Flu Risk," *Saigon Times Daily*, Mar. 13, 2008.

325 **would be unable to keep paying:** See, for example, Jan Slingenbergh, senior officer of the FAO's Animal Production and Health Service in "H5N1 HPAI Pathogenicity Rising, but Situation in Check," FAOAIDE news, Situation Update 55, July 25, 2008.

325 **were both becoming more lethal:** Mary Pantin-Jackwood of the U.S. Department of Agriculture and colleagues from the Viet Nam National Centre for Veterinary Diagnosis in "H5N1 HPAI Pathogenicity Rising." FAOAIDE news, Situation Update 55, July 25, 2008.

325 **"We still have a very serious situation":** "China Needs Better Bird Flu Surveillance Experts," Reuters, Feb. 18, 2009.

326 **the Thai government launched:** Supamit Chunsuttiwat, "Response to Avian Influenza and Preparedness for Pandemic Influenza: Thailand's Experience," *Respirology* 13, suppl. 1 (2008): S36–S40; and Kumnuan Ungchusak, "What Happened When the H5N1 Virus Visited Thailand," lecture at the Asia Medical Forum, Lancet 2006, in Singapore, May 4, 2006.

326 **continued to detect the virus:** Alongkorn Amonsin et al., "Influenza Virus (H5N1) in Live Bird Markets and Food Markets, Thailand," *Emerging Infectious Diseases* 14, no. 11 (Nov. 2008): 1739–42.

326 **"Some commercial producers":** E-mail from Juan Lubroth, Aug. 11, 2006, citing information from the U.S. Agency for International Development.

326 **a baffling transplant:** "New Bird Flu Strain Detected in Nigeria," FAO news release, Aug. 11, 2008. For more discussion see Alice Fusaro et al., "Introduction into Nigeria of a Distinct Genotype," *Emerging Infectious Diseases*, 15, no. 3 (March 2009): 445-47.

326 **"Somewhat surprising":** "Bird Flu Returns to Germany," *Deutsche Welle*, Oct. 9, 2008.

326 **"particularly worrying":** "New Avian Influenza Flare-ups," FAO news release, Jan. 24, 2008.

326 **"a new chapter in the evolution":** "Bird Flu Virus in Europe—a Hidden Danger," FAO news release, Oct. 25, 2007.

327 **"We must never forget":** "Concern Over Flu Pandemic Justified," Address to the Sixty-second World Health Assembly, Geneva, Switzerland, May 18, 2009.

327 **"Do not drop the ball":** "World Is Better Prepared for Influenza Pandemic," Address to the ASEAN+3 Health Ministers' Special Meeting on Influenza A (H1N1), Bangkok, Thailand, via teleconference, May 8, 2009.

327 **"We are certainly aware of the risk":** "Avian Influenza Remains a Global Threat, Says WHO," News release from WHO Western Pacific Region, Nov. 24, 2009.

327 **"Inside China, H5N1 has been existing":** Stefanie McIntyre, "China Expert Warns of Pandemic Flu Mutation," Reuters, Nov. 25, 2009.

328    **For each one of those fatalities:** Juan Lubroth, a senior FAO official, esti-
       mated in June 2008 that 240 million birds had died or been slaughtered. Julia
       Zappei, "Health Experts: Global Fight Against Bird Flu Remains Weak, Can
       Worsen Global Food Crisis," Associated Press, June 20, 2008.

328    **reducing the circulation of these viruses in animals:** The likelihood of
       altogether eliminating the H5N1 virus is at best slim. Animal-health expert Les
       Sims, who first confronted this strain in Hong Kong in 1997 and has followed it
       since, writes, "The prospects of global eradication of H5N1 HPAI viruses circu-
       lating in Asia, Africa and Europe within the next 10 to 20 years is poor. Unless
       the viruses change or there are major shifts in the way poultry are reared, arising
       from accelerated rural development, they may never be eradicated." Les D. Sims,
       "Lessons Learned from Asian H5N1 Outbreak Control," *Avian Diseases* 51, no.
       S1 (2007): 174–81.

328    **relations were strained:** See, for example, the discussion over sample sharing
       between the agencies in Declan Butler, "'Refusal to Share' Leaves Agency Strug-
       gling to Monitor Bird Flu," *Nature* 435 (May 12, 2005): 131.

329    **cash-strapped veterinary counterparts:** Senior officials at the World Orga-
       nization for Animal Health, known by its French initials OIE, offered a similar
       critique: "Although financial resources have been targeted to the human health
       rather than the animal health field under the pressure of a possible human pan-
       demic, the main message . . . remains that the viral load in the environment and
       therewith the risk of a pandemic should be diminished by eradication of the virus
       at its animal source." Christianne Bruschke, Alex Thiermann, and Bernard Val-
       lat, "Implementing Appropriate OIE/FAO Prevention Measures in Different
       Country Contexts," Background Paper at the Technical Meeting on Highly
       Pathogenic Avian Influenza and Human H5N1 Infection, Rome, June 27–29,
       2007.

331    **"The present situation is unique":** "Sharing of Influenza Viruses and Access
       to Vaccines and Other Benefits," Opening Remarks at the Intergovernmental
       Meeting on Pandemic Influenza Preparedness, Geneva, Nov. 20, 2007.

331    **Researchers initially concluded:** Christophe Fraser et al., "Pandemic Poten-
       tial of a Strain of Influenza A (H1N1): Early Findings," *Science*, published online
       before print May 11, 2009, doi: 10.1126/science.1176062.

331    **no greater than that for ordinary flu bugs:** Dr. Daniel Jernigan, deputy di-
       rector of the CDC Influenza Division, at a CDC telebriefing on the investigation
       of human cases of H1N1 flu, May 20, 2009.

331–32 **planning for hospitals and public health systems remains wanting:**
       See, for example, Christopher Lee, "U.S. Flu Outbreak Plan Criticized," *Wash-
       ington Post,* Feb. 2, 2008.

# Index